INTERNATIONAL CENTRE FOR MECHANICAL SCIENCES

COURSES AND LECTURES - No. 328

NONLINEAR ANALYSIS OF SHELLS BY FINITE ELEMENTS

EDITED BY

F.G. RAMMERSTORFER
VIENNA UNIVERSITY OF TECHNOLOGY

SPRINGER - VERLAG WIEN - NEW YORK

Le spese di stampa di questo volume sono in parte coperte da
contributi del Consiglio Nazionale delle Ricerche.

This volume contains 157 illustrations.

Printed in Italy

In order to make this volume available as economically and as rapidly as possible the authors' typescripts have been reproduced in their original forms. This method unfortunately has its typographical limitations but it is hoped that they in no way distract the reader.

ISBN 3-211-82416-2 Springer Verlag Wien-New York
ISBN 0-387-82416-2 Springer Verlag New York-Wien

PREFACE

This monograph is based on the lecture notes of the course "Nonlinear Analysis of Shells by Finite Elements" given at the International Centre for Mechanical Sciences (CISM) in Udine, Italy, from June 24 to 28, 1991 by Professors E.N. Dvorkin, Siderca, Buenos Aires, Argentina, O. Oñate, Universidad Politecnica de Cataluna, Barcelona, Spain, E. Ramm, Univ. Stuttgart, FRG, F.G. Rammerstorfer, Vienna Univ. of Technology, Austria, R.L. Taylor, University of California, USA, W. Wagner, Univ. Hannover, FRG, and W. Wunderlich, München Univ. of Technology, FRG.

Enhanced safety requirements togheter with the demand for reduced weight in the design of mechanical as well as civil engineering structures are leading to the development of new design concepts, to the use of advanced materials or new material combinations and to more accurate calculation methods. In many applications shell structures in combination with composite materials are replacing conventional constructions, and optimization methods become more important. Such weight saving strategies may result in rather flexible structures undergoing large deformations, and the utmost utilization of the strength of the materials requires the consideration of the materials' nonlinear behavior. In the analysis of the structural behavior those aspects can only be treated by nonlinear methods.

Under these aspects, it was the main objective of the course to report on recent developments in the field of stress and deformation analysis as well nonlinear stability and optimization analysis of shells by the finite element method.

The following topics are treated in this monograph:

Alternative shell element formulations in large displacement and rotation analysis comparison of shell theory based elements versus degenerated solid approach, hybrid-mixed formulations, i.e. displacement models with additional assumptions for strains or stresses, assumed strain formulations, assessment of shell elements with respect to locking phenomena, ... Furthermore, implementation and modelling aspects are discussed on the basis of the individual element formulations.

Finite element formulations are described for stiffened shells as well as for composite and sandwich shells under large deformations including some aspects of the material description of composites and concrete. Layered fiber-composite shells and sandwich shells as well as reinforced concrete shells are treated, and algorithms for the computation of failure models and for post-failure analysis for composite shells under mechanical and thermal loads are presented.

With respect to optimization of shell structures this monograph contains optimization strategies, the description of sensitivity analysis, of the design element concept and of shape finding methods of free form shells.

Furthermore, algorithms for the treatment of the nonlinear stability behavior of shell structures (including bifurcation and snap-through buckling) are presented in the book.

The theoretical considerations are accompained by the presentation of numerical examples and practical applications.

It is my pleasure to thank all the colleagues who contributed to the course and this book. I also thank CISM for organizing the course and for the ospitality which the lecturers were provided with during their stay in Udine. The lecturers and authors owe special thank to Professor Sandor Kaliszky, Rector of CISM, for his efforts in supporting this course, and the Professor Carlo Tasso for encouraging the lecturers to write this monograph.

Franz G. Rammerstorfer

CONTENTS

LARGE ROTATIONS IN STRUCTURAL MECHANICS - OVERVIEW

N. Büchter and E. Ramm
University of Stuttgart, Stuttgart, Germany

ABSTRACT

Some basic aspects on the handling of finite rotations in structural mechanics are presented in this overview.

Different possibilities to choose the three independent rotation variables are shown. They are based either on elementary rotations or on rotation vectors.

Contrary to continuum elements the linearization of shell or beam elements with rotational degrees of freedom yields an extra contribution to the stiffness matrix.

1 ROTATION TENSOR AND ROTATIONAL VECTORS

1.1 Preliminary Remarks

Given are 3 linearly independent vectors $\mathbf{a}_1,\mathbf{a}_2,\mathbf{a}_3$ of the 3–dimensional euclidian space $\mathbf{R}^3$. It is characteristic for a rotation $\mathbf{R}$ of the basis $\mathbf{A} = (\mathbf{a}_1,\mathbf{a}_2,\mathbf{a}_3)$ into the new basis

$$\mathbf{A}^* = (\mathbf{a}_1^*, \mathbf{a}_2^*, \mathbf{a}_3^*) = \mathbf{RA} \tag{1.1}$$

that neither the length of the vectors nor the angle between them are changed by this operation. Consequently $\mathbf{R}$ must be orthogonal, because:

$$\mathbf{a}_i^{*T}\mathbf{a}_j^* = \mathbf{a}_i^T\mathbf{a}_j \quad \text{resp.} \quad \mathbf{A}^{*T}\mathbf{A}^* = \mathbf{A}^T\mathbf{R}^T\mathbf{R}\mathbf{A} = \mathbf{A}^T\mathbf{A} \text{ since } \mathbf{R}^T\mathbf{R} = \mathbf{I} \tag{1.2}$$

Eqn. (1.2) means that the 9 components of **R** have to fulfil 6 independent constraints. Consequently each rotation can be determined by 3 parameters. Mathematically spoken a rotation is a linear mapping

$$f : \mathbf{R}^3 \rightarrow \mathbf{R}^3 \quad , \quad \mathbf{x} \rightarrow \mathbf{y} = \mathbf{R}\mathbf{x} \ , \quad \mathbf{R} \in SO(3) \subset \mathbf{R}^9 \tag{1.3}$$

with the non–commutative group

$$SO(3) = \{\mathbf{R} \in \mathbf{R}^9 \ \ | \mathbf{R}^T\mathbf{R} = \mathbf{I}, \det \mathbf{R} = 1\} \tag{1.4}$$

3 dimensional
orthogonal
special ($\det \mathbf{R} = 1$)

The condition $\det \mathbf{R} = 1$ excludes reflections. In the context of a body, a reflection is a self penetration.
Example:

$$\mathbf{R} = \begin{bmatrix} 1 & 0 & 0 \\ 0 & 1 & 0 \\ 0 & 0 & -1 \end{bmatrix}$$

$\det \mathbf{R} = -1$ $\qquad \mathbf{R}^T\mathbf{R} = \mathbf{I}$

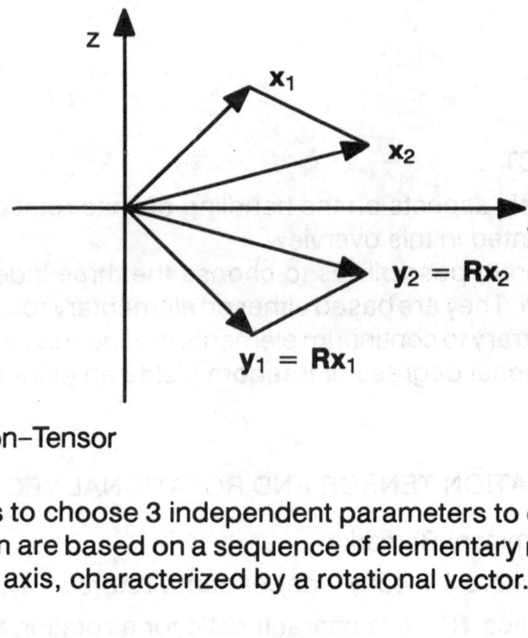

Fig. 1: Example for Reflection–Tensor

There are many possibilities to choose 3 independent parameters to describe a finite rotation. Most of them are based on a sequence of elementary rotations or on a rotation around one axis, characterized by a rotational vector.

1.2 Formulations Based on Rotational Vector

Rotational vectors are directly related to Euler parameters and quaternions. Both formulations start with four parameters to describe finite rotations supplemented of course by an extra constraint.
Euler parameters are based on the idea that each rotation can be described by a rotational axis **e** and its related angle ϕ . A vector parallel to the rotational axis is not changed during the rotation; thus it is an eigenvector of **R** with the eigenvalue $\lambda = 1$.

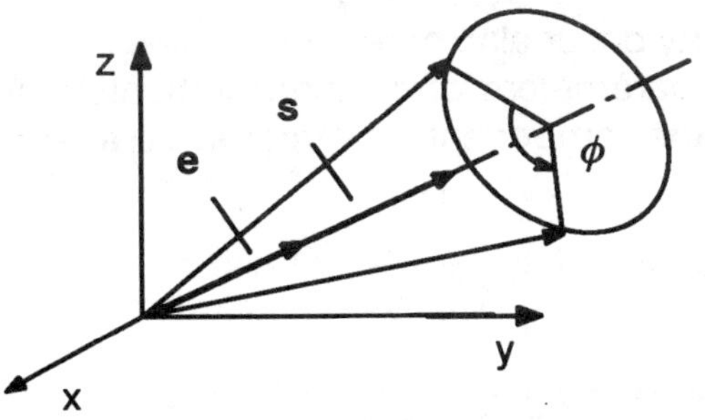

$$\mathbf{R}\ \mathbf{e} = \mathbf{e} \tag{1.5}$$

Fig. 2: Euler parameters, rotational vector

The three components of vector $\mathbf{e}$ and angle ϕ define the Euler parameters. The extra constraint equation is:

$$\mathbf{e}^T\mathbf{e} = 1 \tag{1.6}$$

The rotational tensor $\mathbf{R}$ can be uniquely determined from these parameters without any singularities.

$$\mathbf{R} = \mathbf{I} + \sin\phi\ \hat{\mathbf{e}} + (1 - \cos\phi)\ \hat{\mathbf{e}}\ \hat{\mathbf{e}} \tag{1.7}$$

Matrix $\hat{\mathbf{e}}$ in equation (1.7) operates on a vector like a cross product. For orthogonal cartesian coordinates it follows:

$$\mathbf{e} = e_i\ \mathbf{i}^i \qquad \hat{\mathbf{e}} = \hat{e}_{ij}\ \mathbf{i}^i \otimes \mathbf{i}^j = (\mathbf{e}\ \times)$$

$$[e_i] = \begin{bmatrix} e_1 \\ e_2 \\ e_3 \end{bmatrix} \qquad [\hat{e}_{ij}] = \begin{bmatrix} 0 & -e_3 & e_2 \\ e_3 & 0 & -e_1 \\ -e_2 & e_1 & 0 \end{bmatrix} \tag{1.8}$$

If the constraint equation (1.6) is directly applied in a numerical analysis for example to eliminate $e_3 = \sqrt{1 - e_1^2 - e_2^2}$ leaving ϕ, e_1, e_2 as primary variables, e_3 may become numerically instable.

The four quaternion parameters q_0, $\mathbf{q}$ are in turn closely related to the Euler parameters.

$$q_0 = \cos\frac{\phi}{2} \qquad \mathbf{q} = \sin\frac{\phi}{2}\mathbf{e} \tag{1.9}$$

They lead straightforward to the rotational tensor $\mathbf{R}$.

$$\mathbf{R} = \mathbf{I} + 2q_0\ \hat{\mathbf{q}} + 2\ \hat{\mathbf{q}}\ \hat{\mathbf{q}} \tag{1.10}$$

In finite element analyses the components of $\mathbf{q}$ can be used as primary variables; then q_0 is determined by the constraint equation:

$$q_0 = \sqrt{1 - \mathbf{q}^T\mathbf{q}} \tag{1.11}$$

Again numerical sensitivities may occur also here.
Opposite to Euler or quaternion parameters formulations with rotational vectors only use three parameters, i.e. the components of vector **s**. All alternatives differ only in the length of **s** [6].

1. $$\mathbf{s} = 2\tan\frac{\phi}{2}\,\mathbf{e} \longrightarrow \mathbf{R} = \mathbf{I} + \frac{1}{1+\frac{\mathbf{s}^T\mathbf{s}}{4}}\left(\hat{\mathbf{s}} + \hat{\mathbf{s}}\,\hat{\mathbf{s}}\right) \tag{1.12}$$

2. $$\mathbf{s} = \tan\frac{\phi}{2}\,\mathbf{e} = \frac{\mathbf{q}}{q_0} \longrightarrow \mathbf{R} = \mathbf{I} + \frac{2}{1+\mathbf{s}^T\mathbf{s}}\left(\hat{\mathbf{s}} + \hat{\mathbf{s}}\,\hat{\mathbf{s}}\right) \tag{1.13}$$

3. $$\mathbf{s} = \sin\phi\,\mathbf{e} \longrightarrow \mathbf{R} = \mathbf{I} + \hat{\mathbf{s}} + \frac{1}{2\cos^2\frac{\phi}{2}}\,\hat{\mathbf{s}}\,\hat{\mathbf{s}} \tag{1.14}$$

4. $$\mathbf{s} = \phi\,\mathbf{e} \longrightarrow \mathbf{R} = \mathbf{I} + \frac{\sin\phi}{\phi}\,\hat{\mathbf{s}} + \frac{1-\cos\phi}{\phi^2}\,\hat{\mathbf{s}}\,\hat{\mathbf{s}} = \exp(\hat{\mathbf{s}}) \tag{1.15}$$

Each case has certain pros and cons being more or less relevant. For example cases 1 and 2 lead to simple formulae for the total rotational vector emerging from two subsequent rotations but render an infinite length of the rotational vector at $\phi = (2k-1)\,\pi$. Also the third term of equation (3.14) of case 3 becomes instable at this location. Formulation 4, often attributed to Rodrigues, is the only one without singularities for $\phi \in [0, 2\pi[$ because

$$\lim_{\phi\to 0}\frac{\sin\phi}{\phi} = 1 \qquad \lim_{\phi\to 0}\frac{1-\cos\phi}{\phi^2} = \frac{1}{2} \tag{1.16}$$

1.3 Update of Subsequent Rotations

The total rotation tensor resulting from two subsequent rotations can be calculated by the product of two rotation tensors.

$$\mathbf{R}_3 = \mathbf{R}_2\mathbf{R}_1 \tag{1.17}$$

This operation is not commutative:

$$\mathbf{R}_3 = \mathbf{R}_2\mathbf{R}_1 \neq \mathbf{R}_1\mathbf{R}_2 \tag{1.18}$$

It is important to notice, that simple addition of the rotational vectors belonging to $\mathbf{R}_1$ and $\mathbf{R}_2$ does not result in the total rotational vector of $\mathbf{R}_3$ (see Fig. 3).

$$\mathbf{s}_3 \neq \mathbf{s}_2 + \mathbf{s}_1 \tag{1.19}$$

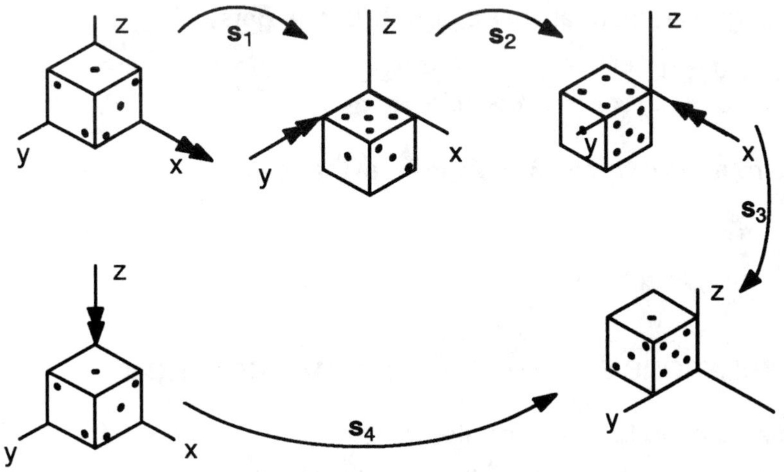

Fig. 3: Subsequent Rotations

$$\mathbf{s}_1 = \begin{bmatrix} +\frac{\pi}{2} \\ 0 \\ 0 \end{bmatrix} \quad \mathbf{s}_2 = \begin{bmatrix} 0 \\ -\frac{\pi}{2} \\ 0 \end{bmatrix} \quad \mathbf{s}_3 = \begin{bmatrix} -\frac{\pi}{2} \\ 0 \\ 0 \end{bmatrix} \quad \mathbf{s}_4 = \begin{bmatrix} 0 \\ 0 \\ -\frac{\pi}{2} \end{bmatrix}$$

$$\mathbf{s}_4 \neq \mathbf{s}_1 + \mathbf{s}_2 + \mathbf{s}_3$$

Update formulae for the rotational vectors can be found by looking at their relation to quaternion parameters. For quaternions we have

$$\begin{aligned} q_{03} &= q_{02}q_{01} - \mathbf{q}_1^T\mathbf{q}_2 \\ \mathbf{q}_3 &= q_{01}\mathbf{q}_2 + q_{02}\mathbf{q}_1 - \mathbf{q}_1 \times \mathbf{q}_2 \end{aligned} \tag{1.20}$$

It follows for Euler parameters:

$$\text{with} \quad q_0 = \cos\frac{\phi}{2} \qquad \mathbf{q} = \sin\frac{\phi}{2}\mathbf{e}$$

$$q_{03} = \cos\frac{\phi_3}{2} = \cos\frac{\phi_2}{2}\cos\frac{\phi_1}{2} - \sin\frac{\phi_1}{2}\sin\frac{\phi_2}{2}\ \mathbf{e}_1^T\mathbf{e}_2$$

$$\mathbf{q}_3 = \sin\frac{\phi_3}{2}\mathbf{e}_3 = \cos\frac{\phi_1}{2}\sin\frac{\phi_2}{2}\mathbf{e}_2 + \cos\frac{\phi_2}{2}\sin\frac{\phi_1}{2}\mathbf{e}_1 - \sin\frac{\phi_1}{2}\sin\frac{\phi_2}{2}\ \mathbf{e}_1 \times \mathbf{e}_2$$

$$\phi_3 = 2\arccos q_{03} \quad \mathbf{e}_3 = \frac{\mathbf{q}_3}{\sin\frac{\phi_3}{2}} \tag{1.21}$$

$$\left(\phi_3 = 2\arcsin\sqrt{\mathbf{q}_3^T\mathbf{q}_3} \quad \text{if } |q_{03}| > 0.984\ (\phi < 20°) \text{ this formula gives more reliable results} \right)$$

For the Rodrigues rotational vector $\mathbf{s} = \phi\ \mathbf{e}$ one gets:

$$\mathbf{s}_3 = \frac{\mathbf{q}_3}{\sqrt{\mathbf{q}_3^T\mathbf{q}_3}} \begin{cases} \text{either} & 2\arccos q_{03} \\ \text{or} & 2\arcsin\sqrt{\mathbf{q}_3^T\mathbf{q}_3} \end{cases} \tag{1.22}$$

and for the rotational vector $\mathbf{s} = 2\tan\frac{\phi}{2}\ \mathbf{e}$ the result is:

$$\mathbf{s}_3 = \frac{\mathbf{s}_1 + \mathbf{s}_2 - \frac{1}{2}\mathbf{s}_1 \times \mathbf{s}_2}{1 - \frac{1}{4}\mathbf{s}_1^T\mathbf{s}_2} \tag{1.23}$$

2 FORMULATIONS BASED ON ELEMENTARY ROTATIONS

2.1 Rotation around fixed and rotated axes

The final rotation is traditionally described by three consecutive individual (so called elementary) rotations, starting with one rotation around a fixed axis followed by two rotations around the new already rotated axes, e.g. for Euler angles

$$\mathbf{R} = \mathbf{R}_z^{**}(\gamma)\ \mathbf{R}_x^{*}(\beta)\ \mathbf{R}_z(\alpha)$$

$\mathbf{R}_x^*$ and $\mathbf{R}_z^{**}$ denote that the rotation is refered to a new axis $\mathbf{x}^*$ or $\mathbf{z}^{**}$, respectively. In contrast to this many references define the final rotation as a sequence of elementary rotations around fixed axes, but in a reversed order, e. g.

$$\mathbf{R} = \mathbf{R}_z(\alpha)\ \mathbf{R}_x(\beta)\ \mathbf{R}_z(\gamma)$$

It is shown below that both formulations lead to the same result.
Let $\mathbf{R}_1$ be the rotation around the fixed axis $\mathbf{e}_1$ with the angle ϕ_1 and $\mathbf{R}_2$ the rotation around the fixed axis $\mathbf{e}_2$ with the angle ϕ_2. The final rotation is $\mathbf{R}_3 = \mathbf{R}_2\mathbf{R}_1$. Now we would like to consider the case that the second axis $\mathbf{e}_2$ has been already rotated to $\mathbf{e}_2^* = \mathbf{R}_1\mathbf{e}_2$ by $\mathbf{R}_1$. We get for the total rotation

$$\mathbf{R}_3 = \mathbf{R}_2^*\mathbf{R}_1 \qquad \text{with} \qquad \mathbf{R}_2^* = \mathbf{R}(\mathbf{s}_2^*) = \mathbf{R}(\mathbf{R}_1\mathbf{s}_2) \tag{2.1}$$

Using the relation

$$\mathbf{R}(\mathbf{Qs}) = \mathbf{Q}\ \mathbf{R}(\mathbf{s})\ \mathbf{Q}^T \qquad \mathbf{Q} \in SO(3) \tag{2.2}$$

(the proof will be given in the appendix) one can write

$$\mathbf{R}_3 = \mathbf{R}_2^*\mathbf{R}_1 = \mathbf{R}_1\mathbf{R}_2\mathbf{R}_1^T\mathbf{R}_1 = \mathbf{R}_1\mathbf{R}_2 \tag{2.3}$$

Eqn.(2.3) means that the same result is obtained if one rotates first around the fixed axis $\mathbf{e}_1$ with ϕ_1 and then with ϕ_2 around $\mathbf{e}_2^* = \mathbf{R}_1\mathbf{e}_2$ or if one rotates first with ϕ_2 around the fixed axis $\mathbf{e}_2$ and then with ϕ_1 around the fixed axis $\mathbf{e}_1$.
The total rotation of three subsequent rotations around rotated axes can be consequently calculated by

$$\mathbf{R}_4 = \mathbf{R}_3^{**}\mathbf{R}_2^{*}\mathbf{R}_1 = (\mathbf{R}_2^{*}\mathbf{R}_1)\mathbf{R}_3(\mathbf{R}_2^{*}\mathbf{R}_1)^{T}(\mathbf{R}_2^{*}\mathbf{R}_1) = (\mathbf{R}_1\mathbf{R}_2)\mathbf{R}_1^{T}\mathbf{R}_1\mathbf{R}_3 = \mathbf{R}_1\mathbf{R}_2\mathbf{R}_3 \qquad (2.4)$$

2.2 Elementary Rotations

Elementary rotations are rotations around one axis.

$$\mathbf{R}_x(a) = \begin{bmatrix} 1 & 0 & 0 \\ 0 & \cos a & -\sin a \\ 0 & \sin a & \cos a \end{bmatrix} \quad \mathbf{R}_y(\beta) = \begin{bmatrix} \cos\beta & 0 & \sin\beta \\ 0 & 1 & 0 \\ -\sin\beta & 0 & \cos\beta \end{bmatrix} \quad \mathbf{R}_z(\gamma) = \begin{bmatrix} \cos\gamma & -\sin\gamma & 0 \\ \sin\gamma & \cos\gamma & 0 \\ 0 & 0 & 1 \end{bmatrix}$$

Among the formulations based on elementary rotations are:

Euler angles: $\mathbf{R} = \mathbf{R}_z(a)\ \mathbf{R}_x(\beta)\ \mathbf{R}_z(\gamma) \ = \mathbf{R}_z^{**}(\gamma)\ \mathbf{R}_x^{*}(\beta)\ \mathbf{R}_z(a)$ (2.5)

and

Cardan angles: $\mathbf{R} = \mathbf{R}_x(a)\ \mathbf{R}_y(\beta)\ \mathbf{R}_z(\gamma) \ = \mathbf{R}_z^{**}(\gamma)\ \mathbf{R}_y^{*}(\beta)\ \mathbf{R}_x(a)$ (2.6)

Unfortunately these formulations are not free of singularities. The uniqueness is lost for Eulerian angles at $\beta = (k-1)\pi$ and for Cardanian angles at $\beta = (2k-1)\pi/2$. In those cases the first and third rotation take place around the same axis. Therefore infinite combinations for α and γ lead to the same rotation tensor **R**.

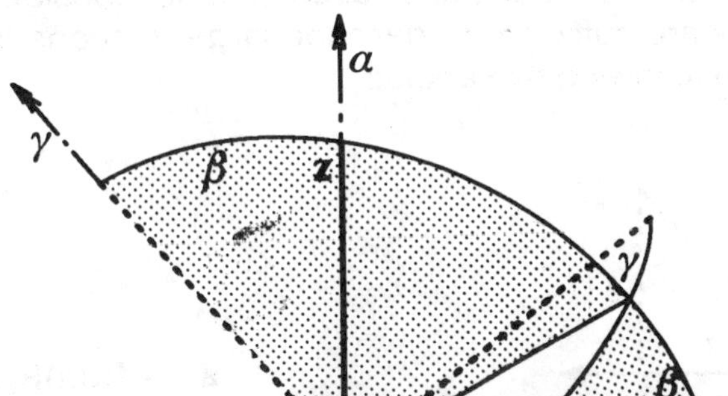

Fig. 4: Euler Angles

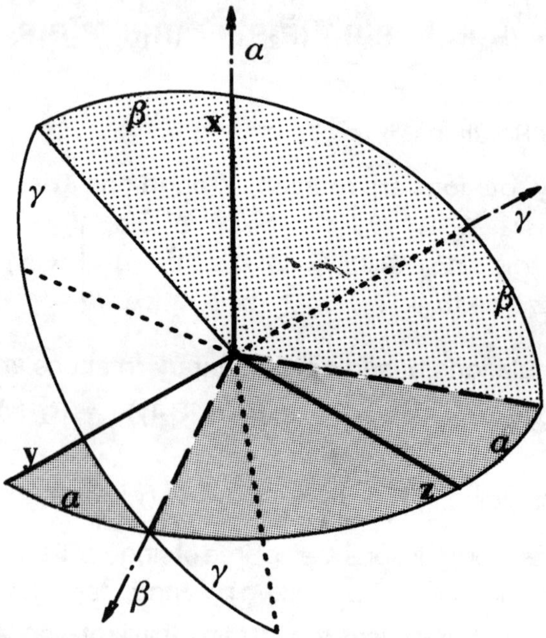

Fig. 5: Cardan Angles

3 LARGE ROTATIONS FOR SHELLS

Usualy the rotation around the shell director is not considered, therefore only two parameters are sufficient to describe large rotations for shells. In [7],[8] Ramm used the following two angles:

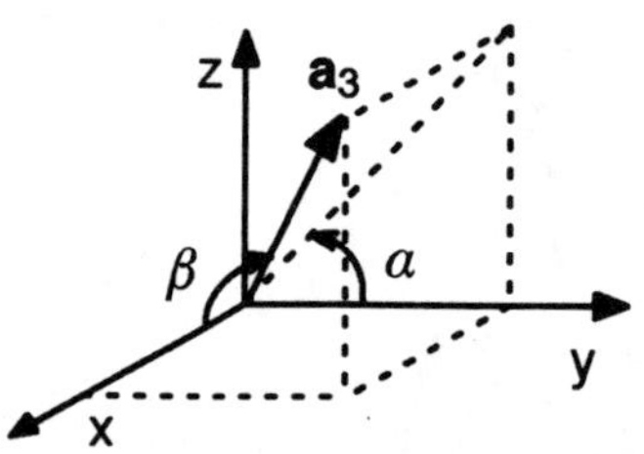

Fig. 6: Rotations chosen in [7],[8]

$$\bar{\mathbf{a}}_3 = \mathbf{R}_x(\alpha)\mathbf{R}_z(\beta-\frac{\pi}{2})\ \mathbf{i}_y = \mathbf{R}_x(\alpha-\frac{\pi}{2})\mathbf{R}_y(-\beta)\ \mathbf{i}_x \tag{3.1}$$

The location of the director $\bar{\mathbf{a}}_3$ with respect to a basis vector $\mathbf{i}_y$ or $\mathbf{i}_x$ can be described in different ways. A singularity occurs for $\beta = (k-1)\pi\ ,\ k \in Z$, where the shell director is parallel to the x–axis and α is not defined. Alternatively it is common to resort to any other of the above described possibilities using for example other combinations of two elementary rotations or the rotational vectors.

Since in shell analysis the rotation around the director is usually not considered a local coordinate system is introduced and the third component is not taken into account. A singularity must occur also for any other choice of the rotation parameters because there is no unique way to describe a rotation of 180° for a single vector. The singularities can be avoided through a multiplicative update [2],[10],[11].

4 LARGE ROTATION – "EXTRA STIFFNESS"

The calculation of nonlinear problems in structural mechanics using the finite element method based on a displacement model usually follows the scheme described below.

4.1 Nonlinear displacement finite element formulation

1. Principle of virtual displacements

$$-\delta W^{int} = \int_V \delta\mathbf{F}\cdot\mathbf{P}dV = \delta W^{ext} \tag{4.1}$$

$$\mathbf{F} = \mathbf{I} + \nabla\mathbf{u} \qquad \delta\mathbf{F} = \nabla\delta\mathbf{u} = \delta\mathbf{u}_{,\mathbf{x}} \qquad \mathbf{F} \text{ – Deformation gradient} \tag{4.2}$$

$$\mathbf{P} = \mathbf{P}(\mathbf{F}) \qquad \mathbf{P} \text{ – PKI stress tensor} \tag{4.3}$$

2. Discretization

$$\mathbf{u} \approx \mathbf{u}_h(N^K(\xi^i), \mathbf{a}^K) \qquad \nabla\mathbf{u}_h \neq \mathbf{0} \tag{4.4}$$

$\mathbf{a}^K$ – nodal parameters
ξ^i – local element coord.
N^K – shape functions

3. Linearization

$$(\ )' = \frac{\partial(\ )}{\partial\mathbf{a}} \qquad \delta(\ ) = (\ )'\delta\mathbf{a} \tag{4.5}$$

$$L(-\delta W^{int}) = \delta\mathbf{a}\int_V \mathbf{F}'_h\cdot\mathbf{P}_h dV + \delta\mathbf{a}\int_V (\mathbf{F}''_h\cdot\mathbf{P}_h + \mathbf{F}'_h\cdot\mathbf{P}'_h)dV\Delta\mathbf{a} \tag{4.6}$$

$\delta\mathbf{a}$ – virtual nodal parameters
$\Delta\mathbf{a}$ – incremental nodal parameters

Tangent stiffness matrix

$$\mathbf{K}_T = \int_V (\underbrace{\mathbf{F}''_h\cdot\mathbf{P}_h}_{\mathbf{K}_g^{II}\ \text{"extra term"}} + \underbrace{\mathbf{F}'_h\cdot\mathbf{P}'_h}_{\mathbf{K}_{e+u} + \mathbf{K}_g^{I}})dV \tag{4.7}$$

$$\mathbf{P} = \mathbf{F}\mathbf{S}$$

$$\mathbf{P}'_h = \mathbf{F}'_h\mathbf{S}_h + \mathbf{F}_h\mathbf{S}'_h$$

$\mathbf{S}$ – PKII stress tensor

4.2 Why "extra term"?

4.2.1 Continuum model

The usual discretization

$$\mathbf{u}_h = \sum_K N^K(\xi^i)\mathbf{u}^K = \mathbf{N}\mathbf{a} \tag{4.8}$$

renders

$$\begin{aligned} \mathbf{F}'_h &= (\mathbf{I} + \nabla \mathbf{u}_h)' = \nabla \mathbf{u}'_h = \nabla \mathbf{N} \\ &=> \mathbf{F}''_h = (\nabla \mathbf{N})' = \mathbf{0} \quad => \mathbf{K}_g^{II} = \mathbf{0} \end{aligned} \tag{4.9}$$

Conclusion: no "extra term" for continuum models.

4.2.2 Shells or beams

The formulations is described for shells; for notation see [3]

shell kinematic: $$\mathbf{u} = \mathbf{v} + \xi^3 \frac{h}{2}(\mathbf{R}(\mathbf{s}) - \mathbf{I})\mathbf{n} \tag{4.10}$$

Discretization for degenerated elements:

$$\mathbf{u}_h = \sum_K N(\xi^\alpha)[\mathbf{v}^K + \xi^3 \frac{h^K}{2}(\mathbf{R}(\mathbf{s}^K) - \mathbf{I})\mathbf{n}^K] \tag{4.11}$$

or for shell theory elements:

$$\mathbf{v} \approx \mathbf{v}_h = \sum_K N^K \mathbf{v}^K \quad ; \quad \mathbf{s} \approx \mathbf{s}_h = \sum_K N^K \mathbf{s}^K \tag{4.12}$$

$\mathbf{R}$ contains trigonometric functions of $\mathbf{s}^K$ or $\mathbf{s}_h$ depending which kind of element formulation is used. Consequently

$$\mathbf{R}'' \neq \mathbf{0} \quad => \quad \mathbf{u}''_h \neq \mathbf{0} \quad => \quad \mathbf{F}'' \neq \mathbf{0} \quad => \quad \mathbf{K}_g^{II} \neq \mathbf{0} \tag{4.13}$$

Conclusion: one more term has to be taken into account in the tangent stiffness matrix [5].

5 EXAMPLE: PINCHED HEMISPHERE [2]

This famous benchmark is presented here to show a large rotation problem. Here a Newton scheme with non-additive update of the rotations has been applied. A quadratic convergence rate is obtained using the correct tangent inculding $\mathbf{K}_g^{II}$.

We used 256 4-node assumed strain elements to calculate one quarter of the hemisphere. The interpolation of the shear strains follows the idea of Dvorkin and Bathe.

Young's modulus E = 6.825E+07
Poisson's ratio $\nu = 0.3$
Thickness h = 0.04
Radius R = 10
Load P = 20*f

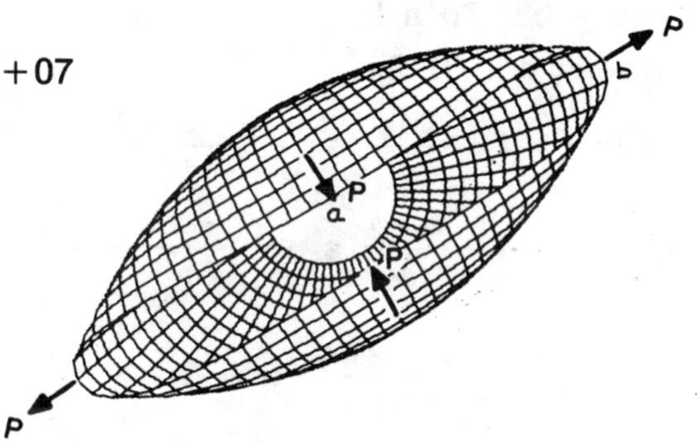

mesh: 16x16 4-node assumed strain elements for one quarter

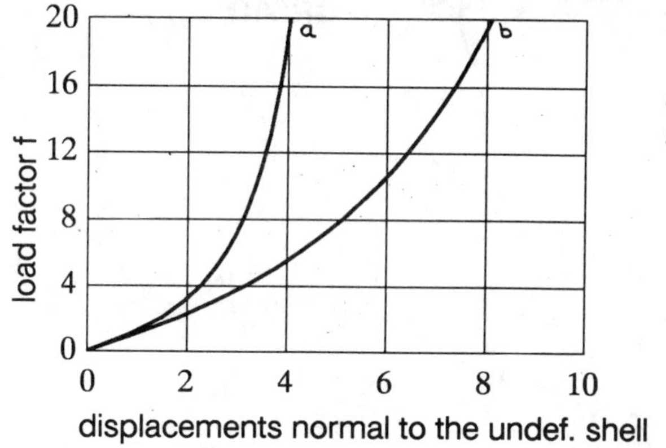

energy norms in
step 10 (f=9.0 -> f=10.0)

3.589E+02
8.163E-03
4.990E-04
4.419E-06
4.016E-10
3.384E-18

Fig. 7: Pinched hemisphere

6 APPENDIX: Proof of eqn. (2.2)

First we proof that

$$\mathbf{Q}^T\hat{\mathbf{s}}^*\mathbf{Q} = \hat{\mathbf{s}} \qquad \text{with} \qquad \mathbf{s}^* = \mathbf{Q}\mathbf{s} \qquad \mathbf{Q} \in SO(3) \qquad (\det \mathbf{Q} = 1) \tag{6.1}$$

Multiplying $\mathbf{Q}^T\hat{\mathbf{s}}^*\mathbf{Q}$ with the base vectors from left and right leads to

$$\mathbf{i}_i^T\mathbf{Q}^T\hat{\mathbf{s}}^*\mathbf{Q}\mathbf{i}_j = \det(\mathbf{i}_i^*, \mathbf{s}^*, \mathbf{i}_j^*) = \det[\mathbf{Q}(\mathbf{i}_i, \mathbf{s}, \mathbf{i}_j)] = \det\mathbf{Q}\det(\mathbf{i}_i, \mathbf{s}, \mathbf{i}_j) = \mathbf{i}_i^T\hat{\mathbf{s}}\mathbf{i}_j$$

Consequently eqn. (6.1) holds. Using the relation

$$\hat{\mathbf{a}}\hat{\mathbf{b}} = \mathbf{b}\mathbf{a}^T - \mathbf{b}^T\mathbf{a}\ \mathbf{I} \tag{6.3}$$

we can verify eqn. (2.2)

$$\begin{aligned}
\mathbf{R}(\mathbf{s}^*) &= \mathbf{I} + \frac{\sin\phi}{\phi}\,\hat{\mathbf{s}}^* + \frac{1-\cos\phi}{\phi^2}\,\hat{\mathbf{s}}^*\,\hat{\mathbf{s}}^* \\
&= \mathbf{I} + \frac{\sin\phi}{\phi}\,\hat{\mathbf{s}}^* + \frac{1-\cos\phi}{\phi^2}\,(\mathbf{s}^*\mathbf{s}^{*T} - \mathbf{s}^{*T}\mathbf{s}^*\mathbf{I}) \\
&= \mathbf{Q}\left(\mathbf{I} + \frac{\sin\phi}{\phi}\,\mathbf{Q}^T\hat{\mathbf{s}}^*\mathbf{Q} + \frac{1-\cos\phi}{\phi^2}\,(\mathbf{s}\mathbf{s}^T - \mathbf{s}^T\mathbf{s}\ \mathbf{I})\right)\mathbf{Q}^T \\
&= \mathbf{Q}\left(\mathbf{I} + \frac{\sin\phi}{\phi}\,\hat{\mathbf{s}} + \frac{1-\cos\phi}{\phi^2}\,\hat{\mathbf{s}}\,\hat{\mathbf{s}}\right)\mathbf{Q}^T \quad = \mathbf{Q}\mathbf{R}(\mathbf{s})\mathbf{Q}^T
\end{aligned} \tag{6.4}$$

7 REFERENCES

1. Büchter, N.: Zusammenführung von Degenerationskonzept und Schalentheorie bei endlichen Rotationen. Dissertation. Bericht Nr. 14, Instituts für Baustatik, Universität Stuttgart, Stuttgart 1992.
2. Büchter, N.; Ramm, E.: Shell Theory versus Degeneration – A Comparison in Large Rotation Finite Element Analysis. International Journal for Numerical Methods in Engineering, Vol. 34 (1992) 39–59.
3. Büchter, N.; Ramm, E.: Comparison of Shell Theory and Degeneration. Nonlinear Analysis of Shells by Finite Elements CISM, Udine, June 24–28, 1991.
4. Cardona, A.; Geradin, M.: A beam finite element non–linear theory with finite rotations. International Journal for Numerical Methods in Engineering, 26, 2403–2438 (1988).
5. Ramm,E.; Matzenmiller, A.: Large deformation shell analyses based on the degeneration concept. State–of–the–Art Texts on 'Finite Element Methods for Plate and Shells Structures', Pineridge Press, Swansea, UK (1986).
6. Pietraszkiewicz, W.; Badur, J.: Finite rotations in the description of continuum deformation. Int. J. Engng. Sci. Vol. 21, No. 9, 1097–1115 (1983).
7. Ramm, E.: Geometrisch nichtlineare Elastostatik und finite Elemente. Habilitation. Bericht Nr. 76–2, Institut für Baustatik, Universität Stuttgart (1976).
8. Ramm, E.: A plate/shell element for large deflections and rotations. US – Germany Symp. on "Formulations and computational algorithms in finite element analysis", MIT 1976, MIT–Press (1977).
9. Schiehlen, W.: Technische Dynamik. Teubner Studienbücher. Stuttgart (1986).
10. Simo, J. C.; Vu–Quoc, L.: On the dynamics of 3–d finite–strain rods. Finite element methods for plate and shell structures, 2: Formulations and algorithms. Pineridge Press, Swansea, U.K., 1–30, (1986).
11. Stanley. G.M.; Park, K.C.; Hughes, T.J.R.: Continuum–based resultant shell elements. Finite element methods for plate and shell structures, ed. T.J.R. Hughes et. al., Pineridge Press, Swansea, pp. 1–45 (1986).

COMPARISON OF SHELL THEORY AND DEGENERATION

N. Büchter and E. Ramm
University of Stuttgart, Stuttgart, Germany

ABSTRACT

The study adresses the controversy 'degenerated solid approach' versus 'shell theory'. It is shown that both formulations differ only in the kind of discretisation if they are based on the same mechanical assumptions. In particular for degenerated shell elements different versions of explicit integration across the thickness are discussed. Among these are the approximation 'jacobian across the thickness is constant' proven to be too restrictive and the series expansion of the inverse jacobian which turns out to be unnecessary although it leads to equations of the same order as those of a 'best first approximation' of a shell theory.

In order to make the differences and identities of the two approaches transparent a notation independent of a specific coordinate system has been utilized; thus transformation between global and local cartesian and curvilinear coordinates are avoided at this stage of the derivation. In addition, the discretization is not yet introduced at this step as it is usually done during the degeneration. For the sake of simplicity only a slight change of the shell thickness is allowed, i.e. the thickness is assumed to be constant.

1 FUNDAMENTAL EQUATIONS

1.1 Shell Theory

Here we restrict the derivation to a material formulation in the (total) Lagrangian way. The starting point are the equations of a 3D–continuum:

- Weak form of the equilibrium equation (principle of virtual work):

$$\iiint_V \mathbf{S}\cdot\delta\mathbf{E}\ dV = \delta W^{ext} \tag{1.1}$$

- kinematic equation: $$\mathbf{F} = \frac{\partial \overline{\mathbf{x}}}{\partial \mathbf{x}} = \overline{\mathbf{x}}_{,\mathbf{x}} = \mathbf{I} + \mathbf{u}_{,\mathbf{x}} \tag{1.2}$$

$$\mathbf{E} = \frac{1}{2}(\mathbf{F}^T\mathbf{F} - \mathbf{I}) = \frac{1}{2}(\mathbf{u}_{,\mathbf{x}}{}^T + \mathbf{u}_{,\mathbf{x}} + \mathbf{u}_{,\mathbf{x}}{}^T\mathbf{u}_{,\mathbf{x}}) \tag{1.3}$$

- linearized form of constitutive equation:

$$\mathbf{S}' = C\cdot\mathbf{E}' \tag{1.4}$$

S, **F**, **E** are the second Piola–Kirchhoff stress tensor, the material deformation gradient and the Green–Lagrange strain tensor, respectively; **x** and $\overline{\mathbf{x}}$ denote the position vectors of an arbitrary point of the shell body in the reference and deformed configuration (Fig. 1); **u** is the displacement field. **r**, $\overline{\mathbf{r}}$ and **v** are the corresponding variables of the reference surface. In view of an incremental formulation the constitutive law is required in a linearized form.

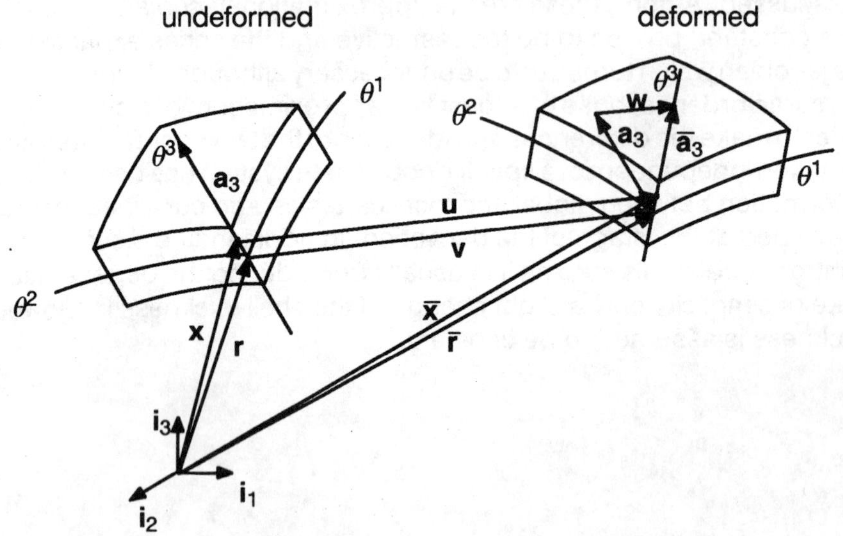

Fig. 1: Deformation of the shell

An assumption for the displacement field across the thickness reduces the 3D–theory to its 2D–equivalent. Small transverse shear deformations justify a linear displacement field:

Assumption A1: $$\mathbf{u} = \bar{\mathbf{x}} - \mathbf{x} = \mathbf{v} + \theta^3\mathbf{w} \tag{1.5}$$

The position vector of the deformed shell is:

$$\bar{\mathbf{x}} = \mathbf{r} + \mathbf{v} + \theta^3(\mathbf{a}_3 + \mathbf{w}) = \bar{\mathbf{r}} + \theta^3\bar{\mathbf{a}}_3 \quad , \quad -\frac{h}{2} \le \theta^3 \le \frac{h}{2} \tag{1.6}$$

h is the shell thickness. The shifter **Z** maps all variables to the metric of the reference surface.

$$\mathbf{J} = \mathbf{Z}\mathbf{J}_0 \tag{1.7}$$

The Jacobians **J** of the shell space and $\mathbf{J}_0$ of the reference surface are derived from:

$$\mathbf{J} = \frac{\partial \mathbf{x}}{\partial \boldsymbol{\theta}} = [\mathbf{g}_1, \mathbf{g}_2, \mathbf{g}_3]_{i^i} \ ; \ \mathbf{J}_0 = \mathbf{J}(\theta^3 = 0) = [\mathbf{a}_1, \mathbf{a}_2, \mathbf{a}_3]_{i^i} \ ; \ \boldsymbol{\theta} = [\theta^1, \theta^2, \theta^3] \tag{1.8}$$

where the column vectors are the covariant base vectors. The contravariant base vectors are obtained as columns of $\mathbf{J}^{-T}$ and $\mathbf{J}_0^{-T}$:

$$\mathbf{Z} = \mathbf{x}_{;r} = (\mathbf{r} + \theta^3\mathbf{a}_3)_{,r} = \mathbf{r}_{;r} + \mathbf{a}_3 \otimes \theta^3{}_{;r} + \theta^3\mathbf{a}_{3;r} = \mathbf{I} - \theta^3\mathbf{B} \tag{1.9}$$

In (1.9) the abbreviation $(\ )_{;r}$ means:

$$(\ )_{;r} = \frac{\partial(\ )}{\partial\theta^i} \otimes \mathbf{a}^i = \left(\frac{\partial(\ )}{\partial\theta^i} \otimes \mathbf{g}^i\right)\mathbf{Z} = (\ )_{,x}\mathbf{Z}$$

I defines the identity tensor; **B** is the symmetrical curvature tensor of the reference surface if $\mathbf{a}_3$ is orthogonal to $\mathbf{a}_1$ and $\mathbf{a}_2$.

$$\mathbf{B} = -\mathbf{a}_{3;r} = -\mathbf{a}_{3,\alpha} \otimes \mathbf{a}^\alpha = b_{\beta\alpha}\ \mathbf{a}^\beta \otimes \mathbf{a}^\alpha \quad \text{with:} \quad b_{\beta\alpha} = -\mathbf{a}_{3,\alpha}\mathbf{a}_\beta \tag{1.10}$$

Greek indices are either 1 or 2; Latin indices run from 1 to 3. The material deformation gradient can be decomposed:.

$$\hat{\mathbf{F}} = \mathbf{F}\mathbf{Z} = \bar{\mathbf{x}}_{,x}\mathbf{x}_{;r} = \bar{\mathbf{x}}_{;r} \tag{1.11}$$

As a consequenze of assumption A1 $\hat{\mathbf{F}}$ consists of a constant and a linear term:

$$\hat{\mathbf{F}} = \bar{\mathbf{x}}_{;r} = \mathbf{x}_{;r} + \mathbf{u}_{;r} = \mathbf{F}_C + \theta^3\ \mathbf{F}_L \tag{1.12}$$

with:

$$\mathbf{F}_C = \mathbf{I} + \mathbf{v}_{;r} + \mathbf{w} \otimes \mathbf{a}_3 = \bar{\mathbf{a}}_i \otimes \mathbf{a}^i \tag{1.13}$$

$$\mathbf{F}_L = \mathbf{w}_{;r} - \mathbf{B} = \bar{\mathbf{a}}_{3;r} = \bar{\mathbf{a}}_{3,\alpha} \otimes \mathbf{a}^\alpha \tag{1.14}$$

Introducing equations (1.9) and (1.11) into equation (1.3) leads to the Green–Lagrange strains:

$$\mathbf{E} = \mathbf{Z}^{-T}\,\frac{1}{2}(\hat{\mathbf{F}}^T\hat{\mathbf{F}} - \mathbf{Z}^T\mathbf{Z})\;\mathbf{Z}^{-1} = \mathbf{Z}^{-T}\hat{\mathbf{E}}\mathbf{Z}^{-1} \tag{1.15}$$

$$= \mathbf{Z}^{-T}\,\frac{1}{2}\Big(\mathbf{u}_{;r}^T\mathbf{Z} + \mathbf{Z}^T\mathbf{u}_{;r} + \mathbf{u}_{;r}^T\mathbf{u}_{;r}\Big)\mathbf{Z}^{-1}$$

$$= \mathbf{Z}^{-T}\,\frac{1}{2}\Big(\mathbf{u}_{;r}^T + \mathbf{u}_{;r} + \mathbf{u}_{;r}^T\mathbf{u}_{;r} + \theta^3(-\,\mathbf{u}_{;r}^T\mathbf{B} - \mathbf{B}\mathbf{u}_{;r})\Big)\mathbf{Z}^{-1}$$

$\hat{\mathbf{E}}$ is quadratic in θ^3.

$$\hat{\mathbf{E}} = \hat{\mathbf{E}}_C + \theta^3\hat{\mathbf{E}}_L + (\theta^3)^2\hat{\mathbf{E}}_Q \tag{1.16}$$

with:

$$\hat{\mathbf{E}}_C = \frac{1}{2}(\mathbf{F}_C^T\mathbf{F}_C - \mathbf{I})$$

$$= \frac{1}{2}\Big(\mathbf{v}_{;r}{}^T + \mathbf{v}_{;r} + \mathbf{w}\otimes\mathbf{a}_3 + \mathbf{a}_3\otimes\mathbf{w}$$
$$\underline{+\,\mathbf{v}_{;r}{}^T\mathbf{v}_{;r} + \mathbf{v}_{;r}{}^T(\mathbf{w}\otimes\mathbf{a}_3) + (\mathbf{a}_3\otimes\mathbf{w})\mathbf{v}_{;r} + (\mathbf{a}_3\otimes\mathbf{w})(\mathbf{w}\otimes\mathbf{a}_3)}\Big) \tag{1.17}$$

$$\hat{\mathbf{E}}_L = \frac{1}{2}(\mathbf{F}_C^T\mathbf{F}_L + \mathbf{F}_L^T\mathbf{F}_C + \mathbf{B}^T + \mathbf{B})$$

$$= \frac{1}{2}\Big(\mathbf{w}_{;r}{}^T + \mathbf{w}_{;r} - \mathbf{v}_{;r}{}^T\mathbf{B} - \mathbf{B}^T\mathbf{v}_{;r} - \mathbf{B}^T(\mathbf{w}\otimes\mathbf{a}_3) - (\mathbf{a}_3\otimes\mathbf{w})\mathbf{B}$$
$$\underline{+\,\mathbf{w}_{;r}{}^T\mathbf{v}_{;r} + \mathbf{v}_{;r}{}^T\mathbf{w}_{;r} + \mathbf{w}_{;r}{}^T(\mathbf{w}\otimes\mathbf{a}_3) + (\mathbf{a}_3\otimes\mathbf{w})\mathbf{w}_{;r}}\Big) \tag{1.18}$$

$$\hat{\mathbf{E}}_Q = \frac{1}{2}(\mathbf{F}_L^T\mathbf{F}_L - \mathbf{B}^T\mathbf{B}) \;=\; \frac{1}{2}\Big(-\,\mathbf{w}_{;r}{}^T\mathbf{B} - \mathbf{B}^T\mathbf{w}_{;r} \;+\; \underline{\mathbf{w}_{;r}{}^T\mathbf{w}_{;r}}\Big) \tag{1.19}$$

Dropping the underlined terms renders a geometrically linear theory.
The Piola–Kirchhoff stress tensor of the second kind is transformed to the reference surface as well $\hat{\mathbf{S}} = \mathbf{Z}^{-1}\mathbf{S}\mathbf{Z}^{-T}$. Equations (1.15) and (1.16) are introduced into the virtual work expression, equation (1.1), denoting the determinant of the shifter by $\mu = \det\mathbf{Z} = \det\mathbf{J}/\det\mathbf{J}_0$:

$$\iiint_V \mathbf{S}\cdot\delta\mathbf{E}\;dV = \iiint_V \mathbf{S}\cdot\mathbf{Z}^{-T}\delta\hat{\mathbf{E}}\mathbf{Z}^{-1}\;dV$$

$$= \iint_A \left[\int_{-h/2}^{h/2}\mu\;\hat{\mathbf{S}}d\theta^3\right]\cdot\delta\hat{\mathbf{E}}_C + \left[\int_{-h/2}^{h/2}\mu\;\hat{\mathbf{S}}\theta^3d\theta^3\right]\cdot\delta\hat{\mathbf{E}}_L + \left[\int_{-h/2}^{h/2}\mu\hat{\mathbf{S}}(\theta^3)^2d\theta^3\right]\cdot\delta\hat{\mathbf{E}}_Q\;dA$$

$$= \iint_A \mathbf{N}_{II}\cdot\delta\hat{\mathbf{E}}_C + \mathbf{M}_{II}\cdot\delta\hat{\mathbf{E}}_L + \mathbf{L}_{II}\cdot\delta\hat{\mathbf{E}}_Q\;dA \tag{1.20}$$

$\mathbf{N}_{II}$, $\mathbf{M}_{II}$ and $\mathbf{L}_{II}$ denote the symmetric pseudo–stress resultant tensors. In view of a Newton–type iteration equation (1.20) is usually linearized; since this process is a conventional process it will not be discussed here further on. Now the constitutive equations for materials varying across the thickness can be defined in rate form for stress resultants.

$$\begin{bmatrix} \mathbf{N}_{II}' \\ \mathbf{M}_{II}' \\ \mathbf{L}_{II}' \end{bmatrix} = \begin{bmatrix} D_0\cdot[\ldots] & D_1\cdot[\ldots] & D_2\cdot[\ldots] \\ D_1\cdot[\ldots] & D_2\cdot[\ldots] & D_3\cdot[\ldots] \\ D_2\cdot[\ldots] & D_3\cdot[\ldots] & D_4\cdot[\ldots] \end{bmatrix} \begin{bmatrix} \hat{\mathbf{E}}_C' \\ \hat{\mathbf{E}}_L' \\ \hat{\mathbf{E}}_Q' \end{bmatrix} \tag{1.21}$$

Here:

$$D_i\cdot[\ldots] = \int_{-h/2}^{h/2} \mu \mathbf{Z}^{-1} C\cdot[\mathbf{Z}^{-T} \ldots \mathbf{Z}^{-1}]\mathbf{Z}^{-T}(\theta^3)^i \, d\theta^3 \tag{1.22}$$

is introduced as abbreviation.

It should be noted that except assumption A1 no further hypotheses have been used. $\mathbf{Z}^{-1} = \mathbf{Z}^{-T}$ and $\mathbf{J}^{-1}$, respectively, can be determined analytically, applying the theorem of Caley–Hamilton. This operation is in general not necessary since the material matrices D_i [...] and the pseudo–stress resultant tensors are integrated numerically.

If a relative error in the order of h/R (h = shell thickness, 1/R = maximal principal curvature = $O(||\mathbf{B}||)$ is tolerated, the above equations may be substantially simplified. (see [4], which refers to the work of Koiter and John).

•Assumption A2: Provided that the gradients of the shear strains are small, the terms of equation (1.16) quadratic in θ^3 can be neglected, i.e. $\hat{\mathbf{E}}_Q = \mathbf{0}$. Because of this the third term in equation (1.20) vanishes.

•Assumption A3: μ is taken as unity and $\mathbf{Z}^{-1}$ as the identity tensor in equation (1.22):

$\mu = \det \mathbf{Z} = \det \mathbf{J} / \det \mathbf{J}_0 \approx 1 \; ; \; \mathbf{Z}^{-1} = \mathbf{Z}^{-T} \approx \mathbf{I}$

This is equivalent to $\hat{\mathbf{S}} \approx \mathbf{S}$; $\hat{\mathbf{E}} \approx \mathbf{E}$, in the energy expression.

•Assumption A4: The normal stresses in thickness direction are neglected, i.e. $\mathbf{a}_3^T \, \mathbf{S} \, \mathbf{a}_3 = S^{33} = 0$. For small strains this assumption allows to introduce an inextensional shell director, that is $|\bar{\mathbf{a}}_3| = 1$. This constraint reduces the 6–parametric theory (components of **v** and **w**) to its 5–parametric subset. The assumption is verified through a condensation of the material law: $C \longrightarrow C^*$. The resulting error is of the order of $\nu h/R$ for Hooke's material law, with ν as Poisson's ratio.

All four assumptions together lead to a first approximation of a geometrically nonlinear shell theory for small strains including transverse shear deformation [5], [13]. Based on curvilinear convective coordinates θ^i the kinematic equations are in vectorial notation:

$$\hat{\mathbf{E}}_C = \frac{1}{2}(\mathbf{F}_C^T\mathbf{F}_C - \mathbf{I}) = \alpha_{ij}\ \mathbf{a}^i \otimes \mathbf{a}^j \quad \text{with: } \alpha_{ij} = \frac{1}{2}(\bar{\mathbf{a}}_i\bar{\mathbf{a}}_j - \mathbf{a}_i\mathbf{a}_j)$$

$$\text{from A4 follows:} \qquad \alpha_{33} = 0 \tag{1.23}$$

$$\hat{\mathbf{E}}_L = \frac{1}{2}(\mathbf{F}_C^T\mathbf{F}_L + \mathbf{F}_L^T\mathbf{F}_C + \mathbf{B}^T + \mathbf{B}) = \beta_{ij}\ \mathbf{a}^i \otimes \mathbf{a}^j$$

$$\text{with:} \qquad \beta_{ij} = \frac{1}{2}(\bar{\mathbf{a}}_i\bar{\mathbf{a}}_{3,j} + \bar{\mathbf{a}}_{3,i}\,\bar{\mathbf{a}}_j - \mathbf{a}_i\mathbf{a}_{3,j} - \mathbf{a}_{3,i}\,\mathbf{a}_j) \tag{1.24}$$

$$\text{from A4 follows:} \quad \bar{\mathbf{a}}_3\bar{\mathbf{a}}_{3,\alpha} = 0;\ \bar{\mathbf{a}}_{3,3} = \mathbf{0} \quad \text{thus:} \quad \beta_{i3} = \beta_{3i} = 0$$

1.2 Concept of Degeneration

a) Numerical integration across thickness

Elements with a implicit (numerical) thickness integration, e.g. [1], [15], start directly from the equations of a continuum and apply only assumption A1 and mostly A4. Insofar, they are from the theoretical point of view at least as rigorous as the above described shell theory.

b) Explicit integration across thickness

- Integration due to Zienkiewicz–Taylor–Too (Z&T&T)

In order to improve the numerical efficiency, in particular for materials with varying properties across the thickness of a shell, the authors in [24] proposed an explicit integration analogous to a shell theory, i.e. equation (1.21). They introduced the assumption:

$$\mathbf{J} = \mathbf{J}_0 \tag{1.25}$$

which is equivalent to $\mathbf{Z}=\mathbf{I}$ and $\mathbf{B}=\mathbf{0}$ in the present notation.

$$\mathbf{E} = \frac{1}{2}(\mathbf{u}_{,x}{}^T + \mathbf{u}_{,x} + \mathbf{u}_{,x}{}^T\mathbf{u}_{,x}) = \frac{1}{2}(\mathbf{Z}^{-T}\mathbf{u}_{;r}{}^T + \mathbf{u}_{;r}\mathbf{Z}^{-1} + \mathbf{Z}^{-T}\mathbf{u}_{;r}{}^T\mathbf{u}_{;r}\mathbf{Z}^{-1})$$

$$\approx \frac{1}{2}(\mathbf{u}_{;r}{}^T + \mathbf{u}_{;r} + \mathbf{u}_{;r}{}^T\mathbf{u}_{;r}) = \mathbf{E}^{\approx} \tag{1.26}$$

This approximation is too restrictive as can be concluded by a comparison with the shell formulation. It follows from assumption A3 and equation (1.15):

$$\mathbf{E} \approx \hat{\mathbf{E}} = \mathbf{E}^{\approx} + \theta^3 \frac{1}{2}\left(-\mathbf{u}_{;r}^T\mathbf{B} - \mathbf{B}\mathbf{u}_{;r}\right) \tag{1.27}$$

It should be pointed out that all missing terms ought to appear already in a geometrically linear theory. The dropped terms may cause unreasonably large errors for rigid body movements.

- Integration due to Milford–Schnobrich (M&S)

Milford and Schnobrich [11] have stressed the significance of the terms with **B**. Belytschko et. al. [6] also emphasized the curvature terms; they demonstrated for the benchmark "pre–twisted beam" that an error up to 70 percent may occur (see example 3.1). The remedy in [11] is a series expansion of $\mathbf{J}^{-1}$ cut off after the linear term. This is equivalent to the approximation:

$$\mathbf{Z}^{-1} \approx \mathbf{I} + \theta^3\mathbf{B} \tag{1.28}$$

so that the strain tensor is:

$$\begin{aligned}\mathbf{E} &\approx \frac{1}{2}\left((\mathbf{I} + \theta^3\mathbf{B})\mathbf{u}_{;r}^T + \mathbf{u}_{;r}(\mathbf{I} + \theta^3\mathbf{B}) + (\mathbf{I} + \theta^3\mathbf{B})\mathbf{u}_{;r}^T\mathbf{u}_{;r}(\mathbf{I} + \theta^3\mathbf{B})\right)\\ &= \frac{1}{2}\left(\mathbf{u}_{;r}^T + \mathbf{u}_{;r} + \mathbf{u}_{;r}^T\mathbf{u}_{;r} + \theta^3(\mathbf{B}\mathbf{u}_{;r}^T + \mathbf{u}_{;r}\mathbf{B})\right.\\ &\qquad \left. + \theta^3(\mathbf{B}\mathbf{u}_{;r}^T\mathbf{u}_{;r} + \mathbf{u}_{;r}^T\mathbf{u}_{;r}\mathbf{B}) + (\theta^3)^2\mathbf{B}\mathbf{u}_{;r}^T\mathbf{u}_{;r}\mathbf{B}\right)\end{aligned} \tag{1.29}$$

Using $\mathbf{u}_{;r} = \mathbf{v}_{;r} + \mathbf{w} \otimes \mathbf{a}_3 + \theta^3\mathbf{w}_{;r}$ and neglecting the terms quadratic and higher in θ^3 we get ($\mathbf{B}\mathbf{a}_3 = \mathbf{0}$):

$$\begin{aligned}\mathbf{E}^{\approx\approx} = \frac{1}{2}\Big(&\mathbf{v}_{;r}^T + \mathbf{v}_{;r} + \mathbf{w} \otimes \mathbf{a}_3 + \mathbf{a}_3 \otimes \mathbf{w}\\ &+ \mathbf{v}_{;r}^T\mathbf{v}_{;r} + \mathbf{v}_{;r}^T(\mathbf{w} \otimes \mathbf{a}_3) + (\mathbf{a}_3 \otimes \mathbf{w})\mathbf{v}_{;r} + (\mathbf{a}_3 \otimes \mathbf{w})(\mathbf{w} \otimes \mathbf{a}_3)\\ &+ \theta^3\left(\mathbf{w}_{;r}^T + \mathbf{w}_{;r} + \mathbf{w}_{;r}^T\mathbf{v}_{;r} + \mathbf{v}_{;r}^T\mathbf{w}_{;r} + \mathbf{w}_{;r}^T(\mathbf{w} \otimes \mathbf{a}_3) + (\mathbf{a}_3 \otimes \mathbf{w})\mathbf{w}_{;r}\right)\\ &+ \theta^3\left(\mathbf{B}\mathbf{v}_{;r}^T + \mathbf{v}_{;r}\mathbf{B} + \mathbf{B}\mathbf{v}_{;r}^T\mathbf{v}_{;r} + \mathbf{B}\mathbf{v}_{;r}^T(\mathbf{w} \otimes \mathbf{a}_3) + \mathbf{v}_{;r}^T\mathbf{v}_{;r}\mathbf{B} + (\mathbf{a}_3 \otimes \mathbf{w})\mathbf{v}_{;r}\mathbf{B}\right)\Big)\end{aligned} \tag{1.30}$$

Since $\mathbf{E}^{\approx}$, $\mathbf{E}^{\approx\approx}$ and $\hat{\mathbf{E}}$ do not differ in the constant terms, the part linear in θ^3 will be closer looked at:

$$\mathbf{E}_L^{\approx} = \frac{1}{2}\left(\mathbf{w}_{;r}^T + \mathbf{w}_{;r} + \mathbf{w}_{;r}^T\mathbf{v}_{;r} + \mathbf{v}_{;r}^T\mathbf{w}_{;r} + \mathbf{w}_{;r}^T(\mathbf{w} \otimes \mathbf{a}_3) + (\mathbf{a}_3 \otimes \mathbf{w})\mathbf{w}_{;r}\right) \tag{1.31}$$

$$\hat{\mathbf{E}}_L = \mathbf{E}_L^{\approx} + \frac{1}{2}\left(-\mathbf{v}_{;r}^T\mathbf{B} - \mathbf{B}\mathbf{v}_{;r} - \mathbf{B}(\mathbf{w} \otimes \mathbf{a}_3) - (\mathbf{a}_3 \otimes \mathbf{w})\mathbf{B}\right) \tag{1.32}$$

$$\mathbf{E}_L^{\approx\approx} = \mathbf{E}_L^{\approx} + \frac{1}{2}\Big(\mathbf{B}\mathbf{v}_{;r}^T + \mathbf{v}_{;r}\mathbf{B} + \mathbf{B}\mathbf{v}_{;r}^T\mathbf{v}_{;r} + \mathbf{B}\mathbf{v}_{;r}^T(\mathbf{w}\otimes\mathbf{a}_3) + \mathbf{v}_{;r}^T\mathbf{v}_{;r}\mathbf{B} + (\mathbf{a}_3\otimes\mathbf{w})\mathbf{v}_{;r}\mathbf{B}\Big) \quad (1.33)$$

Equations (1.32) and (1.33) differ only slightly because:

$$\| \mathbf{E}_L^{\approx\approx} - \hat{\mathbf{E}}_L \| = \| \mathbf{B}\Big(\mathbf{v}_{;r}^T + \mathbf{v}_{;r} + (\mathbf{w}\otimes\mathbf{a}_3) + \mathbf{v}_{;r}^T\mathbf{v}_{;r} + \mathbf{v}_{;r}^T(\mathbf{w}\otimes\mathbf{a}_3)\Big) + \Big(\mathbf{v}_{;r}^T + \mathbf{v}_{;r} + \mathbf{a}_3\otimes\mathbf{w} + \mathbf{v}_{;r}^T\mathbf{v}_{;r} + (\mathbf{a}_3\otimes\mathbf{w})\mathbf{v}_{;r}\Big)\mathbf{B} \|$$

$$= \| \mathbf{B}\hat{\mathbf{E}}_C + \hat{\mathbf{E}}_C\mathbf{B} \| \leq \frac{2}{R}\eta \quad \text{with} \quad O(\| \hat{\mathbf{E}}_C \|) \leq O(\| \hat{\mathbf{E}} \|) = \eta \quad (1.34)$$

Hence, the relative difference between $\mathbf{E}_{C+L}^{\approx\approx}$ and $\hat{\mathbf{E}}_{C+L}$ is of the order of $\theta^3\frac{2}{R} \leq \frac{h}{R}$. For the special case of a geometrically linear theory we have:

$$\hat{\mathbf{E}}_C^{lin} = (\mathbf{v}_{;r} + \mathbf{w}\otimes\mathbf{a}_3)^{sym} = \frac{1}{2}\Big(\mathbf{v}_{;r}{}^T + \mathbf{v}_{;r} + \mathbf{w}\otimes\mathbf{a}_3 + \mathbf{a}_3\otimes\mathbf{w}\Big) \quad (1.35)$$

$\hat{\mathbf{E}}_L$ can now be expressed as:

$$\hat{\mathbf{E}}_L = \mathbf{E}_L^{\approx} - \frac{1}{2}\Big(\mathbf{B}\hat{\mathbf{E}}_C^{lin} + \hat{\mathbf{E}}_C^{lin}\mathbf{B}\Big) - \frac{1}{2}\Big(\mathbf{B}(\mathbf{v}_{;r} + \mathbf{w}\otimes\mathbf{a}_3)^{skew} - (\mathbf{v}_{;r} + \mathbf{w}\otimes\mathbf{a}_3)^{skew}\,\mathbf{B}\Big)$$

$$\approx \mathbf{E}_L^{\approx} - \frac{1}{2}\Big(\mathbf{B}(\mathbf{v}_{;r} + \mathbf{w}\otimes\mathbf{a}_3)^{skew} - (\mathbf{v}_{;r} + \mathbf{w}\otimes\mathbf{a}_3)^{skew}\,\mathbf{B}\Big) \quad (1.36)$$

The skew symmetric part $(\mathbf{v}_{;r} + \mathbf{w}\otimes\mathbf{a}_3)^{skew}$ represents the rigid modes in a linear theory and thus explains the noticeable error, e.g. for the "pre–twisted beam". In order to judge the significance of the extra terms $-\mathbf{B}(\mathbf{w}\otimes\mathbf{a}_3) - (\mathbf{a}_3\otimes\mathbf{w})\mathbf{B}$ in $\hat{\mathbf{E}}_L$ the components of $\mathbf{E}_L^{\approx}$ in a local basis are investigated:

$$\beta_{\alpha\beta}^{\approx} = \mathbf{a}_\alpha\mathbf{E}_L^{\approx}\mathbf{a}_\beta = \frac{1}{2}(\mathbf{w}_{,\alpha}\mathbf{a}_\beta + \mathbf{w}_{,\beta}\mathbf{a}_\alpha + \mathbf{w}_{,\alpha}\mathbf{v}_{,\beta} + \mathbf{w}_{,\beta}\mathbf{v}_{,\beta}) \quad (1.37)$$

$$\beta_{\alpha 3}^{\approx} = \mathbf{a}_\alpha\mathbf{E}_L^{\approx}\mathbf{a}_3 = \mathbf{a}_\alpha\frac{1}{2}\Big(\mathbf{w}_{;r}^T\mathbf{a}_3 + \mathbf{w}_{;r}^T\mathbf{w}\Big) \overset{\text{with A4}}{=} \frac{1}{2}(\mathbf{a}_\alpha\mathbf{B}\mathbf{w}) \neq 0 \;;\; \beta_{33} = 0 \quad (1.38)$$

Although necessarily the curvatures $\beta_{\alpha 3}^{\approx}$ do not vanish despite assumption A4. In a local formulation $\beta_{\alpha 3}^{\approx}$ is a priori not considered. This automatically avoids the above indicated error. Contrary to $\mathbf{E}_L^{\approx}$ $\hat{\beta}_{\alpha 3}$ does vanish in $\hat{\mathbf{E}}_L$ (see also equation (1.24)).

$$\hat{\beta}_{\alpha 3} = \beta_{\alpha 3}^{\approx} + \mathbf{a}_\alpha\frac{1}{2}(-\mathbf{B}\mathbf{w}) = 0 \quad (1.39)$$

If global cartesian coordinates are used $\beta_{\alpha 3}^{\approx}$ cannot be isolated and individually neglected; therefore the extra terms $-\mathbf{B}(\mathbf{w}\otimes\mathbf{a}_3)-(\mathbf{a}_3\otimes\mathbf{w})\mathbf{B}$ in $\hat{\mathbf{E}}_L$ need to be taken into account (see example 3.2).

- Integration due to Irons

Irons derived 1973 [10] the bending strains of his "Semi–Loof" shell element – based on local cartesian coordinates – from the strains at the upper and lower shell surfaces. If the discretization already introduced by Irons at this stage is filtered out, the resulting terms coincide exactly with the bending strains $\hat{\mathbf{E}}_L$ of the shell theory. At that time he obviously was not aware of the relevance of this part because he says: "In the solution to a real problem it is probably an order of magnitude smaller than the orthodox bending terms. It is questionable therefore whether it should be included, ...". And with respect to the derivation he mentions: "There is more engineering intuition than mathematics in this step." Crisfield [9] follows a similar idea to obtain the curvature terms for a 2–dimensional beam element. As a consequence the rigid body movements can be described without defect.

- Integration due to Stanley

Finally, we like to refer to a paper of Stanley [22] in which $\mathbf{Z}^{-1}$ is completely taken into consideration and $\hat{\mathbf{E}}_Q$ is not neglected. In so far the results are identical to those obtained by a numerically integrated version. But it should be noted that not all terms need to be included if a relative error of the order of h/R is tolerated and the analysis is restricted to small strains as well as slightly varying thickness.

1.3 Conclusion for Explicit Integration

It has been shown how the explicit integration across the thickness can be classified under the aspect of a shell theory. The strain tensor $\hat{\mathbf{E}}$ ought to be preferred since consistent shell theory and explicit integration coincide. Then a series expansion of $\mathbf{J}^{-1}$ and $\mathbf{Z}^{-1}$, resp., is unnecessary; the derivation is mathematically consistent, unnessary terms are not carried along.
The only difference between the concept of degeneration and shell theory based finite elements remains in the kind of discretization. The way described by Simo [17]–[20] is also preferred here: It is started from the strain definitions of shell theory, equations (1.23) and (1.24), but a discretization typical for the degeneration concept is introduced (see also [12]). At this point we would like to stress that a 6–parameter theory may be useful in the case of rather thick structures and for large strain problems [20]. An explicit integration in the thickness direction referring also to higher order stress resultants may be computationally advantageous even in that cases, especially for layered structures.

2 COMPARISON OF DISCRETIZATION

2.1 Shell Theory

It is common in shell theory that the initial geometry is analytically defined applying the classical means of differential geometry. However, any other geometrical parametrization is possible, for example the isoparametric interpolation. The often heard argument that the dramatic sensitivity with respect to initial geometrical imperfections in shell stability is very much influenced by this problem is not shared herein. The errors introduced by an approximation of initial and deformed geometry will be of the same order of magnitude. The deformed geometry is indirectly interpolated through the displacement field.

$$\mathbf{u} = \bar{\mathbf{x}} - \mathbf{x} = \mathbf{v} + \theta^3 \mathbf{w} = \mathbf{v} + \theta^3 (\mathbf{R}(\mathbf{s}) - \mathbf{I})\mathbf{a}_3 \qquad (2.1)$$

Usually the components of $\mathbf{v}$ and $\mathbf{s}$ with respect to the local basis $\mathbf{a}_i$ are discretized:

$$\mathbf{v} \approx \mathbf{v}_h = (\sum_{k=1}^{n} N^k \ (v^i)^k\ )\ \mathbf{a}_i \quad ; \quad \mathbf{s} \approx \mathbf{s}_h = (\sum_{k=1}^{n} N^k \ (s^\alpha)^k\ )\ \mathbf{a}_\alpha \qquad (2.2)$$

This kind of interpolation automatically causes difficulties with the rigid body modes of curved structures which, in turn, diminish with a finer discretization. Alternative kinds of interpolation are possible, e.g. with respect to other base vectors.

2.2 Concept of Degeneration

Usually in the degeneration approach initial and deformed geometry are equally interpolated allowing to describe rigid body movements exactly:

$$\mathbf{x} \approx \mathbf{x}_h = \sum_{k=1}^{n} N^k \ (\mathbf{r} + \theta^3 \mathbf{a}_3)^k \quad ; \quad \bar{\mathbf{x}} \approx \bar{\mathbf{x}}_h = \sum_{k=1}^{n} N^k \ (\bar{\mathbf{r}} + \theta^3 \bar{\mathbf{a}}_3)^k \qquad (2.3)$$

The displacement field follows:

$$\mathbf{u} \approx \mathbf{u}_h = \sum_{k=1}^{n} N^k \ (\mathbf{v} + \theta^3 \mathbf{w})^k \qquad (2.4)$$

Assuming constant shell thickness leads to:

$$\mathbf{v} \approx \mathbf{v}_h = (\sum_{k=1}^{n} N^k \ (v^i)^k\ )\ \mathbf{i}_i \quad ; \quad \mathbf{w} \approx \mathbf{w}_h = (\sum_{k=1}^{n} N^k \ (w^i)^k\ )\ \mathbf{i}_i \qquad (2.5)$$

and

$$\bar{\mathbf{r}} \approx \bar{\mathbf{r}}_h = \sum_{k=1}^{n} N^k \ \bar{\mathbf{r}}^k \quad ; \quad \bar{\mathbf{a}}_3 \approx \bar{\mathbf{a}}_{3_h} = \sum_{k=1}^{n} N^k \ \bar{\mathbf{a}}_3 \qquad (2.6)$$

respectively.

This kind of interpolation causes the defect that the assumption $|\bar{\mathbf{a}}_3| = 1$ is violated at each Gauss point. The element turns out to be too flexible, in particular for a coarse mesh and low order elements. Better results are obtained if $\mathbf{s}$ instead of $\mathbf{w} = (\mathbf{R}(\mathbf{s}) - \mathbf{I})\mathbf{a}_3$ is interpolated [7], [19] (see Example 3.3).

$$\mathbf{s} \approx \mathbf{s}_h = (\sum_{k=1}^{n} N^k \ (s^i)^k\ )\ \mathbf{i}_i \tag{2.7}$$

However, the computation of the tangent stiffness matrix will be more expensive. Applying equation (2.7) also leads to the fact that $\mathbf{s}$ is not orthogonal to $\mathbf{a}_3$ at the Gauss points.

Again this defect diminishes when the mesh is refined or higher order elements are chosen.

An alternative scheme is used in [18]: $\bar{\mathbf{a}}_{3h}$ is normalized enforcing the condition $|\bar{\mathbf{a}}_3| = 1$.

3 NUMERICAL EXAMPLES

3.1 Twisted beam [6] – geometrically linear analysis

This often described benchmark was chosen in [6] and is repeated here. It demonstrates the significant error arising if the two terms $-\mathbf{v}_{;r}^T\mathbf{B} - \mathbf{B}\mathbf{v}_{;r}$ are neglected in $\hat{\mathbf{E}}_L$.

Young's modulus $E=29*10^6$
Poisson's ratio $\nu=0.22$
Length = 12 Thickness h
Width = 1.1 Load P
Twist = 90°

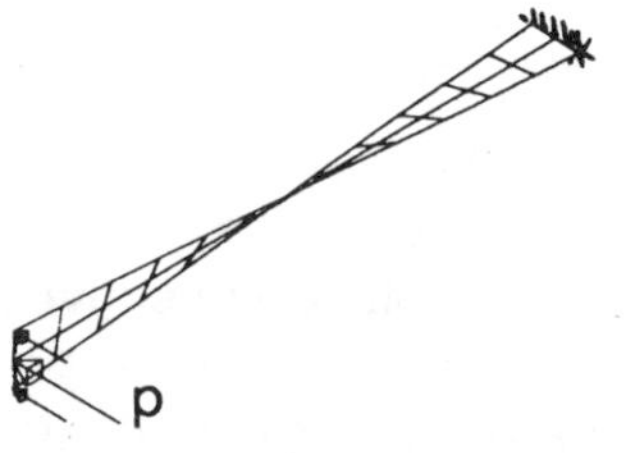

Mesh: 2x12 four–node elements, reduced integrated (1x1)

	h=0.32,P=1	h=0.032,P=10^{-6}
explicit integr. due to Z&T&T (local):	0.2419×10^{-2}	0.2208×10^{-2}
explicit integration with $\hat{\mathbf{E}}$:	0.1887×10^{-2}	0.1281×10^{-2}
exact:	0.1754×10^{-2}	0.1294×10^{-2}

Fig. 2: Twisted beam

3.2 Arch – geometrically linear analysis

This example shows the error if the terms $-\mathbf{B}(\mathbf{w} \otimes \mathbf{a}_3) - (\mathbf{a}_3 \otimes \mathbf{w})\mathbf{B}$ are missing in eqn. (6.35). As mentioned above, this defect is automatically corrected in most local formulations, but does not vanish in formulations referring stresses and strains to global cartesian coordinates.

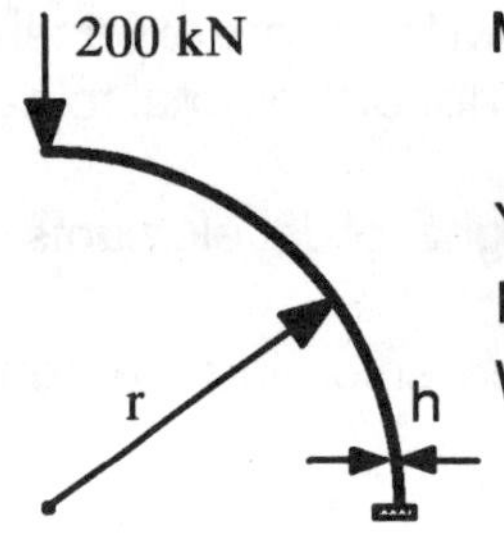

Mesh: 4 nine–node elements, reduced integrated (2x2)
(nodes in width direction are coupled)

Young's modulus $E = 2.1\ 10^5\ \text{N/mm}^2$
Poisson's ration $\nu = 0$
Width $b = 1000$ mm ; Radius $r = 5000$ mm
Thickness $h = 50$ mm

displacement in load direction	
explicit integration due to Z&T&T (global):	3.834 mm
explicit integration with $\hat{\mathbf{E}}$:	5.714 mm
exact:	5.714 mm

Fig. 3: Arch

3.3 Cantilever beam – geometrically nonlinear analysis

This simple example demonstrates too weak behavior if the director or respectively **w** is interpolated, so that the inextensibility requirement of the director is violated. An interpolation of **s** gives much better results [7], [19], but leads to a more elaborate formulation. 1-point reduced integrated elements are used. The nodes in width direction are coupled.
Using Hooke's material law between the Biot stresses and engineering strains, the applied moment leads to a tip rotation of 360.00° [7]. However if the material law uses Green–Lagrange strains and 2 nd Piola–Kirchhoff stresses the exact result is 360.96° for these properties.

Young's modulus E = 21000
Poisson's ratio ν=0.0
Length l = 100
Width b = 1.0
Thickness h = 2.0
Moment M = 879.6456

tip rotation:	**s** interpolated	**w** interpolated
10 four–node elements	361.94°	392.07°
20 four–node elements	361.94°	368.30°
exact:	360.96°	

Fig. 4: Cantilever beam

4 APPENDIX

A.1 Additional bending terms in local, orthogonal cartesian coordinates

The local base vectors at the reference surface $\theta^3 = 0$ are denoted by $\mathbf{e}_1,\mathbf{e}_2,\mathbf{e}_3$ and the coordinates by y^1,y^2,y^3. $\mathbf{e}_3$ is parallel to $\mathbf{a}_3$.

$$\hat{\beta}_{AB} = \mathbf{e}_A^T \hat{\mathbf{E}}_L \mathbf{e}_B = \beta_{\widetilde{\widetilde{AB}}} + \frac{1}{2}\left(\frac{\partial \mathbf{v}}{\partial y^A}\frac{\partial \mathbf{a}_3}{\partial y^B} + \frac{\partial \mathbf{a}_3}{\partial y^A}\frac{\partial \mathbf{v}}{y^B}\right) \qquad A,B = 1,2$$

A.2 Additional bending terms in global, orthogonal cartesian coordinates

$$\hat{\beta}_{ij} = \mathbf{i}_i^T \hat{\mathbf{E}}_L \mathbf{i}_j = \beta_{\widetilde{\widetilde{ij}}} + \frac{1}{2}\left[\frac{\partial \mathbf{v}}{\partial x^i}\frac{\partial \mathbf{a}_3}{\partial x^j} + \frac{\partial \mathbf{a}_3}{\partial x^i}\frac{\partial \mathbf{v}}{\partial x^j} + \frac{\partial \mathbf{a}_3}{\partial x^i}\mathbf{w}\frac{\partial \theta^3}{\partial x^j} + \frac{\partial \mathbf{a}_3}{\partial x^j}\mathbf{w}\frac{\partial \theta^3}{\partial x^i}\right]_{\theta^3 = 0}$$

5 NOTATION

Symbols

scalar	a
vector	$\mathbf{a} = a^i \mathbf{g}_i = a_i \mathbf{g}^i$
tensor of second order or matrix	$\mathbf{A} = A^{ij}\mathbf{g}_i \otimes \mathbf{g}_j = \ldots.$
tensor of fourth order	$\mathcal{A} = A^{ijkl}\mathbf{g}_i \otimes \mathbf{g}_j \otimes \mathbf{g}_k \otimes \mathbf{g}_l = \ldots.$

indices: $\alpha \in [1,2]$ $\quad$ $i \in [1,2,3]$

Multiplication:

scalar product :	$\mathbf{B} = a\mathbf{A}$	$B^{ij} = aA^{ij}$
Multiplication with summation over the last index of the first factor and the first index of the second factor.	$c = \mathbf{ab}$ $\mathbf{b} = \mathbf{Aa}$ $\mathcal{B} = \mathbf{A}\mathcal{C}$	$c = a^i b_j = \ldots$ $b^i = A^{ij}a_j = \ldots$ $B^{ijkl} = A^{im}C_m^{\cdot jkl} = \ldots$
Multiplication with summation over the last two indices of the first factor and the first two indices of the second factor.	$c = \mathbf{A} \cdot \mathbf{B}$ $\mathbf{B} = \mathcal{C} \cdot \mathbf{A}$ $\mathbf{B} = \mathbf{A} \cdot \mathcal{C}$	$c = A^{ij}B_{ij} = \ldots$ $B^{ij} = C^{ijkl}A_{kl} = \ldots$ $B_{ij} = A^{nm}C_{nmij} = \ldots$
Multliplication without summation.	$\mathbf{B} = \mathbf{a} \otimes \mathbf{b}$ $\mathcal{C} = \mathbf{A} \otimes \mathbf{B}$	$B^{ij} = a^i b^j = \ldots$ $C^{ijkl} = A^{ij}B^{kl} = \ldots$

Differentiation:

partial derivative with respect to θ^i $\quad (\,)_{,i} = \dfrac{\partial(\,)}{\partial \theta^i}$

covariant derivative $\quad (\,)|_i$

$$
\begin{aligned}
\mathbf{a}_{,\mathbf{x}} = \frac{\partial \mathbf{a}}{\partial x^j} \otimes \mathbf{i}^j &= \frac{\partial \mathbf{a}}{\partial \theta^k}\frac{\partial \theta^k}{\partial x^j} \otimes \mathbf{i}^j = \mathbf{a}_{,k} \otimes \mathbf{g}^k \\
&= (a_m \mathbf{g}^m)_{,k} \otimes \mathbf{g}^k \\
&= (a_{m,k}\,\mathbf{g}^m + a_m \mathbf{g}^m{}_{,k}) \otimes \mathbf{g}^k \\
&= (a_{m,k} - a_n \Gamma^n_{mk})\, \mathbf{g}^m \otimes \mathbf{g}^k \\
&= a_m|_k\, \mathbf{g}^m \otimes \mathbf{g}^k \\
&= a_m|_k\, \mathbf{g}^m \otimes \mathbf{g}^k \quad = \nabla_{\mathbf{x}}\mathbf{a}
\end{aligned}
$$

Example: $\quad \mathbf{w} = \mathbf{w}(\theta^1, \theta^2) \qquad (\theta^3 \mathbf{w})_{;\mathbf{r}} = ?$

$$
\begin{aligned}
\nabla_{\mathbf{x}}(\theta^3 \mathbf{w}) = (\theta^3 \mathbf{w})_{,\mathbf{x}} = \frac{\partial(\theta^3 \mathbf{w})}{\partial x^j} \otimes \mathbf{i}^j &= \frac{\partial \theta^3}{\partial x^j}\mathbf{w} \otimes \mathbf{i}^j + \theta^3 \frac{\partial \mathbf{w}}{\partial x^j} \otimes \mathbf{i}^j \\
&= \mathbf{w} \otimes \mathbf{g}^3 + \theta^3 \mathbf{w}_{,\mathbf{x}} \\
&= \mathbf{w} \otimes \mathbf{g}^3 + \theta^3 \mathbf{w}_{,\alpha} \otimes \mathbf{g}^\alpha
\end{aligned}
$$

$$
\begin{aligned}
(\theta^3 \mathbf{w})_{;\mathbf{r}} &= (\theta^3 \mathbf{w})_{,\mathbf{x}}\, \mathbf{x}_{;\mathbf{r}} = (\mathbf{w} \otimes \mathbf{g}^3 + \theta^3 \mathbf{w}_{,\alpha} \otimes \mathbf{g}^\alpha)\mathbf{Z} \\
&= \mathbf{w} \otimes \mathbf{a}^3 + \theta^3 \mathbf{w}_{,\alpha} \otimes \mathbf{a}^\alpha = \mathbf{w} \otimes \mathbf{a}^3 + \theta^3 \mathbf{w}_{;\mathbf{r}}
\end{aligned}
$$

6 REFERENCES

1. Ahmad, S.; Irons, B.M.; Zięnkiewicz, O.C.:Curved thick shell and membrane elements with particular reference to axi-symmetric problems. Proc. 2nd Conf. Matrix Methods in Structural Mechanics.Wright-Patterson A.F. Base, Ohio 1968.
2. Ahmad, S.; Irons, B.M.; Zienkiewicz, O.C.: Analysis of thick and thin shell structures by curved finite elements. Int. J. Num. Meth. Eng., 2 (1970) 419-451.
3. Basar, Y.: Eine konsistente Theorie für Flächentragwerke endlicher Verformungen und deren Operatordarstellung auf variationstheoretischer Grundlage. ZAMM 66, 7 (1986) 297-308.
4. Basar, Y.; Krätzig, W.B.: Mechanik der Flächentragwerke. Friedr. Vieweg & Sohn, Braunschweig 1985.
5. Basar, Y.; Ding, Y.: Theory and finite element formulation for shell structures undergoing finite rotations. Advances in the theory of plates and shells (ed. Voyiadjis, G.Z. and Karamanlidis, D.) Amsterdam 1990.
6. Belytschko, T.; Wong, B. L.: Assumed strain stabilisation procedure for the 9-node Lagrange shell element. Int. J. Num. Meth. Eng., 28 (1989) 385-414.
7. Büchter, N.: Ein nichtlineares degeneriertes Balkenelement (2-D) unter Verwendung von Biotspannungen. Mitteilung Nr. 8 des Instituts für Baustatik, Universität Stuttgart 1989.
8. Büchter, N.: Zusammenführung von Degenerationskonzept und Schalentheorie bei endlichen Rotationen. Dissertation. Bericht Nr. 14, Instituts für Baustatik, Universität Stuttgart, Stuttgart 1992.
9. Crisfield, M. A.: Explicit integration and the isoparametric arch and shell elements. Communications in applied numerical methods., 2 (1986) 181-187.
10. Irons, B.M.: The semiloof shell element. Lecture notes. Int . Res. Seminar on Theory and Application of Finite Elements. Univ. of Calgary, Canada 1973.
11. Milford, R.V.; Schnobrich, W. C.: Degenerated isoparametric finite elements using explicit integration. Int. J. Num. Meth. Eng., 23, (1986) 133-154.
12. Parisch, H.: An investigation of a finite rotation four node shell element. Int. J. Num. Meth. Eng., 31 (1991) 127-150.
13. Pietraszkiewicz, W.: Geometrically nonlinear theories of thin elastic shells. Advances in Mechanics, Vol. 12 No. 1 (1989).
14. Ramm, E.: Geometrisch nichtlineare Elastostatik und finite Elemente. Habilitation. Bericht Nr. 76-2, Institut für Baustatik, Universität Stuttgart 1976.

15. Ramm, E.: A plate/shell element for large deflections and rotations. US – Germany Symp. on "Formulations and computational algorithms in finite element analysis", MIT 1976, MIT–Press (1977).
16. Reissner, E.: On one–dimensional finite–strain beam theory: the plane problem. ZAMP, 23 (1972).
17. Simo, J.C.; Fox, D. D.: On a stress resultant geometrically exact shell model. Part I: Formulation and optimal parametrization. Comp. Meth. Appl. Mech. Eng. 72 (1989) 267–304.
18. Simo, J.C.; Fox, D.D.; Rifai, M.S.: On a stress resultant geometrically exact shell model. Part II: The linear theory. Comp. Meth. Appl. Mech. Eng., 73 (1989) 53–62.
19. Simo, J.C.; Fox, D.D.; Rifai, M.S.: On a stress resultant geometrically exact shell model. Part III: Computational aspects of the nonlinear theory. Comp. Meth. Appl. Mech. Eng. 79 (1990) 21–70.
20. Simo, J.C.; Rifai, M.S.; Fox, D.D.: On a stress resultant geometrically exact shell model. Part IV: Variable thickness shells with through–the–thickness stretching. Comp. Meth. Appl. Mech. Eng., 81 (1990) 53–91.
21. Stander, N.; Matzenmiller, A.; Ramm, E.: An assessment of assumed strain methods in finite rotation shell analysis. Engineering Computations, 6 (1989) 57–66.
22. Stanley, G.M.: Continuum- based shell elements. PH. D. Thesis, Stanford Univ. 1985.
23. Stanley. G.M.; Park, K.C.; Hughes, T.J.R.: Continuum–based resultant shell elements. Finite element methods for plate and shell structures, ed. T.J.R. Hughes et. al., Pineridge Press, Swansea 1986, 1–45.
24. Zienkiewicz, O.C,; Taylor, R.L.; Too, J.M.: Reduced integration technique in general analysis of plates and shells. Int. J. Num. Meth. Eng., 3 (1971) 275–290.

AN ASSESSMENT OF HYBRID-MIXED FOUR-NODE SHELL ELEMENTS

U. Andelfinger and E. Ramm
University of Stuttgart, Stuttgart, Germany

1 INTRODUCTION

The most popular shell finite elements are shear flexible C^0–continuous displacement models. These elements which are easy to formulate usually give displacements that are too small and in the extreme case (for instance thin plate or thin shell) exhibit severe stiffening, known as locking.

Locking can be avoided with hybrid–mixed formulations in which additional assumptions for strains or stresses are made. These extra variables are discontinuous from element to element, so that the independent degrees of freedom can be eliminated on the element level. Therefore, a hybrid–mixed formulation always leads to an element stiffness matrix.

When assuming strain or stress fields, however care must be taken to avoid instabilities. Necessary conditions (ellipticity– and Babuska–Brezzi–conditions) to get stable elements will not be given here; they are discussed for instance in [2] or [4].

After a careful inspection of the different locking phenomena, three hybrid–mixed formulations are summarized, the Hellinger – Reissner stress – displacement formulation, the so–called assumed strain method [5], [6], [7], [8] and the enhanced assumed strain method [12]. Special versions of four–node shell elements are presented in Chapter 4. The shell formulation is derived from the three–dimensional degenerated continuum using a local coordinate formulation. In Chapter 5 the elements are compared to each other through numerical tests.

2 THE LOCKING PHENOMENA

A severe problem for most finite elements is the locking which can appear as shear, membrane or volumetric locking. The three types of locking are listed in Table 2.1. The condition which leads to a high stiffness ratio is given in the second column, while in the third one the corresponding constraints to avoid locking are listed. The table clearly shows the similarity of the different locking phenomena.

typ of locking	condition for high stiffness ratio	constraint to avoid locking
shear locking	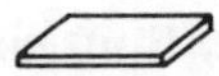 thin plate $\frac{k_{shear}}{k_{bend}}$ is high	for bending modes $\gamma_i = 0$
membrane locking	 thin shell $\frac{k_{memb.}}{k_{bend.}}$ is high	for bending modes $\varepsilon_{ij} = 0$
volumetric locking	ν close to 0,5 (plane strain, 3D, axisymm.) $\frac{k_{vol.}}{k_{dev}}$ is high	for deviatoric modes $\varepsilon_{ii} = 0$

Table 2.1: The locking phenomena

Shear locking can occur in thin plates where the shear stiffnesses are a lot higher than the bending stiffnesses (for an element with length L and thickness t, the ratio $\mathbf{k}_{shear}/\mathbf{k}_{bend.}$ is proportional to L^2/t^2). If the bending modes contain parasitic shear strains, they behave just as the shear modes, that means far too stiff if compared to their desired performance. To avoid shear locking, the constraint $\gamma_\xi = \gamma_\eta = 0$ has to be fulfilled for the bending modes.
The same arguments are valid for membrane and volumetric locking. Membrane locking can occur in thin shells, where the membrane stiffnesses are a lot higher than the bending stiffnesses. Again, if for the bending modes the constraint $\epsilon_{ij} = 0$ is not satisfied, leading to parasitic membrane strains, these bending modes behave as stiff as the membrane modes which in turn for bending dominated problems results in displacements that are far too small.
Volumetric locking can appear in three-dimensional, in plane strain or axisymmetric elements (not in plane stress elements), where a Poisson ratio close to 0,5 leads to a nearly incompressible situation, resulting in very high volumetric stiffnesses compared to the deviatoric ones[1]. If for the deviatoric modes, the constraint $\epsilon_{ii} = 0$ (no volumetric strains in the deviatoric modes) is not fulfilled, the deviatoric modes start to lock.
In all three cases we are facing a high stiffness ratio, either due to geometric conditions or because of an (almost) incompressible material. If the soft modes (bending or deviatoric modes) are spoiled by parasitic strains, they behave far too stiff. This

1. See also the flow situation in J_2-plasticity

always happens, if the strain interpolation is too high, compared to the chosen displacement field. To avoid locking, we need balanced elements, where the strain fields fit to the displacement fields, so that the displacement modes are not overloaded with strains.
It is important to note that the discussed types of locking have nothing to do with ill-conditioning (sometimes called numerical locking, which however is a misleading term). Shear, membrane and volumetric locking would occur even if a computer had an infinite number of digits.

3 HYBRID-MIXED FORMULATIONS

In order to avoid locking one has to leave the one-field principle of potential energy. To derive balanced elements, three possibilities are discussed. The first one is based on the Hellinger – Reissner functional where stresses and displacements are interpolated. Next we discuss the so-called assumed natural strain method. Starting from the principle of Hu – Washizu, locking-free elements are derived through a reduction of the compatible strain fields. In Chapter 3.3 the enhanced assumed strain method is considered which again is based on the Hu – Washizu functional but uses enriched strain fields instead.

3.1 Hybrid-mixed stress-displacement approximation

Hybrid-mixed stress-displacement elements are based on the principle of Hellinger and Reissner (HR). Considering as usual that the total internal energy is the sum of the contributions of all elements, the internal energy of one element is given through

$$U_e = \int_{\Omega_e} -\frac{1}{2}\boldsymbol{\sigma}^T \mathbf{C}^{-1}\boldsymbol{\sigma} + \boldsymbol{\sigma}^T \overset{D}{\mathbf{B}}\mathbf{u}\; d\Omega \tag{3.1}$$

In (3.1) $\boldsymbol{\sigma}$ stands for the stresses, $\mathbf{C}^{-1}$ is the inverse constitutive tensor, $\overset{D}{\mathbf{B}}$ is the differential operator linking the displacements $\mathbf{u}$ to the strains and Ω_e is the volume of one element. While the displacements $\mathbf{u}$ are approximated continuously as

$$\mathbf{u} = \mathbf{N}\mathbf{d} \tag{3.2}$$

the stress fields are assumed discontinuously:

$$\boldsymbol{\sigma} = \mathbf{P}\boldsymbol{\beta} \tag{3.3}$$

Hereby $\mathbf{N}$ are the usual C^0-continuous shape functions, $\mathbf{P}$ contains the polynomial terms for the stress approximation and $\mathbf{d}$ and $\boldsymbol{\beta}$ are the vectors of the discrete unknowns. Using (3.2) and (3.3) in (3.1) gives

$$U_e = -\frac{1}{2}\boldsymbol{\beta}^T \mathbf{H}\boldsymbol{\beta} + \boldsymbol{\beta}^T \mathbf{G}\mathbf{d} \tag{3.4}$$

where

$$\mathbf{H} = \int_{\Omega_e} \mathbf{P}^T \mathbf{C}^{-1} \mathbf{P} \, d\Omega \tag{3.5}$$

$$\mathbf{G} = \int_{\Omega_e} \mathbf{P}^T \overset{D}{\mathbf{B}} \mathbf{N} \, d\Omega \;=\; \int_{\Omega_e} \mathbf{P}^T \mathbf{B} \, d\Omega \tag{3.6}$$

Variation with respect to the unknowns $\boldsymbol{\beta}$ and **d** leads to the following equations:

$$-\mathbf{H}\boldsymbol{\beta} + \mathbf{G}\mathbf{d} = \mathbf{0} \tag{3.7}$$

$$\mathbf{G}^T\boldsymbol{\beta} = \mathbf{R} \tag{3.8}$$

where **R** stands for the applied loads. Provided that **H** is positive definite, the stress parameters $\boldsymbol{\beta}$ can be eliminated through

$$\boldsymbol{\beta} = \mathbf{H}^{-1}\mathbf{G}\mathbf{d} \tag{3.9}$$

which ends in an element stiffness matrix

$$\mathbf{k} = \mathbf{G}^T\mathbf{H}^{-1}\mathbf{G} \tag{3.10}$$

3.1.1 Example

A well-known example for a Hellinger – Reissner element is the element of Pian and Sumihara [9]. For the four-node membrane element the displacements are assumed bilinear, while the stress approximation follows a five-parameter assumption. The five stress fields correspond to the five deformation modes and, therefore, lead to a locking-free element (no shear locking, no volumetric locking).

3.2 The assumed natural strain method

In the so-called assumed natural strain method (ANS) the idea to avoid locking is to reduce the strain fields of a displacement model. The new strain fields are defined through carefully selected sampling points, so that the method could be interpreted as sort of an underintegration. The number of sampling points however is high enough, so that no zero-energy-modes show up.

In contradiction to the Hellinger-Reissner or the enhanced assumed strain method, the additionally assumed fields of the ANS-method do not introduce independent unkowns. The fields are rather linked directly to the fields of the displacement derivatives. The resulting efficiency (only displacement degrees of freedom) makes this method in particular appealing and popular.

In [11] it is shown that the underlying variational basis is the principle of Hu - Washizu. The internal energy for one element is written as follows:

$$U_e = \int_{\Omega_e} \frac{1}{2}\boldsymbol{\epsilon}^T\mathbf{C}\boldsymbol{\epsilon} - \boldsymbol{\sigma}^T\boldsymbol{\epsilon} + \boldsymbol{\sigma}^T\overset{D}{\mathbf{B}}\mathbf{u} \, d\Omega \tag{3.11}$$

In order to clarify the concept, the strain fields $\epsilon^{(u)} = \overset{D}{\mathbf{B}}\mathbf{u}$ which result from the displacement derivatives are split into three parts, a constant part ϵ^c, a higher order part ϵ^l and a part ϵ^s that causes stiffening:

$$\epsilon^{(u)} = \overset{D}{\mathbf{B}}\mathbf{u} = \epsilon^c + \epsilon^l + \epsilon^s \tag{3.12}$$

If the strain fields ϵ in (3.11) are chosen to be

$$\epsilon = \epsilon^c + \epsilon^l \tag{3.13}$$

which means that the parasitic strain fields are omitted, (3.11) gives

$$U_e = \int_{\Omega_e} \frac{1}{2}(\epsilon^c + \epsilon^l)^T C(\epsilon^c + \epsilon^l) - \sigma^T(\epsilon^c + \epsilon^l) + \sigma^T(\epsilon^c + \epsilon^l + \epsilon^s)\, d\Omega \tag{3.14}$$

or

$$U_e = \int_{\Omega_e} \frac{1}{2}(\epsilon^c + \epsilon^l)^T C(\epsilon^c + \epsilon^l) + \sigma^T \epsilon^s\, d\Omega \tag{3.15}$$

If the stress fields are chosen orthogonal to ϵ^s, the last term in (3.15) drops and we end up with some sort of a modified potential energy:

$$U_e = \int_{\Omega_e} \frac{1}{2}(\epsilon^c + \epsilon^l)^T C(\epsilon^c + \epsilon^l)\, d\Omega \tag{3.16}$$

The above equations show that the strain fields of a displacement model can be reduced. If the stress fields are orthogonal to the strains ϵ^s which are omitted, the variational basis of these "strain-reduced" displacement models is the principle of Hu-Washizu.

3.2.1 Example

As an example we want to mention the plate elements of Bathe and Dvorkin [3], [5] or Huang and Hinton [6], where the fields of the two transverse shear strains are reduced through selected sampling points. For instance for the four-node plate element the original strain fields γ_ξ and γ_η are bilinear (which causes locking), whereas in the ANS-model the strain fields are reduced to constant-linear fields due to the two sampling points (Fig. 3.1).

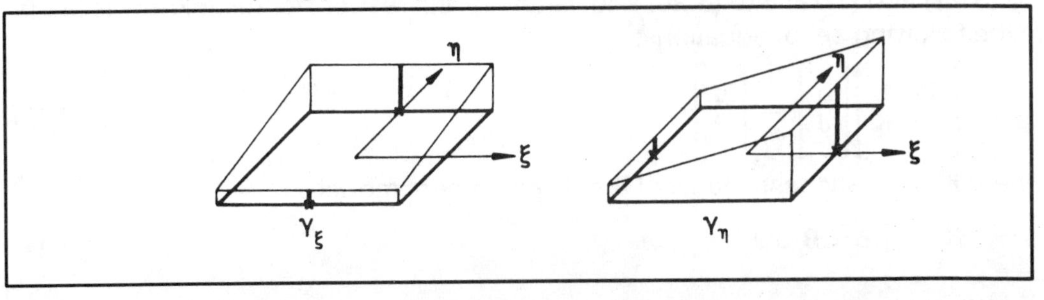

Fig. 3.1: Reduced transverse shear strain fields of the four-node ANS-plate model

3.3 The enhanced assumed strain method

In the HR–approximation or the ANS–method stiffening is avoided by reducing the stress or strain fields of the displacement model. In the enhanced assumed strain method (EAS–method) [12], the strain fields are enriched which again can lead to a balanced, stiffening–free element. In [12] it is shown that the method encompasses the classical method of incompatible displacements [13], [15]. Starting with the principle of Hu – Washizu, we again write

$$U_e = \int_{\Omega_e} \frac{1}{2}\boldsymbol{\epsilon}^T\mathbf{C}\boldsymbol{\epsilon} - \boldsymbol{\sigma}^T\boldsymbol{\epsilon} + \boldsymbol{\sigma}^T\overset{D}{\mathbf{B}}\mathbf{u}\ d\Omega \tag{3.17}$$

Opposite to (3.13) the strain fields are chosen to be

$$\boldsymbol{\epsilon} = \overset{D}{\mathbf{B}}\mathbf{u} + \tilde{\boldsymbol{\epsilon}} \tag{3.18}$$

which means the assumption consists of the compatible part $\boldsymbol{\epsilon}^{(u)} = \overset{D}{\mathbf{B}}\mathbf{u}$ plus an additional part $\tilde{\boldsymbol{\epsilon}}$. Introducing (3.18) into (3.17) gives

$$U_e = \int_{\Omega_e} \frac{1}{2}(\overset{D}{\mathbf{B}}\mathbf{u} + \tilde{\boldsymbol{\epsilon}})^T\mathbf{C}(\overset{D}{\mathbf{B}}\mathbf{u} + \tilde{\boldsymbol{\epsilon}}) - \boldsymbol{\sigma}^T\tilde{\boldsymbol{\epsilon}}\ d\Omega \tag{3.19}$$

Similarly to (3.15) the stress fields are chosen to be orthogonal to the additional strain fields $\tilde{\boldsymbol{\epsilon}}$. This leads to

$$U_e = \int_{\Omega_e} \frac{1}{2}(\overset{D}{\mathbf{B}}\mathbf{u} + \tilde{\boldsymbol{\epsilon}})^T\mathbf{C}(\overset{D}{\mathbf{B}}\mathbf{u} + \tilde{\boldsymbol{\epsilon}})\ \ d\Omega \tag{3.20}$$

In contradiction to (3.13) the EAS–method uses additional fields on top of the compatible strains $\boldsymbol{\epsilon}^{(u)} = \overset{D}{\mathbf{B}}\mathbf{u}$, however rendering extra unknowns, the parameters for the enhanced strain fields. Besides the displacement assumption

$$\mathbf{u} = \mathbf{N}\mathbf{d} \tag{3.21}$$

we use a discontinuous assumption for $\tilde{\boldsymbol{\epsilon}}$:

$$\tilde{\boldsymbol{\epsilon}} = \mathbf{M}\,\boldsymbol{\alpha} \tag{3.22}$$

where **M** contains polynomials and α is the vector of discrete unknowns. Introducing (3.21) and (3.22) into (3.20) and variation with respect to the unknowns leads to the following set of equations:

$$\begin{bmatrix} \mathbf{D} & \mathbf{L} \\ \mathbf{L}^T & \mathbf{K} \end{bmatrix} \begin{bmatrix} \boldsymbol{\alpha} \\ \mathbf{d} \end{bmatrix} = \begin{bmatrix} \mathbf{0} \\ \mathbf{R} \end{bmatrix} \tag{3.23}$$

where **K** is the stiffness matrix of the displacement model

$$\mathbf{K} = \int_{\Omega_e} \mathbf{B}^T\mathbf{C}\mathbf{B}\ d\Omega \tag{3.24}$$

and

$$\mathbf{D} = \int_{\Omega_e} \mathbf{M}^T\mathbf{C}\mathbf{M}\ d\Omega \tag{3.25}$$

$$\mathbf{L} = \int_{\Omega_e} \mathbf{M}^T\mathbf{C}\mathbf{B}\ d\Omega \tag{3.26}$$

Due to discontinuous approximations for the enhanced strains $\tilde{\boldsymbol{\epsilon}}$, the parameters $\boldsymbol{\alpha}$ can again be eliminated and the element stiffness matrix is given by

$$\mathbf{k}_e = \mathbf{K} - \mathbf{L}^T\mathbf{D}^{-1}\mathbf{L} \tag{3.27}$$

Considering equations (3.20) to (3.27), the similarity to the method of incompatible displacements is evident, especially if one identifies the strain fields $\tilde{\boldsymbol{\epsilon}}$ with the derivatives of incompatible displacements $\mathbf{u}_\lambda$:

$$\tilde{\boldsymbol{\epsilon}} = \overset{D}{\mathbf{B}}\mathbf{u}_\lambda \tag{3.28}$$

However the EAS–method firstly shows that the underlying variational basis is the principle of Hu–Washizu and secondly makes clear that the load vector remains unchanged, a fact that was uncertain when using incompatible displacements.

4 FOUR–NODE SHELL ELEMENTS

In Chapter 3 three hybrid–mixed formulations are discussed, the HR–, the ANS– and the EAS–method.

Considering a four–node shell element, which consists of a membrane, a bending and a transverse shear part, the question arises which formulation can be used to get an efficient and stiffening–free element.

The membrane and bending parts can be treated in the same way. This is possible due to the similar differential relations and advisable because they are coupled sometimes (eccentric stiffeners, unsymmetrically layered models, etc.).

For the membrane and bending parts, the HR– and the EAS–formulation lead to optimal elements with high coarse mesh accuracy. The reason that the ANS–concept is less suited is that, due to Poisson's ratio, the membrane and bending strains cannot be reduced as it is done for the transverse shear strains.

For the transverse shear part, the HR– and EAS–formulations only work for elements with parallelogram geometries. For arbitrarily distorted elements the formulations cannot lead to a locking–free element unless additional projection methods [14] are used. Therefore, the ANS–concept developed in [5], [7], [8] seems to be unbeatable, especially due to its efficiency.

Therefore, two elements are formulated (see Table 4.1). The element HR–ANS has a five–parameter Hellinger – Reissner approximation (according to [9]) for membrane and bending and an ANS–formulated transverse shear part. For the element EAS–ANS a four–parameter EAS–assumption is chosen for membrane and bending [2].

For comparison, we use the fully integrated pure displacement model (DISP), the uniformly reduced (URI), the selectively reduced integrated shell elements (SRI) and the element of Bathe and Dvorkin, which uses a pure displacement formulation for membrane and bending.

	URI	DISP	SRI	B/D	HR–ANS	EAS–ANS
membrane	1x1	DISP	DISP	DISP	HR–5	EAS–4
bending	1x1	DISP	DISP	DISP	HR–5	EAS–4
tr. shear	1x1	DISP	1x1	ANS	ANS	ANS

Table 4.1: Four–node shell elements with different membrane, bending and transverse shear formulations

5 NUMERICAL EXAMPLES

5.1 Eigenvalue analysis of different four–node plate elements

For the 2x2–integrated displacement model DISP, the reduced integrated models URI and SRI as well as for the elements B/D and HR–ANS, the stiffnesses of the 12 modes of a rectangular plate element are investigated through an eigenvalue analysis. (Due to the rectangular geometry, the element EAS–ANS is identical to HR–ANS). The tested element has equal side lengths 1.0 and a thickness of 0.1. The eigenvalues and the corresponding strains are shown in table 5.1. The results of the fully integrated displacement model DISP are given in the second column. Due to parasitic shear strains γ_ξ^{lin} and γ_η^{lin}, the eigenvalues of the curvature modes $\kappa_{\xi\xi}^c$ and $\kappa_{\eta\eta}^c$ are too large (0.118056 instead of 0.083333). The reduced integrated models URI and SRI show the exact value which means that the constant bending modes are locking free. However due to underintegration, the element URI has four and SRI two additional zero–energy–modes.

The element B/D passes the Patch test because it has no zero–energy–modes and the bending modes $\kappa_{\xi\xi}^c$ and $\kappa_{\eta\eta}^c$ are exact as well. The element HR–ANS has linear bending modes which are improved compared to B/D or SRI (0.027777 instead of 0.041667) because $\kappa_{\xi\eta}$ is assumed to be constant.

A decoupling of the linear shear modes and the constant drilling mode $\kappa_{\xi\eta}^c$ is not possible. This coupling however, which does not appear in the reduced integrated elements URI and SRI (because they do not contain linear shear modes), is only leading to locking, when the elements are warped (see Chapter 5.4).

URI (1 x 1)	DISP (2 x 2)	SRI	B/D	HR-ANS
0	0	0	0	0
0	0	0	0	0
0	0	0	0	0
0	0.0347222 $\gamma^{lin}_{\xi}, \gamma^{lin}_{\eta}$	0	0.0347222 $\gamma^{lin}_{\xi}, \gamma^{lin}_{\eta}$	0.027777 $\kappa^{lin}_{\xi\xi}$
0	0.0532407 $\kappa^{lin}_{\xi\xi}, \kappa^{lin}_{\xi\eta}, \gamma^{lin}_{\xi}$	0	0.0416667 $\kappa^{lin}_{\xi\xi}, \kappa^{lin}_{\xi\eta}$	0.027777 $\kappa^{lin}_{\eta\eta}$
0	0.0532407 $\kappa^{lin}_{\eta\eta}, \kappa^{lin}_{\xi\eta}, \gamma^{lin}_{\eta}$	0.0416667 $\kappa^{lin}_{\xi\xi}, \kappa^{lin}_{\xi\eta}$	0.0416667 $\kappa^{lin}_{\eta\eta}, \kappa^{lin}_{\xi\eta}$	0.0347222 $\gamma^{lin}_{\xi}, \gamma^{lin}_{\eta}$
0	0.0713352 $\kappa^{c}_{\xi\eta}, (\gamma^{lin}_{\xi}, \gamma^{lin}_{\eta})$	0.0416667 $\kappa^{lin}_{\eta\eta}, \kappa^{lin}_{\xi\eta}$	0.0713352 $\kappa^{c}_{\xi\eta}, (\gamma^{lin}_{\xi}, \gamma^{lin}_{\eta})$	0.0713352 $\kappa^{c}_{\xi\eta}, (\gamma^{lin}_{\xi}, \gamma^{lin}_{\eta})$
0.083333 $\kappa^{c}_{\xi\xi}$	0.118056 $\kappa^{c}_{\xi\xi}, \gamma^{lin}_{\xi}$	0.083333 $\kappa^{c}_{\xi\xi}$	0.083333 $\kappa^{c}_{\xi\xi}$	0.083333 $\kappa^{c}_{\xi\xi}$
0.083333 $\kappa^{c}_{\eta\eta}$	0.118056 $\kappa^{c}_{\eta\eta}, \gamma^{lin}_{\eta}$	0.083333 $\kappa^{c}_{\eta\eta}$	0.083333 $\kappa^{c}_{\eta\eta}$	0.083333 $\kappa^{c}_{\eta\eta}$
0.083333 $\kappa^{c}_{\xi\eta}$	0.324498 $\gamma^{lin}_{\xi}, \gamma^{lin}_{\eta}, (\kappa^{c}_{\xi\eta})$	0.083333 $\kappa^{c}_{\xi\eta}$	0.324498 $\gamma^{lin}_{\xi}, \gamma^{lin}_{\eta}, (\kappa^{c}_{\xi\eta})$	0.324498 $\gamma^{lin}_{\xi}, \gamma^{lin}_{\eta}, (\kappa^{c}_{\xi\eta})$
0.520833 γ^{c}_{ξ}	0.520833 γ^{c}_{ξ}	0.520833 γ^{c}_{ξ}	0.520833 γ^{c}_{ξ}	0.520833 γ^{c}_{ξ}
0.520833 γ^{c}_{η}	0.520833 γ^{c}_{η}	0.520833 γ^{c}_{η}	0.520833 γ^{c}_{η}	0.520833 γ^{c}_{η}

Table 5.1: Eigenvalues and corresponding strains of different four-node plate models

5.2 Morley's skew plate

In the second example Morley's skew plate (length L = 100, thickness t = 1, $E = 10^5$, $\nu = 0.3$, uniform pressure load q = 1) is analyzed. The boundary conditions of the simply-supported plate are idealized as soft support (SS1). In table 5.2 the centerpoint deflection (the analytical solution is 4,455 [12]) is given for different element meshes. Due to the parallelogram geometry of all elements, HR-ANS and EAS-ANS are again identical. For all meshes they show a clear improvement over the element B/D.

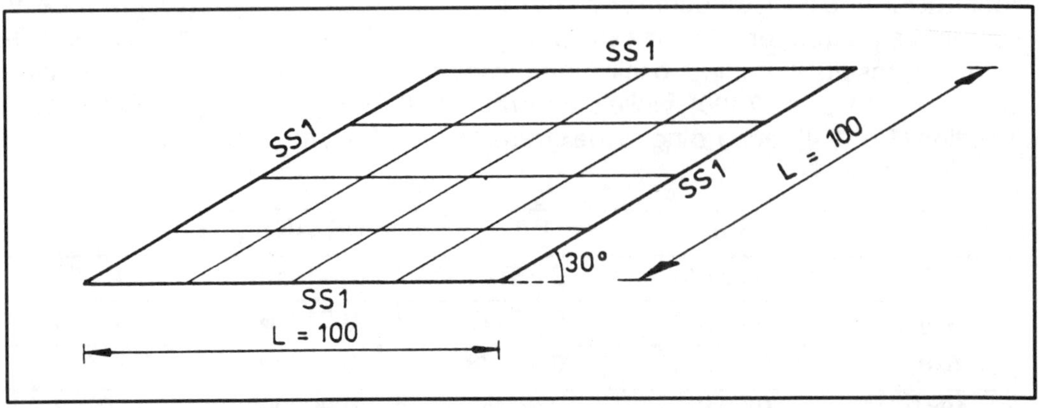

Fig. 5.1: Morley's skew plate

mesh	URI	SRI	B/D	HR–ANS	EAS–ANS
4x4	4,8568	4,5085	3,9182	4,2122	4,2122
6x6	4,9492	4,4853	3,8873	4,2234	4,2234
8x8	4,8542	4,4378	3,8991	4,2239	4,2239
10x10	4,7776	4,4387	3,9752	4,2591	4,2591
12x12	4,7293	4,4511	4,0589	4,3032	4,3032
14x14	4,6996	4,4665	4,1299	4,3424	4,3424
16x16	4,6812	4,4820	4,1875	4,3738	4,3738

Table 5.2: Centerpoint deflection of Morley's skew plate

5.3 Scordelis – Lo roof

The geometry of the Scordelis–Lo roof, a cylindrical shell under self weight, is given in figure 5.2. The length of the structure which is supported through rigid end diaphragms is $L=50$, the radius is $R=25$ and the thickness is $h=0{,}25$. Due to symmetry conditions only one quarter of the structure is idealized. In Table 5.3 the vertical deflection w_B is reported for different finite element meshes. A "deep shell solution" is given in [10] as 0,3008.
For the 2x2 mesh the elements B/D and SRI show better results than the elements HR–ANS and EAS–ANS. The reason is that the elements converge from above, so that the membrane and bending parts of B/D and SRI, which are too stiff, have a positive influence. For fine meshes the elements HR–ANS and EAS–ANS show excellent results.
The advantage of the elements HR–ANS or EAS–ANS is even more obvious if one looks at the stress distributions. For the membrane shear forces n_{rs} for instance (the forces are plotted for a 8x8 mesh for a quarter of the structure), the element B/D (Fig. 5.3) results in distinct oscillations, especially at point B (the results oscillate from – 17400 to + 17400), while the element HR–ANS gives a qualitatively correct solution (Fig. 5.4); according to the stress assumption, the distribution is elementwise constant.

mesh	SRI	B/D	HR–ANS	EAS–ANS
2x2	0,37778	0,35728	0,41952	0,41952
4x4	0,28894	0,28320	0,31469	0,31470
8x8	0,29678	0,29428	0,30428	0,30429
16x16	0,30164	0,30040	0,30308	0,30308

Table 5.3: Vertical deflection w_B of the Scordelis – Lo roof

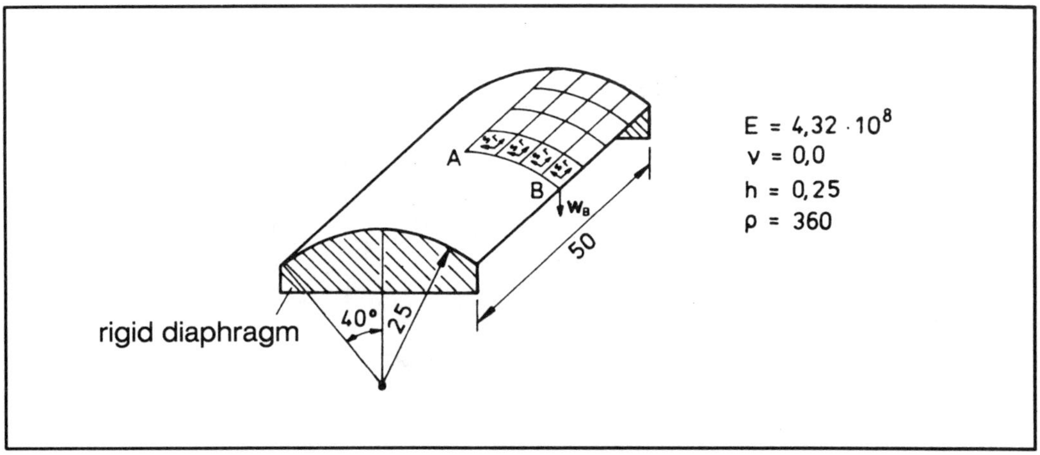

Fig. 5.2: Scordelis – Lo roof

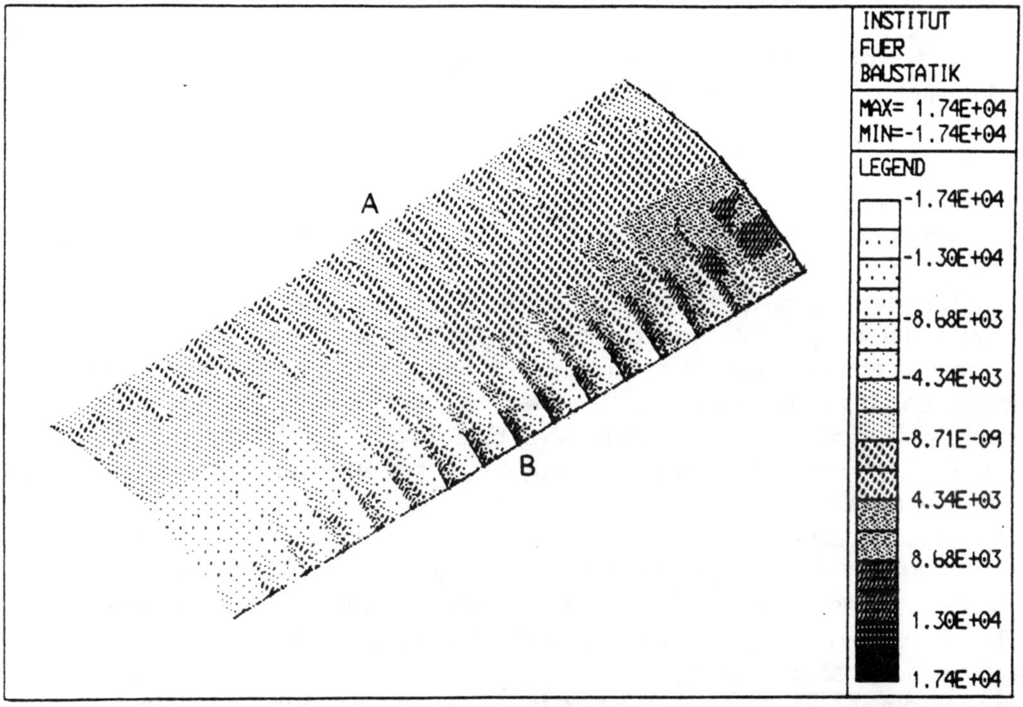

Fig. 5.3: Membrane shear forces n_{rs} of the Scordelis–Lo roof, using the element B/D (half of the structure is shown)

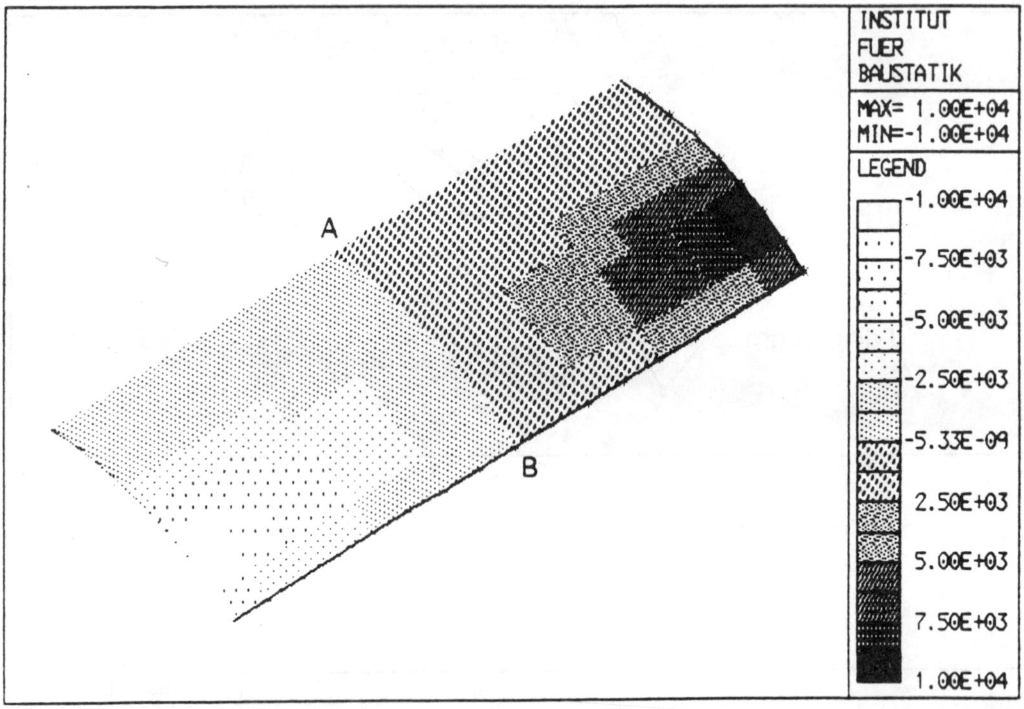

Fig. 5.4: Membrane shear forces n_{rs} of the Scordelis–Lo roof, using the element HR–ANS (half of the structure is shown)

5.4 Locking test of a cylindrical panel

In Fig. 5.5 a cylindrical panel with 90^0 is shown. It is clamped on one side and on the other side a constant moment is applied, so that the problem is in a pure bending mode. To test the locking of the finite elements, the thickness of the structure is reduced successively by a factor of 10, increasing the displacement each time by a factor of 10^3. For the regular mesh (Fig. 5.6) all elements show the correct displacements and exact stress resultants (constant moment, membrane and shear forces are zero). However, if the mesh is distorted in such a way that the elements are warped, the elements HR–ANS, EAS–ANS and B/D exhibit severe shear locking – displacements increase only by a factor of 10 – with high oscillations for the transverse shear stresses.

In order to exclude membrane locking, the element 1x1–ANS (membrane and bending are 1x1–integrated, the transverse shear part is formulated with the ANS–concept) was tested as well. However, the only element that does not lock is SRI. This is due to the underintegrated transverse shear part. Therefore, the locking can be identified as shear locking.

The example shows that care must be taken, when four–node shell elements are warped because the constant drilling curvature $\kappa_{\xi\eta}$ is spoiled with linear transverse shear strains. (The reason why the elements do not lock for the pretwisted beam [2] is because for this example $\kappa_{\xi\eta}$ is negligible.)

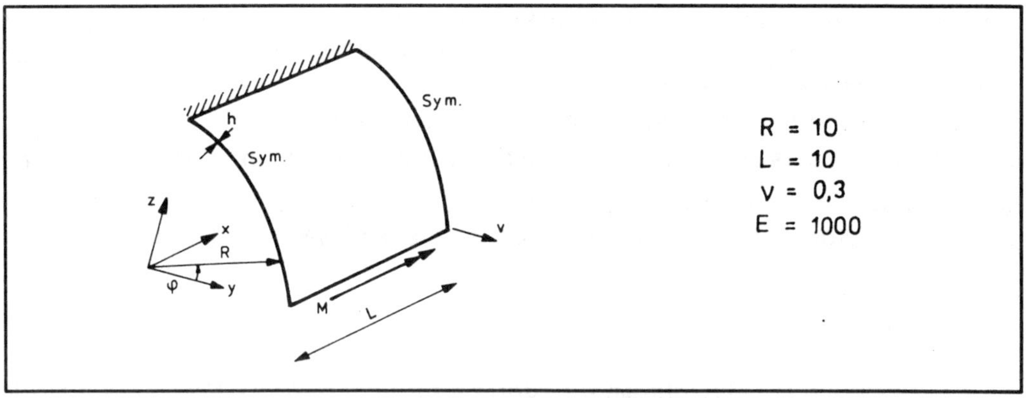

Fig. 5.5: 90^0 – cylindrical panel under constant bending

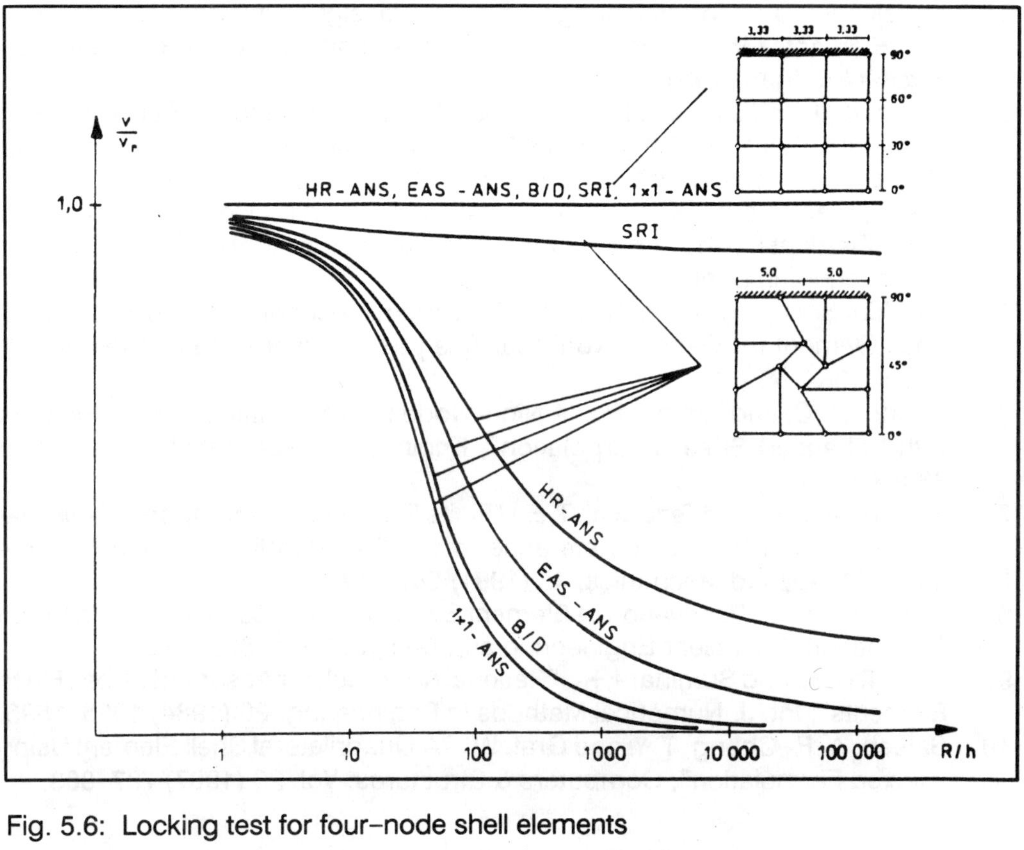

Fig. 5.6: Locking test for four–node shell elements

6 CONCLUSIONS

It was shown that locking occurs whenever constant bending modes are spoiled by parasitic shear or membrane strains (or when deviatoric modes are spoiled by volumetric strains). To avoid such a stiffening hybrid-mixed formulations are well suited. For materially nonlinear applications, EAS-ANS seems to be the best suited element, because it is not based on stress assumptions.
A locking-phenomena that is almost unknown in the literature is the shear-locking of warped elements. The locking is due to a coupling of constant drilling $\kappa^c_{\xi\eta}$ and linear transverse shear strains. It probably can be solved by using a modified shear correction factor, which reduces the shear stiffness of warped elements.

7 REFERENCES

1. Andelfinger, U., Matzenmiller, A. and Ramm, E., "Hybrid-Mixed Four-Node Shell Elements, Different Two-Field Assumptions and a Stability Test", in: Analytical and Computational Models of Shells (A.K. Noor et al., eds.), ASME Special Publication, 1989.
2. Andelfinger, U.: "Untersuchungen zur Zuverlässigkeit hybrid-gemischter Finiter Elemente für Flächentragwerke", Dissertation, Institut für Baustatik, Universität Stuttgart, 1991.
3. Bathe, K. J. and Dvorkin, E. N., "A Four-Node Plate Bending Element Based on Mindlin/Reissner Theory and a Mixed Interpolation", Int. J. for Numerical Methods in Engineering, 21 (1985), 367–383.
4. Brezzi, F. and Bathe, K. J., "A Discourse on the Stability Conditions for Mixed Finite Element Formulations", Computer Methods in Applied Mechanics and Engineering (in print).
5. Dvorkin, E. N. and Bathe, K. J., "A Continuum Mechanics Based Four-Node Shell Element for General Nonlinear Analysis", Engineering Computations, No. 1 (1984) 77–88.
6. Huang, H. C. and Hinton, E., "A Nine-Node Lagrange Mindlin Plate Element with Enhanced Shear Interpolation", Engineering Computations, 1 (1984) 369–379.
7. Hughes, T. J. R. and Tezduyar, T. E., "Finite Elements Based Upon Mindlin Plate Theory with Particular Reference to the Four-Node Isoparametric Element", J. Applied Mechanics, 48 (1981) 587–596.
8. MacNeal, R. H., "Derivation of Element Stiffness Matrices by Assumed Strain Distributions", Nuclear Engineering and Design, 70 (1982) 3–12.
9. Pian, T. H. H. and Sumihara, K., "Rational Approach for Assumed Stress Finite Elements", Int. J. Numerical Methods in Engineering, 20 (1984) 1685–1695.
10. Saleeb, A. F., Chang, T. Y. and Graf, W., "A Quadrilateral Shell Element Using a Mixed Formulation", Computers & Structures, Vol. 26 (1987) 787–803.

11. Simo, J. C. and Hughes, T. J. R., "On Variational Foundations of Assumed Strain Methods", J. Applied Mechanics, 53 (1986) 51–54.
12. Simo, J. C. and Rifai, M.S.,. "A Class of Mixed Assumed Strain Methods and the Method of Incompatible Modes", Int. J. Numerical Methods in Engineering, 29 (1990) 1595–1638.
13. Taylor, R. L., Beresford, P. J. and Wilson, E. L., "A Non-Conforming Element for Stress Analysis", Int. J. Numerical Methods in Engineering, 10 (1976) 1211–1220.
14. Wang, X. J. and Belytschko, T., "A Study of Stabilization and Projection in the 4-Node Mindlin Plate Element", Int. J. Numerical Methods in Engineering, 28 (1989) 2223–2238.
15. Wilson, E. L., Taylor, R. L. Doherty, W. P. and Ghaboussi, J., "Incompatible Displacement Models", in: "Numerical and Computational Methods in Structural Mechanics" (S. T. Fenves et al., eds.), Academic Press, 1973, 43–57.

ON NONLINEAR ANALYSIS OF SHELLS USING FINITE ELEMENTS BASED ON MIXED INTERPOLATION OF TENSORIAL COMPONENTS

E. N. Dvorkin

Center for Industrial Research, Fudetec, Buenos Aires, Argentina

ABSTRACT

Most of the research work developed in the area of nonlinear finite element analysis of shells since 1970 has been done on elements that while being based on the Ahmad-Irons-Zienkiewicz element overcome the locking problem. In particular, the elements based on mixed interpolation of tensorial components belong to the above mentioned set. In this Chapter, the formulation of these elements is reviewed. Its implementation in general purpose nonlinear finite element codes is examined and some modelling considerations are discussed.

1 . INTRODUCTION

In 1970 Ahmad, Irons and Zienkiewicz in a now classical paper [1] presented an isoparametric linear shell element with independent interpolations for displacements and rotations (interpolations that require only C^0 continuity). This shell element formulation is obtained by introducing some kinematic assumptions in the general 3D continuum formulation, therefore it is also referred as the degenerated continuum formulation.

The extension of the A-I-Z shell element to nonlinear analyses is quite natural and was developed by Ramm [2] in 1977 and by Kråkeland[3] also in 1977.

In our presentation we will follow the formulation developed in Ref.[4] by Bathe and Bolourchi (see also Ref.[5]).

The A-I-Z shell element uses interpolation functions with C^0 continuity at the price of introducing the shear deformations in the analysis (this elements are therefore also referred as Reissner/Mindlin elements). And we put it in this way ("at the price") because the main objective is to have C^0 continuity and the shear deformations are a by-product that while being desirable for the analysis of moderately thick shells introduce the main numerical difficulty of these elements: the locking problem.

Most of the research work developed in the area of finite element analysis of shells since 1970 has been devoted to elements that while being based on the A-I-Z element overcome the locking problem.

2 . THE AHMAD-IRONS-ZIENKIEWICZ SHELL ELEMENT

In Fig. 1 we present a general A-I-Z element. In order to define its geometry we use [5]:

- The coordinates of the mid-surface nodes.
- Director vectors at the nodes that approximate the shell normals at those points.

At any time (load level) t, the position vector of any point inside the shell element, with natural coordinates (r_1,r_2,r_3) is given by:

$$ {}^t\underline{x} = h_k \; {}^t\underline{x}^k + \frac{r_3}{2} \, h_k \, a_k \; {}^t\underline{V}_n^k \tag{1} $$

(t=0 represents the reference undeformed configuration)

In Eqn.(1) the summation convention is used and:

h_k : 2D isoparametric interpolation function [5] corresponding to the k-th. node.

${}^t\underline{x}^k$: position vector of the k-th. mid-surface node at time t.

a_k: thickness at the k-th. node.

$|{}^t\underline{V}_n^k| = 1$

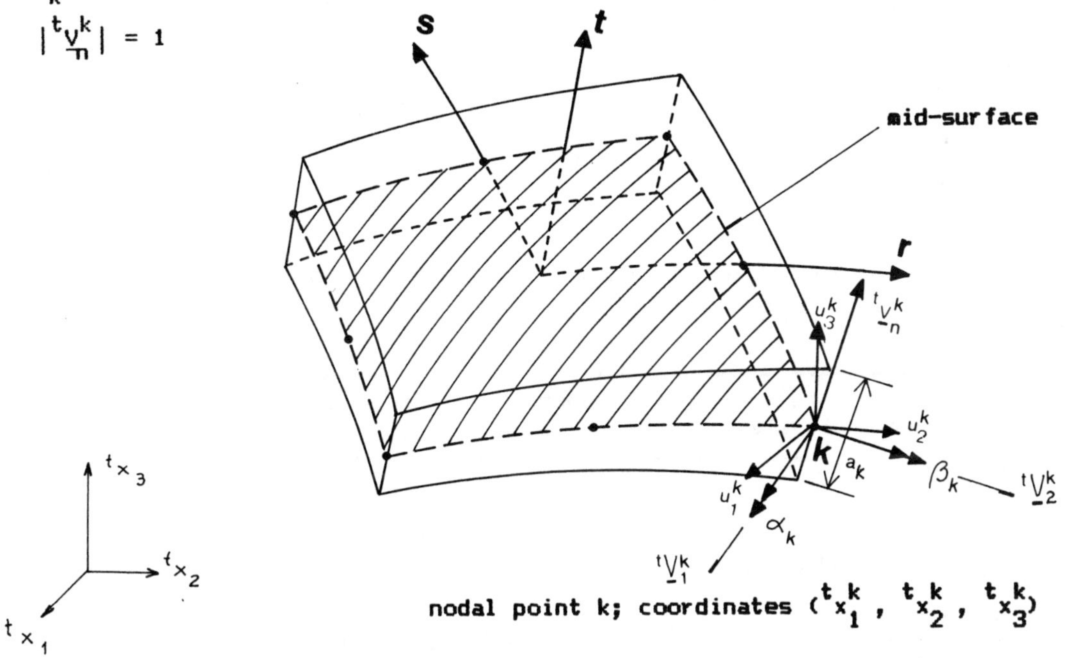

FIGURE 1 - Ahmad-Irons-Zienkiewicz shell element

This element can be used to model variable thickness shells. The following kinematic assumptions are introduced:

1 - The director vectors remain straight during deformation
2 - The thickness (a_k) remains constant during deformation (only valid for small strain problems)

For the point (r_1,r_2,r_3) the incremental displacement $\underline{u}$ can be therefore written as:

$$\underline{u} = h_k\, \underline{u}_k + \frac{r_3}{2}\, h_k\, a_k\, (-\alpha_k\, {}^t\underline{V}_2^k + \beta_k\, {}^t\underline{V}_1^k) \qquad (2)$$

In the above:

$\underline{u}_k$: nodal incremental displacement

α_k, β_k : incremental rotations

$$ {}^t\underline{V}_1^k = \frac{\underline{e}_y \times {}^t\underline{V}_n^k}{|\underline{e}_y \times {}^t\underline{V}_n^k|} \qquad (3.a) $$

$$ {}^t\underline{V}_2^k = {}^t\underline{V}_n^k \times {}^t\underline{V}_1^k \qquad (3.b) $$

A special definition of ${}^t\underline{V}_1^k$ and ${}^t\underline{V}_2^k$ is used for the case when ${}^t\underline{V}_n^k$ is parallel to the y-axis [5]

> Please note that only 5 degrees of freedom are used for each shell element node

3 . THE LOCKING PROBLEM

In order to introduce the shear locking problem, in what follows we resource to a very simple example. Let us consider the two-node beam element shown in Fig. 2.a, this beam element is also based on independent interpolations for the transverse displacements (w) and the rotation angles (θ) (Timoshenko beam element).

The above mentioned interpolations are:

$$ w = h_1 \, w_1 + h_2 \, w_2 \qquad (4.a) $$

$$ \theta = h_1 \, \theta_1 + h_2 \, \theta_2 \qquad (4.b) $$

w and θ are both interpolated using linear interpolation functions. It is easy to see that the Bernoulli condition

$$ \gamma = \frac{dw}{dx} - \theta = 0 \qquad (5) $$

leads to θ=const., and therefore in order to satisfy the boundary conditions at node 1 we must have θ=0 all over the beam: locking. This locking (shear locking) is due to the fact that the interpolation functions cannot satisfy the condition of zero shear strain all over the element.

We used a simple beam element to present a picture of shear locking but in general plate / shell elements it is also possible to very clearly identify the shear locking. The potential energy of a plate element can be written as [5,6],

$$\pi = \frac{h^3}{2}\left[\int_A \underline{\chi}^T \underline{C}_b \underline{\chi}\, dA + \alpha \int_A \underline{\gamma}^T \underline{C}_s \underline{\gamma}\, dA\right] - \mathbb{V} \qquad (6.a)$$

where:

h = plate thickness

$$\underline{\chi} = \begin{bmatrix} \dfrac{\partial\alpha}{\partial x} \\ \dfrac{-\partial\beta}{\partial y} \\ \dfrac{\partial\alpha}{\partial y} - \dfrac{\partial\beta}{\partial x} \end{bmatrix} \quad ; \quad \underline{\gamma} = \frac{1}{L}\begin{bmatrix} \dfrac{\partial u_3}{\partial y} - \beta \\ \dfrac{\partial u_3}{\partial x} + \alpha \end{bmatrix}$$

$$\underline{C}_b = \frac{E}{12(1-\nu^2)}\begin{bmatrix} 1 & \nu & 0 \\ \nu & 1 & 0 \\ 0 & 0 & (1-\nu)/2 \end{bmatrix}$$

$$\underline{C}_s = \frac{k\,E}{2\,(1+\nu)}\begin{bmatrix} 1 & 0 \\ 0 & 1 \end{bmatrix}$$

and:

E : Young's modulus

ν : Poisson ratio

k : shear correction factor [5]

$$\alpha = \frac{L}{h} \xrightarrow[h \to 0]{} \infty$$

$\mathbb{V}$ = potential of external loads

Therefore if the trial functions cannot exactly represent $\underline{\gamma} \equiv \underline{0}$ (Kirchhoff-Love hypothesis) α will amplify any error in $\underline{\gamma}$ when $h \rightarrow 0$ (thin shells) leading to a locking behavior.

In order to introduce the membrane/shear locking problem [7] we again resource to a very simple example. Let us consider the curved three-node Timoshenko beam element shown in Fig. 2.b, the functional of total potential energy is:

$$\Pi = \frac{EI}{2}\left[\int_0^L \theta_{,s}^2\, ds + \alpha_m \int_0^L u_{s,s}^2\, ds + \alpha_s \int_0^L (u_{n,s} - \theta)^2 ds\right] - \mathbb{V} \tag{6.b}$$

where L is the length of the circular beam measured along its axis, the s-direction is tangential to its axis and the n-direction is normal to its axis.

Since $\alpha_m = 12/h^2$ and $\alpha_s = 12Gk/h^2$, it is evident that when the beam becomes very thin, the second and third integrals act as a penalty to impose the conditions

$$\varepsilon_{ss} = u_{s,s} = 0 \tag{6.c}$$

$$\gamma_{ns} = u_{n,s} - \theta = 0 \tag{6.d}$$

that is to say zero membrane and transverse shear strains.

If in the three-node Timoshenko beam in Fig. 2.b we prescribe the three rotations $\theta_i =$ (i=1,2,3) corresponding to the analytical solution and, from Eqs.(6.c) and (6.d) imposed at the Gauss points we try to calculate the displacements, we get a system of 6 equations with 4 unknowns which in general cannot be solved, demonstrating the combined shear/membrane locking.

If the shear locking is removed by interpolating the transverse shear strains through two points we still have a system of 5 equations with 4 unknowns which cannot be solved, demonstrating the membrane locking.

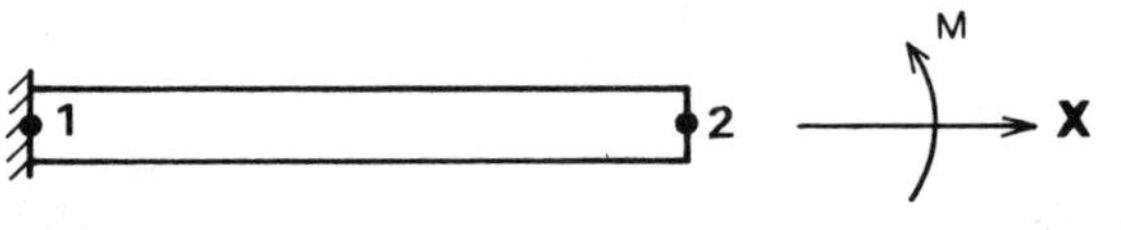

(a) Two-node Timoshenko beam element

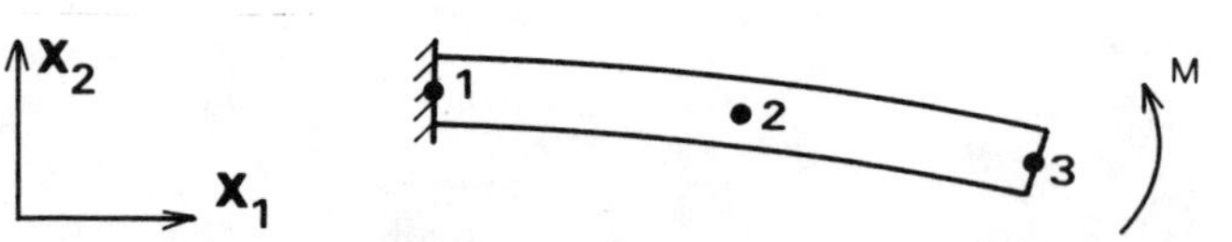

(b) Three-node Timoshenko beam element (curved)

FIGURE 2

In Figs. 3 and 4 we present some solutions showing shear locking (Fig. 3) and combined shear and membrane locking (Fig. 4). From those results we can conclude that locking is an element property related to the element ratio (L/h).

uniform pressure q
thickness h

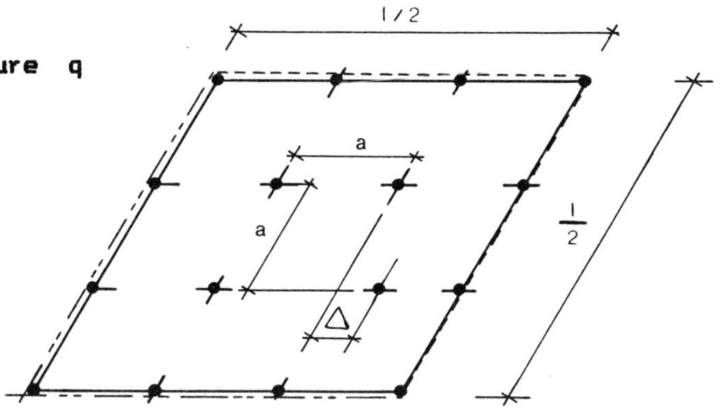

CASE	Δ / a	INTEGRATION
1	0	4 x 4 x 2
2	1 / 50	4 x 4 x 2
3	1 / 20	4 x 4 x 2
4	1 / 20	3 x 3 x 2

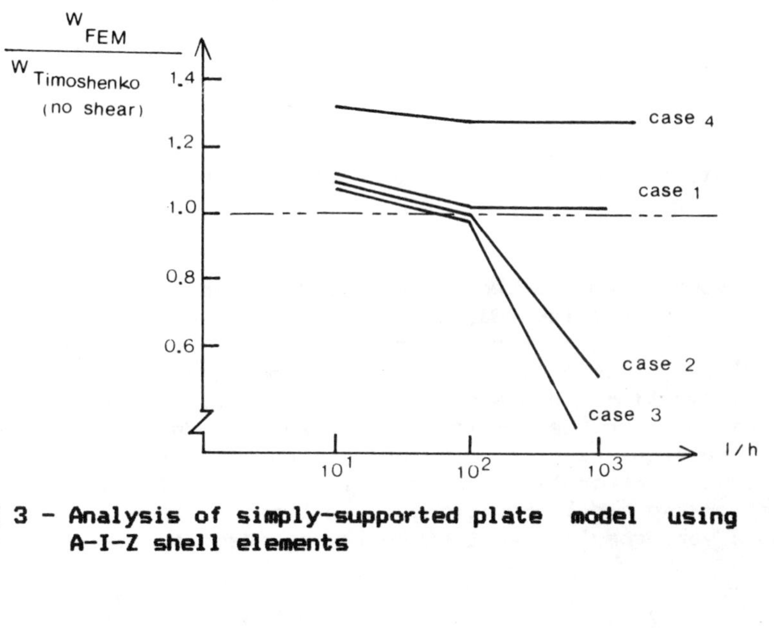

FIGURE 3 - Analysis of simply-supported plate model using standard A-I-Z shell elements

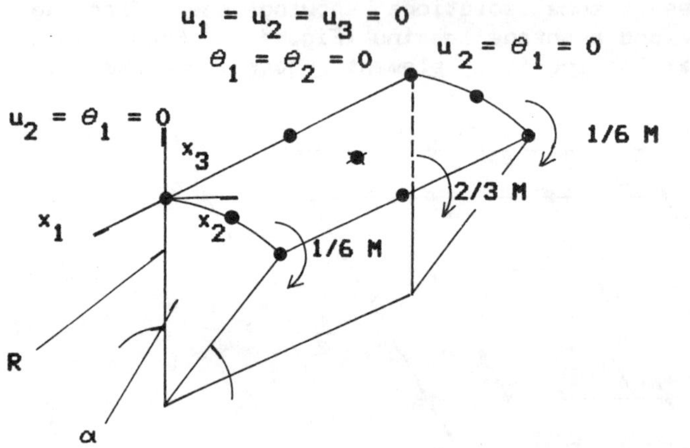

R = 20
h (thickness) = 0.2
E = 2.1 x 10^6
ν = 0.3
Int. = 4x4x2

α	$\theta_{FE} / \theta_{TH}$
30°	0.02
20°	0.09
10°	0.62
5°	0.96

3 x 10° elements $\theta_{FE} / \theta_{TH}$ = 0.61
6 x 5° elements $\theta_{FE} / \theta_{TH}$ = 0.96

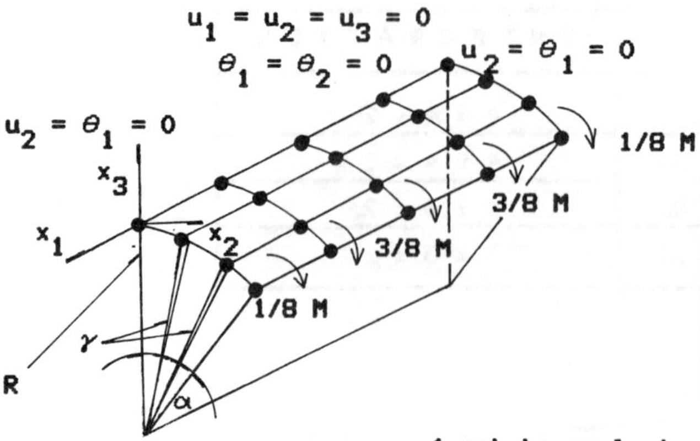

R = 20
h (thickness) = 0.2
E = 2.1 x 10^6
ν = 0.3
Int. = 4x4x2

α	γ	$\theta_{FE} / \theta_{TH}$
30°	0°	0.91
30°	4°	0.01

(not to scale)

FIGURE 4 - Analysis of curved cantilever model using standard A-I-Z shell elements

Although in some cases reduced or selective integration can be a remedy for the locking problem [8], the introduction of spurious zero energy modes [5,9] lowers the reliability of this results for engineering analyses, specially in the case of nonlinear analyses.

As a very straightforward example, designed to illustrate the above statement let us consider the simple finite element model shown in Fig. 5.a [9]; the analysis results obtained using for the 2D element complete and reduced integration are shown in Fig 5.b. It is clear that the reduced integrated model predicts an unrealistic collapse.

It is also a very complicated task to try to identify the rigid body modes in order to constraint them because they depend on the element geometry, as it can be seen in Fig. 6.

Therefore it is desirable to produce a shell element formulation that overcomes the locking problem without incorporating spurious rigid body modes.

(a)

truss elements : elastic-perfectly plastic
2D elements (plane stress) : elastic

(b)

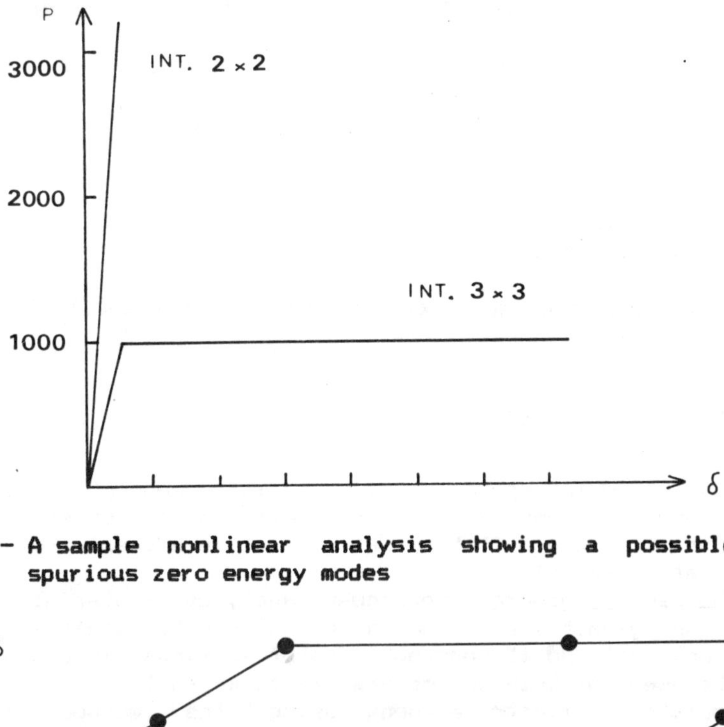

FIGURE 5 - A sample nonlinear analysis showing a possible effect of spurious zero energy modes

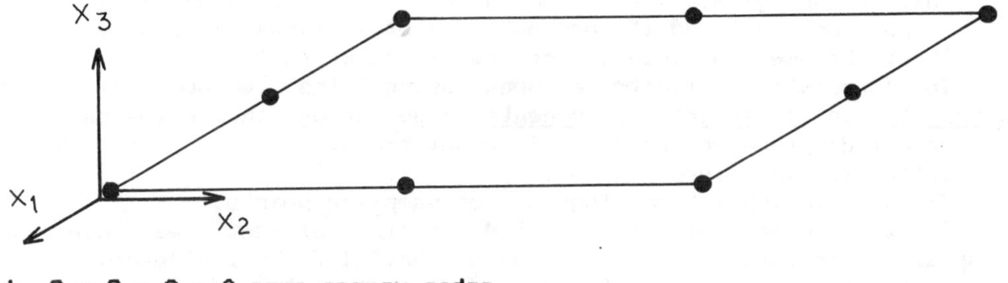

Int. 3 x 3 x 2 : 0 zero energy modes
Int. 2 x 2 x 2 : 2 zero energy modes
(a) No Distortion

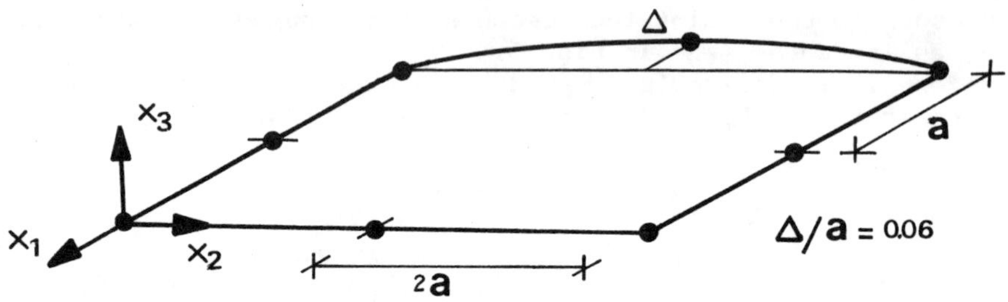

Int. 2 x 2 x 2 : 2 zero energy modes
(b) In-plane Distortion

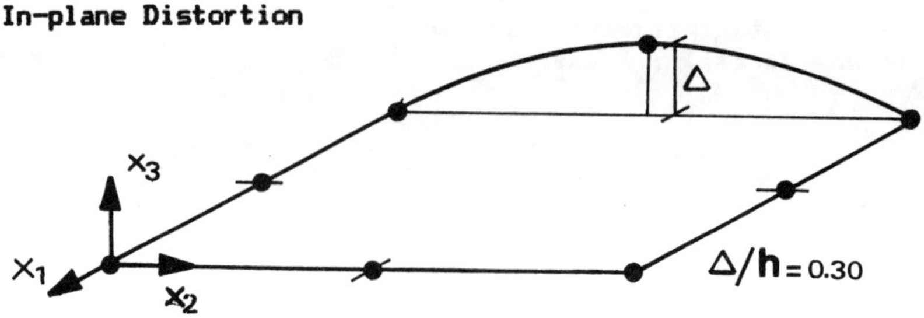

Int. 2 x 2 x 2 : 0 zero energy modes
(c) Out-of-Plane Distortion

FIGURE 6 - Spurious zero energy modes in 8-node shell element

4 . THE MITC4 SHELL ELEMENT

The MITC4 (mixed interpolation of tensorial components, four nodes) is a general shell element [10] with the following features:

- It can be used in non-flat geometries (it is a shell element rather than a plate element)
- It can be used in general nonlinear analyses (material nonlinear analyses and geometrically nonlinear analyses but small strains)
- It does not lock and it does not present spurious rigid body modes
- It can be used for thin and moderately thick shells

To formulate a finite element using the method of mixed interpolation of tensorial components we go through the following steps:

a) Select displacement/rotation interpolations.
b) Select strain interpolations.
c) Tie both interpolations together at sampling points.

In the case of the MITC4 shell element we use the displacement/rotation interpolations of the A-I-Z shell element.

Since it is convenient to use different interpolations for the in-layer strains (corresponding to the plane stress conditions at r_3=const.) and for the transverse shear strains, we formulate the element in its natural coordinate system (general curvilinear coordinates).

4.1 LINEAR ANALYSIS

In Fig. 7 we present the element description. In that figure $\underline{g}_1, \underline{g}_2$ and $\underline{g}_3$ are the usual covariant base vectors [11]

$$\underline{g}_i = \frac{\partial \underline{x}}{\partial r_i} \tag{7}$$

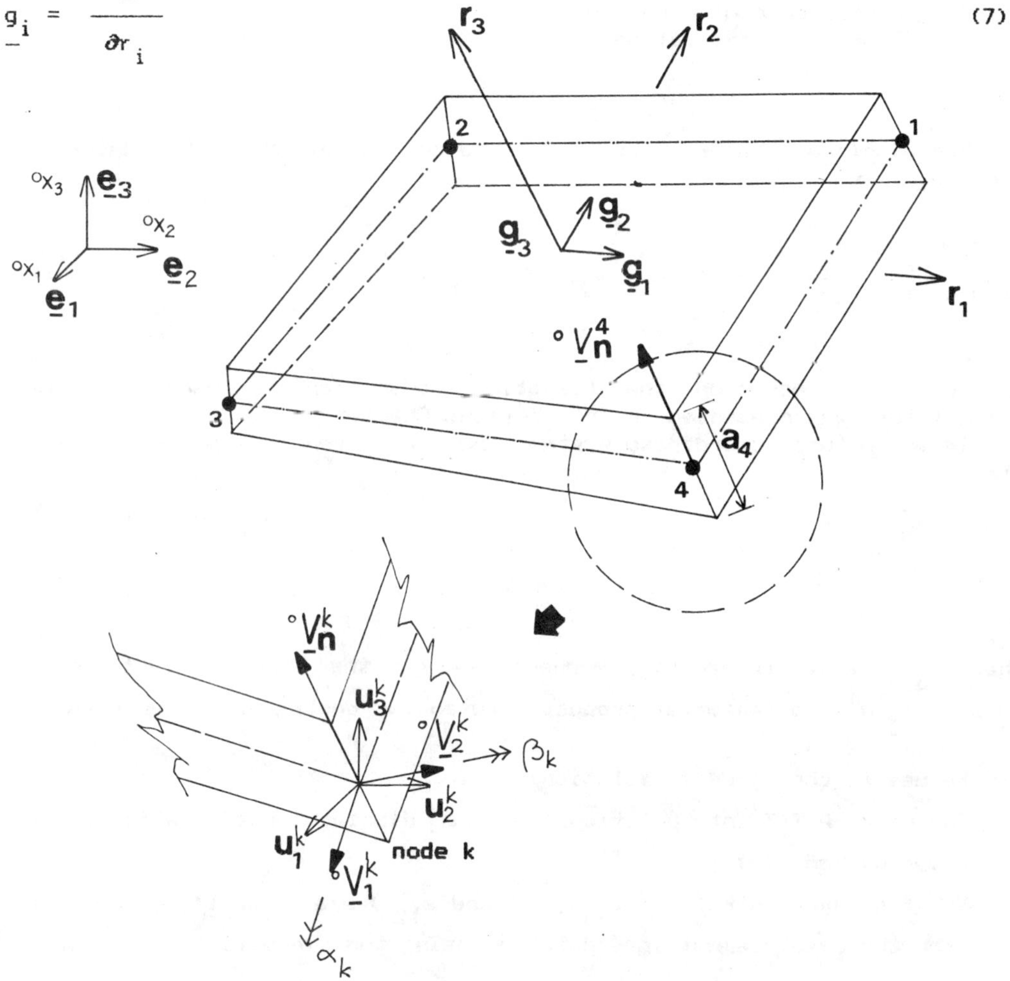

FIGURE 7 - Four-node shell element

Following the A-I-Z element description for any point with natural coordinates (r_1,r_2,r_3) the Cartesian coordinates at a time (load level) "l" are,

$$ {}^{l}x_i = h_k\ {}^{l}x_i^k + \frac{r_3}{2}\ h_k\ a_k\ {}^{l}V_{ni} \qquad (8) $$

l = 0 undeformed configuration
l = t deformed configuration

The displacements from the undeformed configuration to the deformed one are,

$$ u_i = h_k\ u_i^k + \frac{r_3}{2}\ h_k\ a_k\ (-\alpha_k\ {}^{\circ}V_{2i}^k + \beta_k\ {}^{\circ}V_{1i}^k) \qquad (9) $$

in the above equation the summation convention is used and the nomenclature was presented in the previous Chapter.

In the natural coordinate system, the strain tensor can be written as:

$$ \underline{\underline{\varepsilon}} = \tilde{\varepsilon}_{ij}\ \underline{g}^i\ \underline{g}^j \qquad (10) $$

where $\tilde{\varepsilon}_{ij}$ are the covariant components and $\underline{g}^i$ the contravariant base vectors. ($\underline{g}^i\underline{g}^j$ is a tensorial product between two contravariant vectors)

We use in the MITC4 formulation:

- The usual A-I-Z interpolations for the displacements/rotations, as given by Eqn. (9)
- The "in-layer" strains ($\tilde{\varepsilon}_{11}$, $\tilde{\varepsilon}_{22}$ and $\tilde{\varepsilon}_{12}$) are directly calculated from the displacement interpolation using the kinematic relations [5]
- The transverse shear strains $\tilde{\varepsilon}_{13}$ and $\tilde{\varepsilon}_{23}$ are interpolated using the interpolations shown in Fig. 8

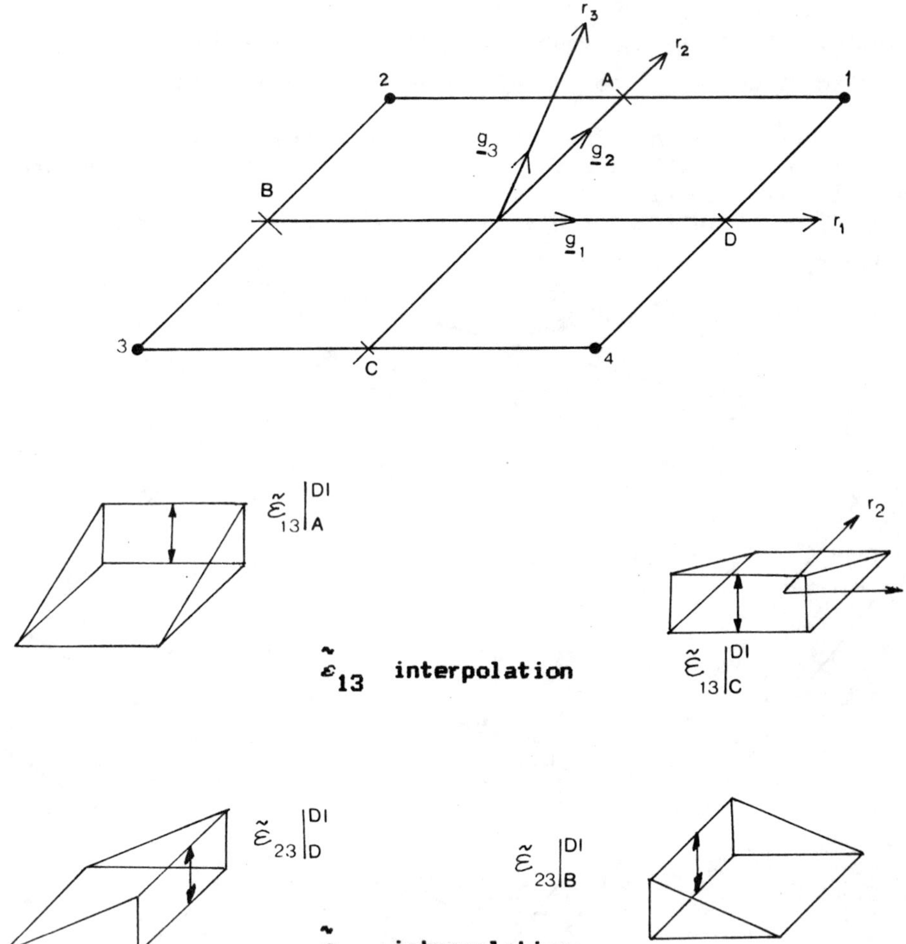

FIGURE 8 - Transverse strain interpolation functions used in the MITC4 formulation

$$\tilde{\varepsilon}_{13} = \frac{1}{2}(1 + r_2)\, \tilde{\varepsilon}_{13}\Big|_A^{DI} + \frac{1}{2}(1 - r_2)\, \tilde{\varepsilon}_{13}\Big|_C^{DI} \tag{11.a}$$

$$\tilde{\varepsilon}_{23} = \frac{1}{2}(1 + r_1)\, \tilde{\varepsilon}_{23}\Big|_D^{DI} + \frac{1}{2}(1 - r_1)\, \tilde{\varepsilon}_{23}\Big|_B^{DI} \tag{11.b}$$

In the above $\tilde{\varepsilon}_{ij}\Big|_P^{DI}$ are the covariant strain components $\tilde{\varepsilon}_{ij}$ calculated at point P from the displacement interpolation.

Equations (11) provide non-compatible shear strains, therefore Irons' Patch Test will have to be satisfied in order to assure the element convergence [12,13].

At any point inside the element a local Cartesian system $(\hat{e}_1, \hat{e}_2, \hat{e}_3)$ is defined, as shown in Fig. 9.

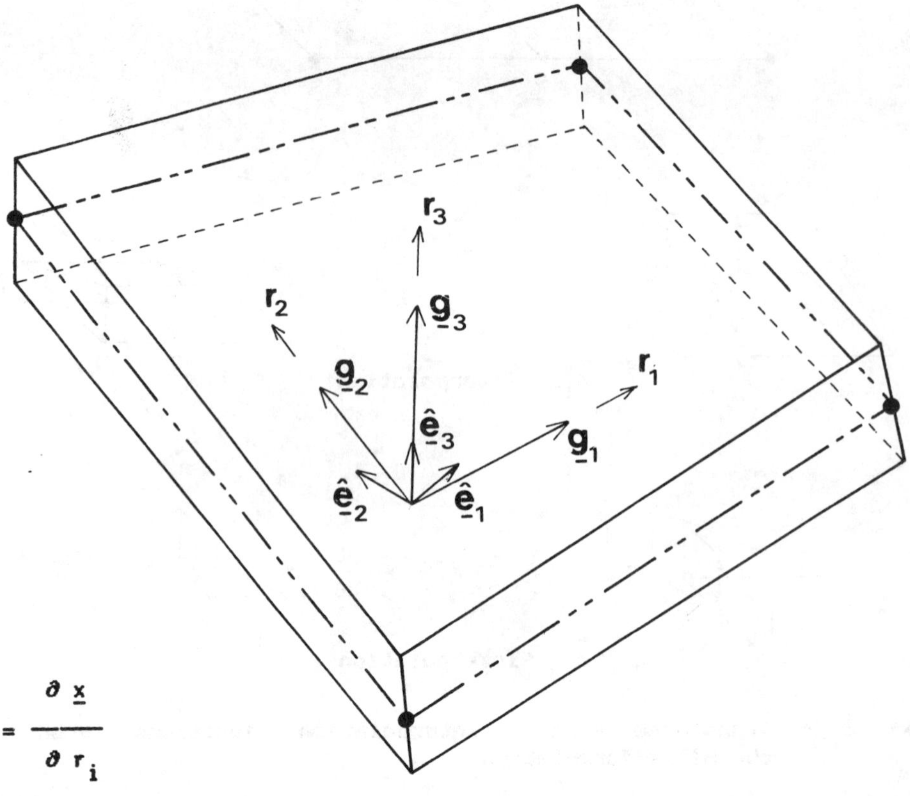

$$\underline{g}_i = \frac{\partial \underline{x}}{\partial r_i}$$

$$\hat{\underline{e}}_3 = \frac{\underline{g}_3}{|\underline{g}_3|} \quad ; \quad \hat{\underline{e}}_1 = \frac{\underline{g}_2 \times \hat{\underline{e}}_3}{|\underline{g}_2 \times \hat{\underline{e}}_3|} \quad ; \quad \hat{\underline{e}}_2 = \hat{\underline{e}}_3 \times \hat{\underline{e}}_1$$

FIGURE 9 - Local Cartesian coordinate system

At the local Cartesian system the constitutive fourth-order tensor $\hat{C}^{mnop}$ is defined, degenerating the usual 3D tensor so as to have $\tau_{33}=0$ (also $\hat{\varepsilon}_{33}=0$) [5]. In the curvilinear system,

$$\tilde{C}^{ijkl} = (\underline{g}^i \cdot \hat{\underline{e}}_m)(\underline{g}^j \cdot \hat{\underline{e}}_n)(\underline{g}^k \cdot \hat{\underline{e}}_o)(\underline{g}^l \cdot \hat{\underline{e}}_p)\hat{C}^{mnop} \tag{12}$$

hence,

$$\tilde{\tau}^{ij} = \tilde{C}^{ijkl}\,\tilde{\varepsilon}_{kl} \tag{13}$$

4.2 NONLINEAR ANALYSIS

We formulate an incremental analysis: being known the configuration at time (load level) t we search for the configuration at time (load level) $t+\Delta t$, using the Total Lagrangian Formulation [5]. We consider in what follows only small strain problems.

For an equilibrium configuration at time $t+\Delta t$ the Principle of Virtual Work, can be stated as,

$$\int_{{}^{\circ}V} {}^{t+\Delta t}_{\;\;\circ}\tilde{S}^{ij}\;\delta\,{}^{t+\Delta t}_{\;\;\circ}\tilde{\varepsilon}_{ij}\;{}^{\circ}dV = {}^{t+\Delta t}\mathcal{R} \tag{14}$$

(Bathe's notation [5])

In the above:

${}^{t+\Delta t}_{\;\;\circ}\tilde{S}^{ij}$: contravariant components, in the natural coordinate system, of the 2nd. Piola-Kirchhoff stress tensor at time $t+\Delta t$, referred to the undeformed configuration (t=0)

${}^{t+\Delta t}_{\;\;\circ}\tilde{\varepsilon}_{ij}$: covariant components, in the natural coordinate system, of the Green-Lagrange strain tensor at time $t+\Delta t$, referred to the undeformed configuration

${}^{\circ}V$: volume at t=o

${}^{t+\Delta}\mathcal{R}$: virtual work of the external forces acting at $t+\Delta t$

In incremental form

$${}^{t+\Delta t}_{\;\;\circ}\tilde{S}^{ij} = {}^{t}_{\circ}\tilde{S}^{ij} + \underbrace{{}_{\circ}\tilde{S}^{ij}}_{\text{increment}} \tag{15.a}$$

$${}^{t+\Delta t}_{\;\;\circ}\tilde{\varepsilon}_{ij} = {}^{t}_{\circ}\tilde{\varepsilon}_{ij} + \underbrace{{}_{\circ}\tilde{\varepsilon}_{ij}}_{\text{increment}} \tag{15.b}$$

and,

$$ {}_{\circ}\tilde{\varepsilon}_{ij} = \underbrace{{}_{\circ}\tilde{e}_{ij}}_{\text{linear terms}} + \underbrace{{}_{\circ}\tilde{\eta}_{ij}}_{\text{nonlinear terms}} \tag{15.c} $$

For the linearized step we use,

$$ {}_{\circ}\overset{\circ}{\tilde{S}}{}^{ij} = {}_{\circ}\tilde{C}^{ijkl} \; {}_{\circ}\tilde{e}_{kl} \tag{16.a} $$

$$ \delta\,{}_{\circ}\tilde{\varepsilon}_{ij} = \delta\,{}_{\circ}\tilde{e}_{ij} \tag{16.b} $$

In Eqn. (16.a) ${}_{\circ}\tilde{C}^{ijkl}$ is the tangent constitutive tensor

The linearized equation of motion is:

$$ \int_{{}^{\circ}V} {}_{\circ}\tilde{C}^{ijkl}\,{}_{\circ}\tilde{e}_{kl}\;\delta\,{}_{\circ}\tilde{e}_{ij}\,d^{\circ}V + \int_{{}^{\circ}V} {}^{t}_{\circ}\tilde{S}^{ij}\;\delta\,{}_{\circ}\tilde{\eta}_{ij}\,d^{\circ}V = {}^{t+\Delta t}\mathcal{R} - \int_{{}^{\circ}V} {}^{t}_{\circ}\tilde{S}^{ij}\;\delta\,{}_{\circ}\tilde{e}_{ij}\,d^{\circ}V \tag{17} $$

For our element we interpolate the incremental displacements / rotations as usual (Eqn.(9)); the in-layer incremental strains ${}_{\circ}\tilde{e}_{11}$, ${}_{\circ}\tilde{e}_{22}$, ${}_{\circ}\tilde{e}_{12}$, ${}_{\circ}\tilde{\eta}_{11}$, ${}_{\circ}\tilde{\eta}_{22}$, and ${}_{\circ}\tilde{\eta}_{12}$ are derived from the displacement interpolations and the terms ${}_{\circ}\tilde{e}_{13}$, ${}_{\circ}\tilde{e}_{23}$, ${}_{\circ}\tilde{\eta}_{13}$ and ${}_{\circ}\tilde{\eta}_{23}$ are interpolated using the functions shown in Fig. 8.

In this way, an equation of the form:

$$ ({}^{t}_{\circ}\underline{K}_{L} + {}^{t}_{\circ}\underline{K}_{NL})\;\underline{U} = {}^{t+\Delta t}\underline{R} - {}^{t}_{\circ}\underline{F} \tag{18} $$

is obtained for the linearized incremental step. In Eq. (18):

${}^{t}_{\circ}\underline{K}_{L}$: linear stiffness matrix

${}^{t}_{\circ}\underline{K}_{NL}$: nonlinear stiffness matrix [14]

$\underline{U}$: vector of generalized incemental displacements

${}^{t+\Delta t}\underline{R}$: vector of nodal forces equivalent (in the sense of the Principle of Virtual Work) to the external loads acting at t+Δt

${}^{t}_{\circ}\underline{F}$: vector of nodal forces equivalent to the stresses acting at time t.

The solution of Eq. (18) provides a linearized approximation to the step from t to t+Δt.
An iterative scheme [5,15,16] is used until the equation

$$ {}^{t+\Delta t}\underline{R} - {}^{t+\Delta t}_{\circ}\underline{F} = \underline{0} \tag{19} $$

is satisfied within certain computational tolerances.

4.3 NUMERICAL EXPERIMENTATION

Our purpose, in the present Chapter is to report upon numerical experimentation with the MITC4 element [9,10,18,19], in an organized way such that the following objectives are achieved:

- (i) We show that the MITC4 converges, implying by this [17] that it is stable and consistent.
- (ii) We examine the solutions it provides for some elementary problems so that we can verify that the element behaves as we expected (e.g. ability to approximate the Kirchhoff-Love hypothesis for thin shells under different geometrical configurations), and at the same time gain more insight about the element performance.
- (iii) For some benchmark linear problems we compare its rate of convergence against other existing elements.
- (iv) We show via some simple problems that the element can be reliably used for general nonlinear analysis.
- (v) We compare the solutions it provides in nonlinear analyses with experimental, analytical or numerical solutions obtained using other nonlinear elements. In this way we will be able to judge on the element performance (accuracy plus efficiency) in nonlinear analyses.

The numerical solutions are all obtained using 2x2 Gauss integration in (r_1, r_2) and 2 and 4 Gauss points in r_3 for elastic and elasto-plastic analyses respectively.

4.3.1 CONVERGENCE

First the stability issue will be discussed. The eigenvalues of the stiffness matrices of undistorted and distorted elements were calculated. In all cases, as expected, the element displayed the six rigid body modes and no spurious zero energy modes.

In order to check for the consistency of the formulation the Patch

Tests shown in Fig.10 were performed. In all cases the rotations, transverse displacements and stresses exactly agreed with the analytical results. An analytical proof that the MITC4 satisfies the bending Patch Test was given in [19].

Finally it should be noted that the Patch Test is of course passed for the three membrane strain states.

$E = 2.1 \times 10^6$
$\nu = 0.3$
thickness = { 1 / 0.001

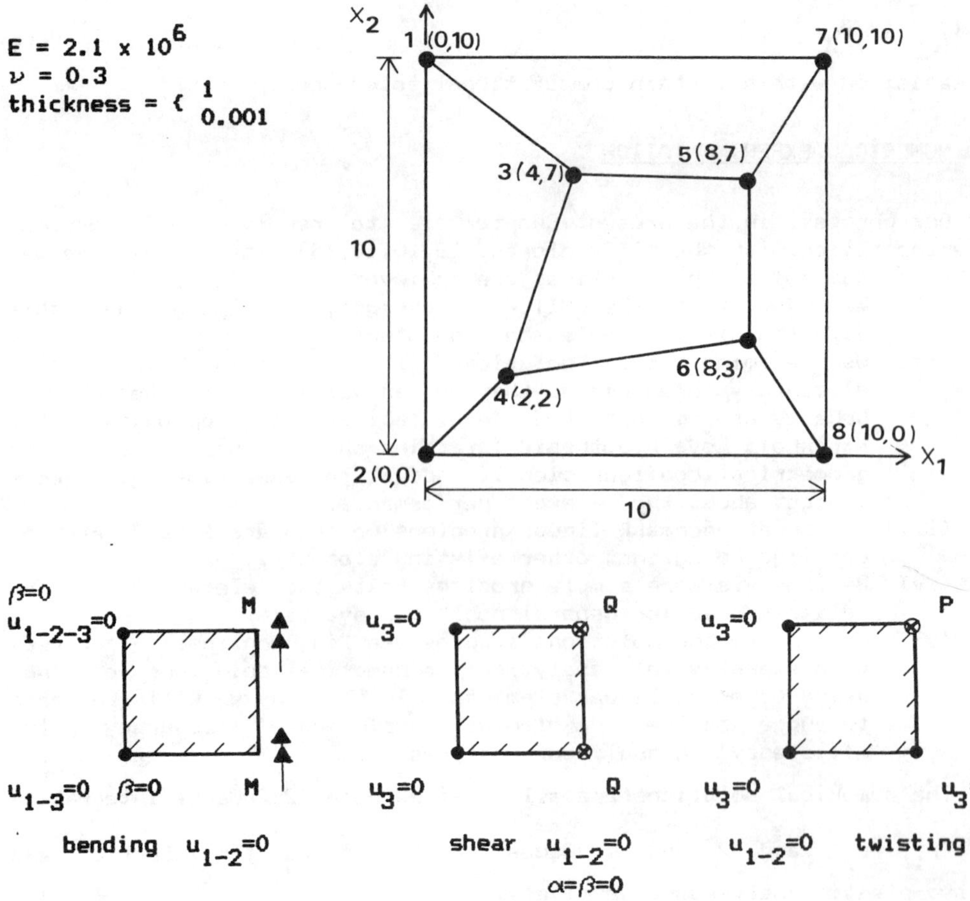

FIGURE 10 - Patch Tests for the MITC4 element

4.3.2 LINEAR ANALYSES

In Fig.11 we show our results for the classical Scordelis-Lo shell. The MITC4 results are compared against the 16-node A-I-Z shell element results and also against the DKT element [20] results.

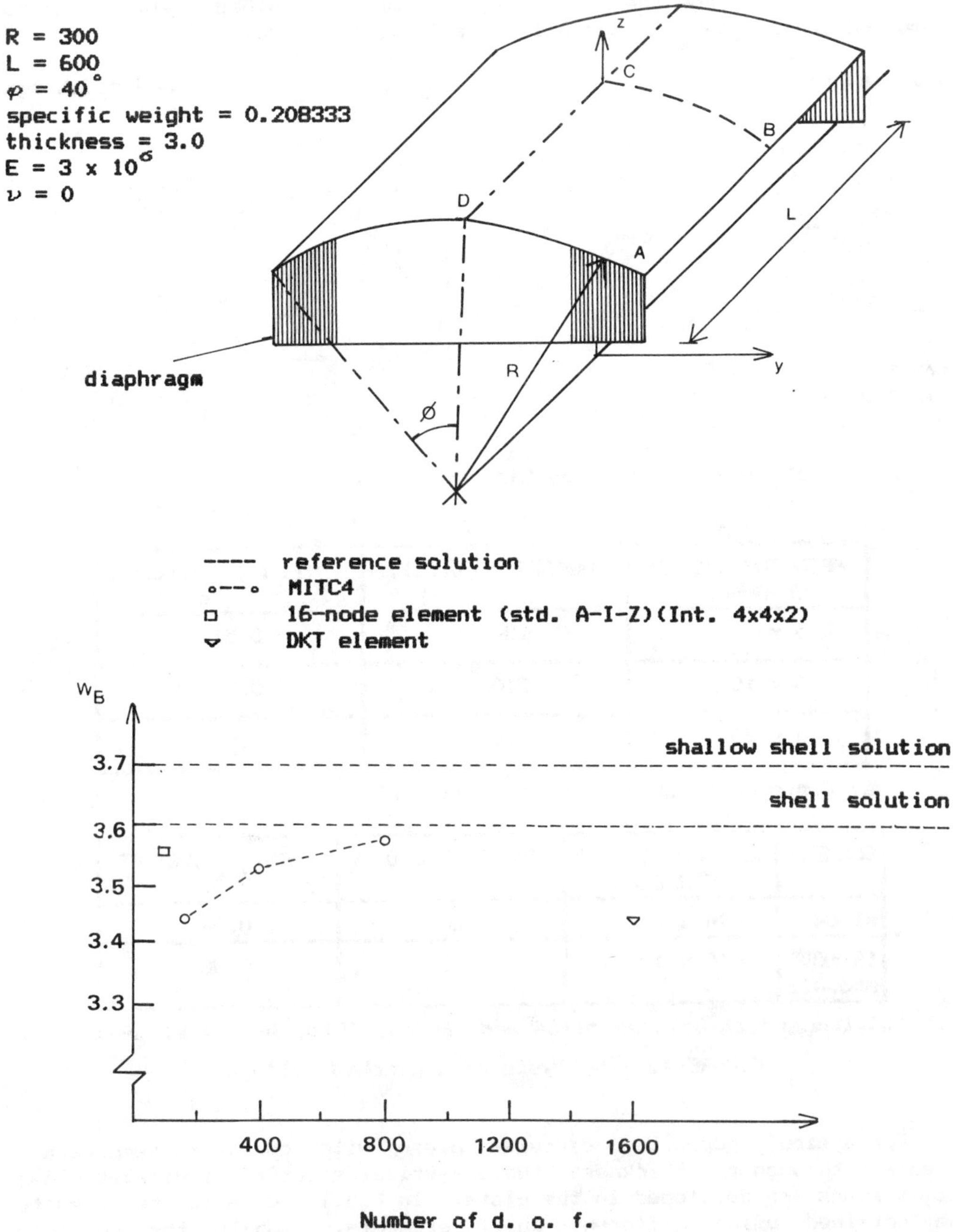

FIGURE 11 - Analysis of the Scordelis-Lo shell

In Fig.12 we show the MITC4 results for a pinched cylinder also compared against the 16-node A-I-Z shell element results.

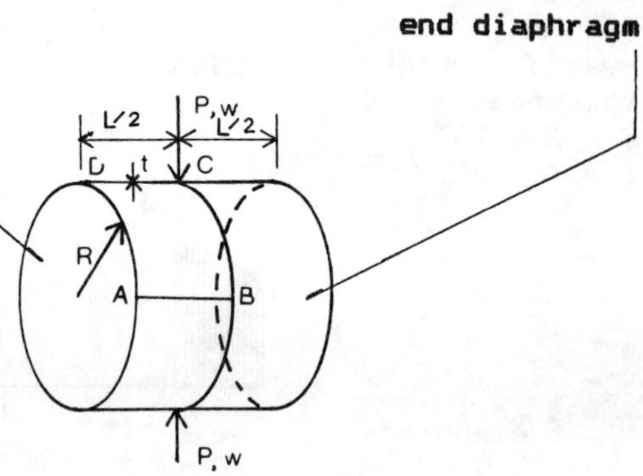

R/t = 100
L/R = 2
ν = 0.3

$$\hat{w}_c = \frac{w_c E t}{P}$$

$\hat{w}_c$ series solution = -164.24 by Lindberg et. al.

MESH FOR 1/8 TH OF SHELL	NUMBER OF D.O.F.	$\hat{w}_c$ FEM / $\hat{w}_c$ ANALYT.
5 x 5	130	0.51
10 x 10	510	0.83
20 x 20	2020	0.96

a) Convergence study for MITC4 element

ELEMENT	MESH FOR 1/8 TH OF SHELL	NUMBER OF D.O.F.	$\hat{w}_c$ FEM / $\hat{w}_c$ ANALYT.
MITC4	20 x 20	2020	0.96
16-NODE std.AIZ	10 x 10	4530	0.98

b) Comparison between MITC4 and 16-node (Std. A-I-Z) elements

FIGURE 12 - Analysis of a pinched cylinder

For a simply supported circular plate with constant temperature gradient through the thickness, the analytical solution indicates that no stresses are developed in the plate. In Fig.13 we show the results we obtained using different shell elements. While the complete integrated A-I-Z elements predict "parasitic" stresses, the MITC4 element predicts exactly zero stresses.

	F.E.M. MODEL	$\lvert\varepsilon_w\rvert = \frac{\lvert W_{FE}-W_{AN}\rvert}{\lvert W_{AN}\rvert}$	$\varepsilon_\tau = \frac{\lvert \tau_{pp}^{max} \rvert}{\lvert E\alpha\theta \rvert}$
STANDARD A-I-Z ELEMENTS	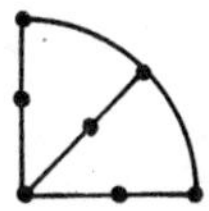	0.26	1.10
		0.07	0.54
	INT.3x3x2 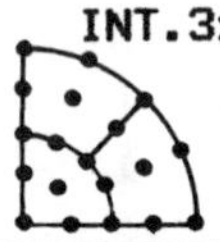	0.008	CENTRAL ELEM. 0.62 OUTER ELEM. 0.22
	SAME MODEL INT.2x2x2	0.014	0.02
	INT.4x4x2 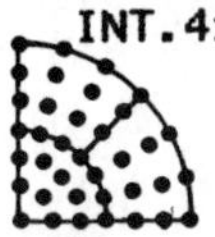	0.003	0.26
	SAME MODEL INT.3x3x2	0.0001	0.04
MITC4	INT.2x2x2 	0	0

FIGURE 13 - Analysis of a circular plate with constant temperature gradient through the thickness

4.3.3 NONLINEAR ANALYSES

In Fig.14 some results corresponding to a large displacements/rotations analysis of a cantilever are presented. Please note that model II (A-I-Z) yields an accurate response solution in linear analysis but locks once the element is curved in the nonlinear response solution.

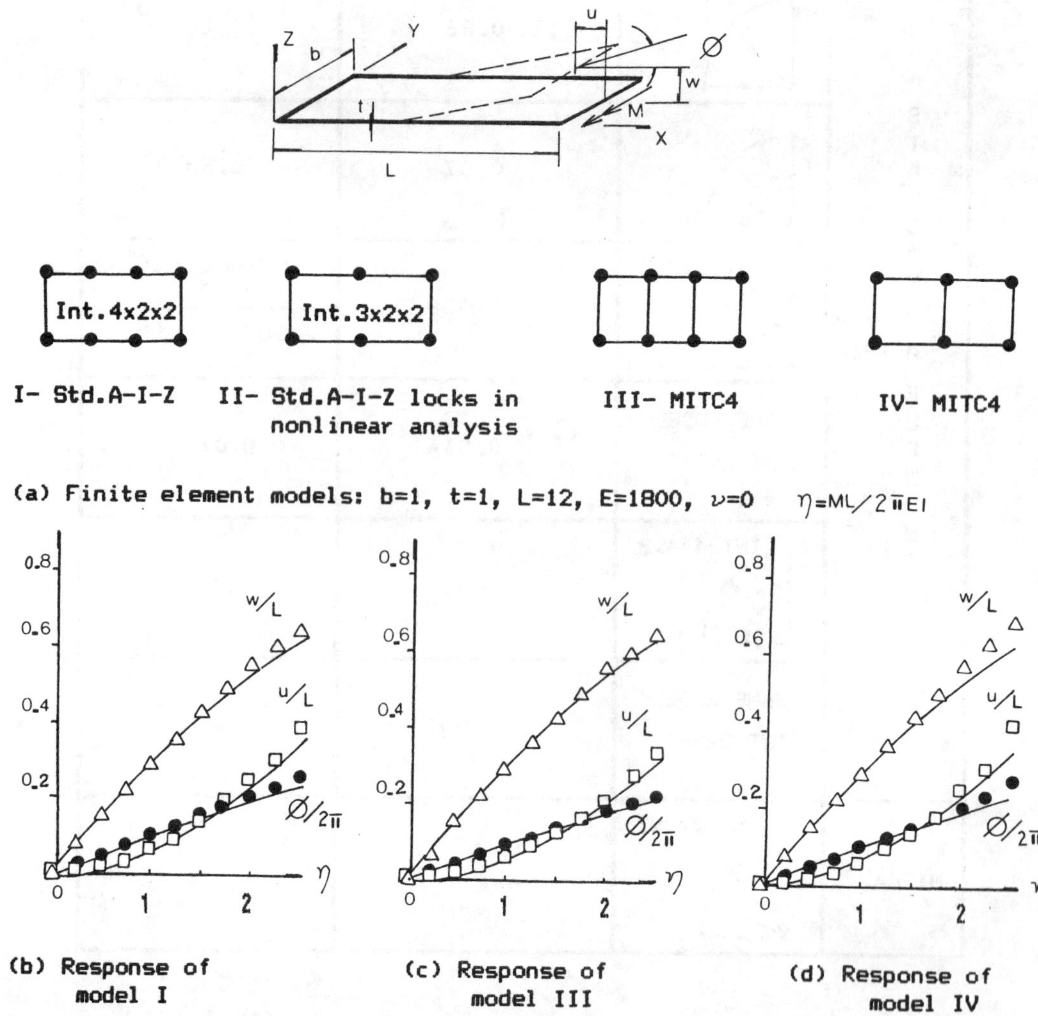

FIGURE 14 - Large displacement/rotation analysis of a cantilever using the MITC4 element

In Figs. 15 and 16 we present some other MITC4 solutions for nonlinear shell problems that are usually used in the literature as benchmarks.

$R_1 = R_2 = 2540$
a = 784.90
h = 99.45
E = 68.95
ν = 0.3
All edges are hinged and immovable

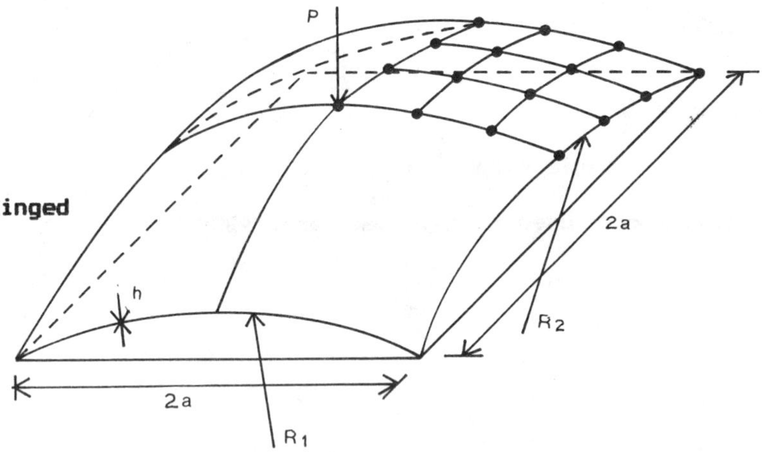

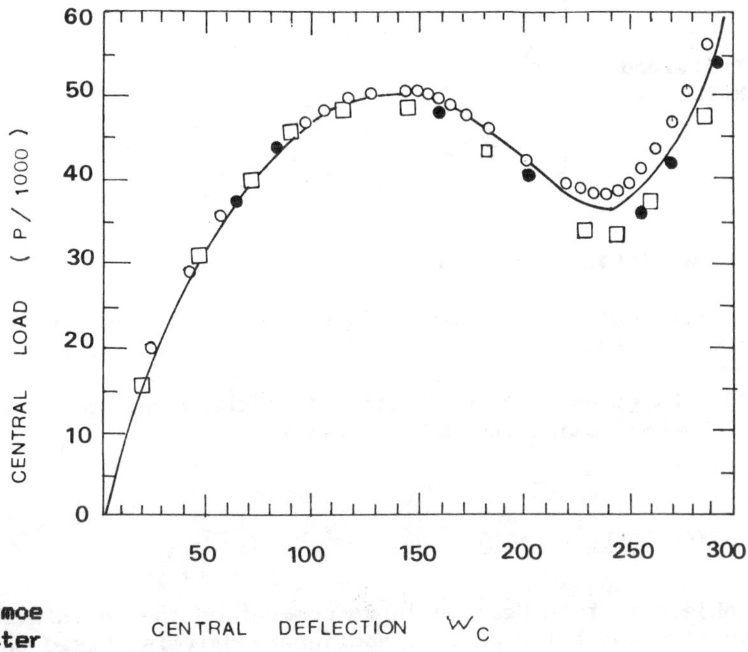

○ Horrigmoe
— Leicester
● Nine 4-node elements
□ One 16-node element (Std. A-I-Z) Int. 4x4x2

FIGURE 15 - Geometrical nonlinear response of a spherical shell

E = 21000
ν = 0
E_T = 0
σ_y = 4.2
L = 15200
R = 7600
ϕ = 40°
thickness = 76

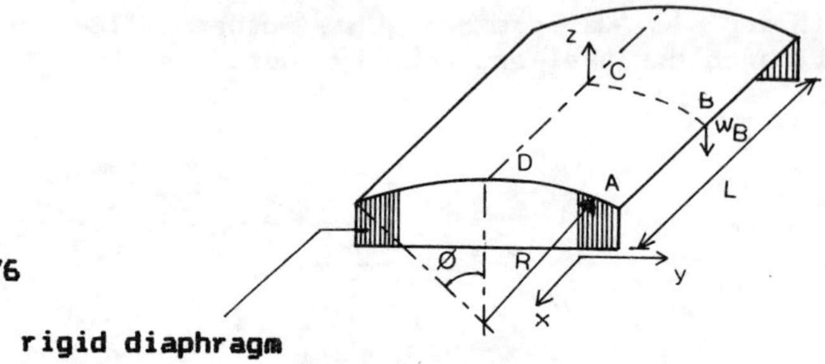

9x9 uniform mesh used to represent area ABCD

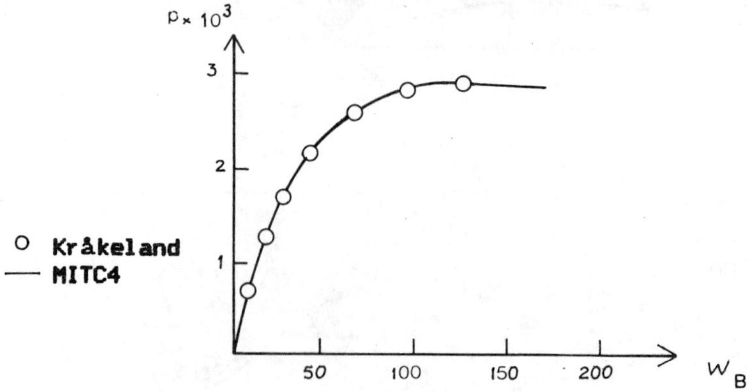

4 point Gauss integration through element thickness

p is the vertical pressure per unit of projected area on the horizontal plane

FIGURE 16 - Large deflection elastic-plastic analysis of a cylindrical shell using the MITC4 element

5. THE MITC8 SHELL ELEMENT

Our object in this Section is to comment on the formulation of an 8-node shell element for general nonlinear analysis, based on the method of mixed interpolation of tensorial components: the MITC8 [19].

When developing the MITC8 element we used the following design criteria:

i) The element must not present either shear locking or membrane locking.
ii) The element must not have spurious zero energy modes.
iii) The element must satisfy Irons' Patch Test.
iv) The element must have low sensitivity to distortions.
v) The computer implementation must be as effective as possible.

To construct this new shell element we used our accumulated insight into the behavior of elements to satisfy the above design criteria as closely as possible. As a basic tool to test our ideas we used the Patch Tests in Fig.17, after which we measured the order of convergence of our elements by solving some well-established plate and shell problems.

$E = 2.1 \times 10^6$
$\nu = 0.3$
thickness = 0.01

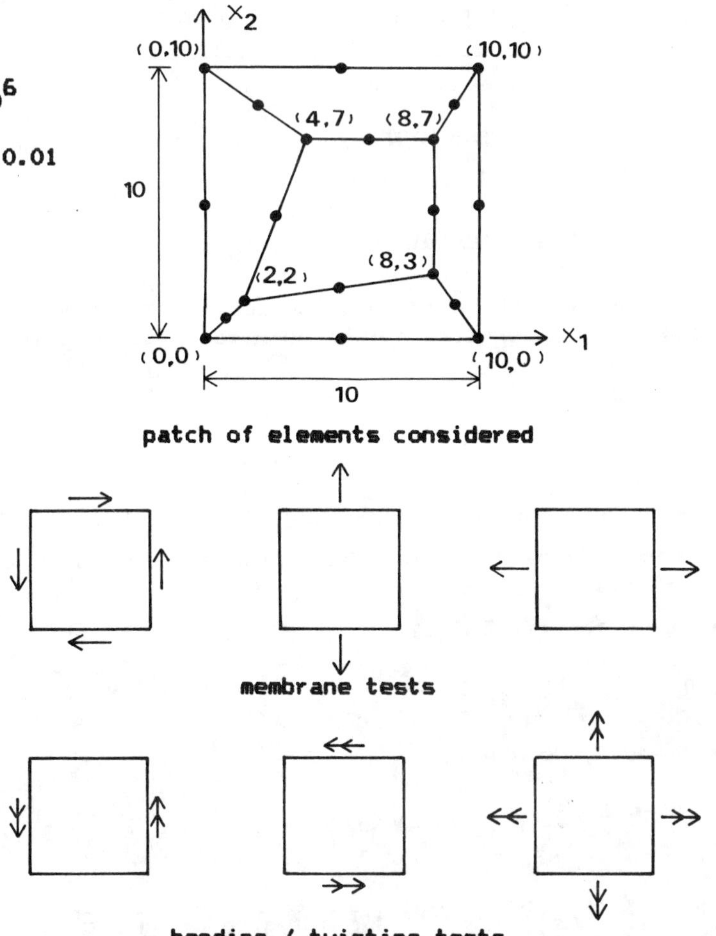

FIGURE 17 - Patch Tests for the MITC8 element

5.1 THE MITC8 FORMULATION

The element geometry and displacement/rotation interpolations follow the A-I-Z element (see Fig.18).

The strain tensor at any point inside the element is written in the natural coordinate system of the element as:

$$\underset{\text{in-layer strains}}{\underline{\underline{\varepsilon}} = \underbrace{\tilde{\varepsilon}_{rr}\,\underline{g}^r\,\underline{g}^r + \tilde{\varepsilon}_{ss}\,\underline{g}^s\,\underline{g}^s + \tilde{\varepsilon}_{rs}\,(\underline{g}^r\,\underline{g}^s + \underline{g}^s\,\underline{g}^r)}}$$

$$+ \underbrace{\tilde{\varepsilon}_{rt}\,(\underline{g}^r\,\underline{g}^t + \underline{g}^t\,\underline{g}^r) + \tilde{\varepsilon}_{st}\,(\underline{g}^s\,\underline{g}^t + \underline{g}^t\,\underline{g}^s)}_{\text{transverse strains}} \tag{20}$$

In the formulation of the MITC8 shell element we use the following strain interpolations:

5.1.1 In-layer strain interpolation

To avoid membrane locking and have no spurious zero energy modes, we use the following in-layer strain interpolation, (see Fig.18).

$$\underline{\underline{\varepsilon}} = \sum_{i=1}^{8} h_i^{IS}\,\underline{\underline{\varepsilon}}\big|_i \tag{21}$$

where the h_i^{IS} are obtained from the isoparametric interpolation functions in Ref.[5] by replacing the variable r with r/a, and the variable s with s/a; $a = 1/\sqrt{3}$.

Also, we have for i=1,2,3 and 4,

$$\underline{\underline{\varepsilon}}\big|_i = \tilde{\varepsilon}_{rr}\,\underline{g}^r\,\underline{g}^r\big|_i^{DI} + \tilde{\varepsilon}_{ss}\,\underline{g}^s\,\underline{g}^s\big|_i^{DI} + \tilde{\varepsilon}_{rs}\,(\underline{g}^r\,\underline{g}^s + \underline{g}^s\,\underline{g}^r)\big|_i^{DI} \tag{22}$$

For i = 5 and 7, we use

$$\underline{\underline{\varepsilon}}\big|_5 = \tilde{\varepsilon}_{ss}\,\bar{\underline{g}}^s\,\bar{\underline{g}}^s\big|_5^{DI} + \{\bar{\underline{g}}_r.[\tfrac{1}{2}(\underline{\underline{\varepsilon}}\big|_1^{DI} + \underline{\underline{\varepsilon}}\big|_2^{DI})].\bar{\underline{g}}_r\}\;\bar{\underline{g}}^r\,\bar{\underline{g}}^r\big|_5^{DI}$$

$$+ \{\bar{\underline{g}}_r.[\tfrac{1}{2}(\underline{\underline{\varepsilon}}\big|_1^{DI} + \underline{\underline{\varepsilon}}\big|_2^{DI})].\bar{\underline{g}}_s\}\,(\bar{\underline{g}}^r\,\bar{\underline{g}}^s + \bar{\underline{g}}^s\,\bar{\underline{g}}^r)\big|_5^{DI} \tag{23}$$

$$\underline{\underline{\varepsilon}}\Big|_7 = \tilde{\varepsilon}_{ss}\, \underline{\bar{g}}^s\, \underline{\bar{g}}^s \Big|_7^{DI} + \left\{ \underline{\bar{g}}_r \cdot \left[\frac{1}{2} \left(\underline{\underline{\varepsilon}}\Big|_3^{DI} + \underline{\underline{\varepsilon}}\Big|_4^{DI} \right) \right] \cdot \underline{\bar{g}}_r \right\} \underline{\bar{g}}^r\, \underline{\bar{g}}^r \Big|_7^{DI}$$

$$+ \left\{ \underline{\bar{g}}_r \cdot \left[\frac{1}{2} \left(\underline{\underline{\varepsilon}}\Big|_3^{DI} + \underline{\underline{\varepsilon}}\Big|_4^{DI} \right) \right] \cdot \underline{\bar{g}}_s \right\} \left(\underline{\bar{g}}^r\, \underline{\bar{g}}^s + \underline{\bar{g}}^s\, \underline{\bar{g}}^r \right) \Big|_7^{DI} \tag{24}$$

where

$$\underline{\bar{g}}_s \equiv \underline{g}_s \quad ; \quad \underline{\bar{g}}_t \equiv \underline{g}_t$$

$$\underline{\bar{g}}_r = \underline{g}_r - \alpha\, \underline{g}_s \quad ; \qquad \alpha = \frac{g_{rs}}{g_{ss}} \tag{25}$$

For $i = 6$ and 8, we use

$$\underline{\underline{\varepsilon}}\Big|_6 = \tilde{\varepsilon}_{rr}\, \underline{\bar{g}}^r\, \underline{\bar{g}}^r \Big|_6^{DI} + \left\{ \underline{\bar{g}}_s \cdot \left[\frac{1}{2} \left(\underline{\underline{\varepsilon}}\Big|_2^{DI} + \underline{\underline{\varepsilon}}\Big|_3^{DI} \right) \right] \cdot \underline{\bar{g}}_s \right\} \underline{\bar{g}}^s\, \underline{\bar{g}}^s \Big|_6^{DI}$$

$$+ \left\{ \underline{\bar{g}}_r \cdot \left[\frac{1}{2} \left(\underline{\underline{\varepsilon}}\Big|_2^{DI} + \underline{\underline{\varepsilon}}\Big|_3^{DI} \right) \right] \cdot \underline{\bar{g}}_s \right\} \left(\underline{\bar{g}}^r\, \underline{\bar{g}}^s + \underline{\bar{g}}^s\, \underline{\bar{g}}^r \right) \Big|_6^{DI} \tag{26}$$

$$\underline{\underline{\varepsilon}}\Big|_8 = \tilde{\varepsilon}_{rr}\, \underline{\bar{g}}^r\, \underline{\bar{g}}^r \Big|_8^{DI} + \left\{ \underline{\bar{g}}_s \cdot \left[\frac{1}{2} \left(\underline{\underline{\varepsilon}}\Big|_1^{DI} + \underline{\underline{\varepsilon}}\Big|_4^{DI} \right) \right] \cdot \underline{\bar{g}}_s \right\} \underline{\bar{g}}^s\, \underline{\bar{g}}^s \Big|_8^{DI}$$

$$+ \left\{ \underline{\bar{g}}_r \cdot \left[\frac{1}{2} \left(\underline{\underline{\varepsilon}}\Big|_1^{DI} + \underline{\underline{\varepsilon}}\Big|_4^{DI} \right) \right] \cdot \underline{\bar{g}}_s \right\} \left(\underline{\bar{g}}^r\, \underline{\bar{g}}^s + \underline{\bar{g}}^s\, \underline{\bar{g}}^r \right) \Big|_8^{DI} \tag{27}$$

where

$$\underline{\bar{g}}_r \equiv \underline{g}_r \quad ; \quad \underline{\bar{g}}_t \equiv \underline{g}_t$$

$$\underline{\bar{g}}_s = \underline{g}_s - \beta\, \underline{g}_r \quad ; \quad \beta = \frac{g_{rs}}{g_{rr}} \tag{28}$$

Equation (21) corresponds to a mixed interpolation of the tensorial components such that the membrane patch test is passed (for straight-sided elements) and, in non-flat elements, membrane locking is avoided.

5.1.2 Transverse shear strain interpolation

The transverse shear strain interpolation is selected to avoid shear locking and, as for the in-layer strain interpolation, no spurious zero energy modes must be introduced. We use the following interpolation for $\tilde{\varepsilon}_{rt}\ \underline{g}^r\ \underline{g}^t$ (see Fig.18).

$$\tilde{\varepsilon}_{rt}\underline{g}^r\underline{g}^t = \sum_{i=1}^{4} h_i^{RT}\tilde{\varepsilon}_{rt}\underline{g}^r\underline{g}^t\Big|_i^{DI} + h_5^{RT}\left[\frac{1}{2}\left(\tilde{\varepsilon}_{rt}\Big|_{RA}^{DI} + \tilde{\varepsilon}_{rt}\Big|_{RB}^{DI}\right)\right]\underline{g}^r\underline{g}^t\Big|_5^{DI} \tag{29}$$

where

$$\begin{aligned}
h_1^{RT} &= \frac{1}{4}\left[1 + \frac{r}{a}\right](1 + s) - \frac{1}{4}h_5^{RT} \\
h_2^{RT} &= \frac{1}{4}\left[1 - \frac{r}{a}\right](1 + s) - \frac{1}{4}h_5^{RT} \\
h_3^{RT} &= \frac{1}{4}\left[1 - \frac{r}{a}\right](1 - s) - \frac{1}{4}h_5^{RT} \\
h_4^{RT} &= \frac{1}{4}\left[1 + \frac{r}{a}\right](1 - s) - \frac{1}{4}h_5^{RT} \\
h_5^{RT} &= \left[1 - \left(\frac{r}{a}\right)^2\right](1 - s^2)
\end{aligned} \tag{30}$$

where $a = 1/\sqrt{3}$. We note that in equation (29) we have replaced $\tilde{\varepsilon}_{rt}\big|_5^{DI}$ by the mean of the components at points RA and RB.

Similarly, we use the following interpolation for $\tilde{\varepsilon}_{st}\underline{g}^s\underline{g}^t$

$$\tilde{\varepsilon}_{st}\underline{g}^s\underline{g}^t = \sum_{i=1}^{4} h_i^{ST}\ \tilde{\varepsilon}_{st}\underline{g}^s\underline{g}^t\Big|_i^{DI} + h_5^{ST}\left[\frac{1}{2}\left(\tilde{\varepsilon}_{st}\Big|_{SA}^{DI} + \tilde{\varepsilon}_{st}\Big|_{SB}^{DI}\right)\right]\underline{g}^s\underline{g}^t\Big|_5^{DI} \tag{31}$$

where

$$
\begin{aligned}
h_1^{ST} &= \frac{1}{4}\left[1+\frac{s}{a}\right](1+r)-\frac{1}{4}h_5^{ST}\\
h_2^{ST} &= \frac{1}{4}\left[1+\frac{s}{a}\right](1-r)-\frac{1}{4}h_5^{ST}\\
h_3^{ST} &= \frac{1}{4}\left[1-\frac{s}{a}\right](1-r)-\frac{1}{4}h_5^{ST}\\
h_4^{ST} &= \frac{1}{4}\left[1-\frac{s}{a}\right](1+r)-\frac{1}{4}h_5^{ST}\\
h_5^{ST} &= \left[1-\left(\frac{s}{a}\right)^2\right](1-r^2)
\end{aligned}
\tag{32}
$$

where $a = 1/\sqrt{3}$.

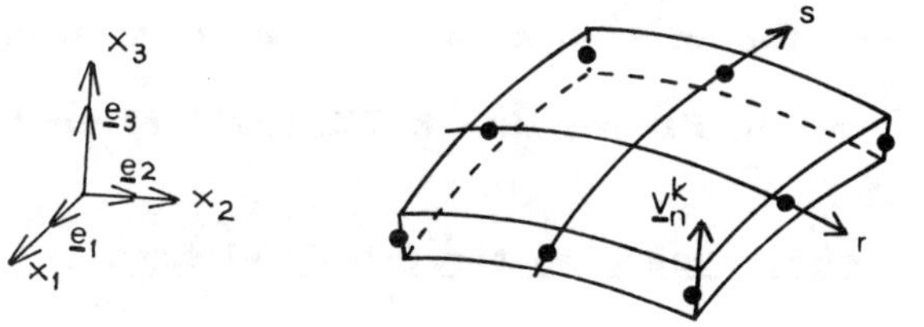

(a) Element geometry

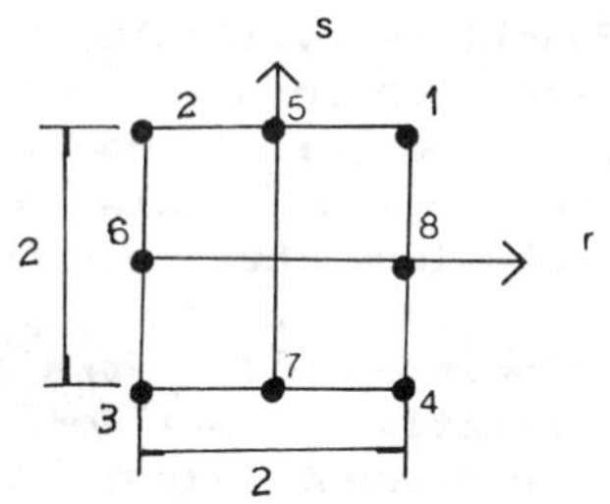

(b) Nodes used for interpolation of displacements and rotations

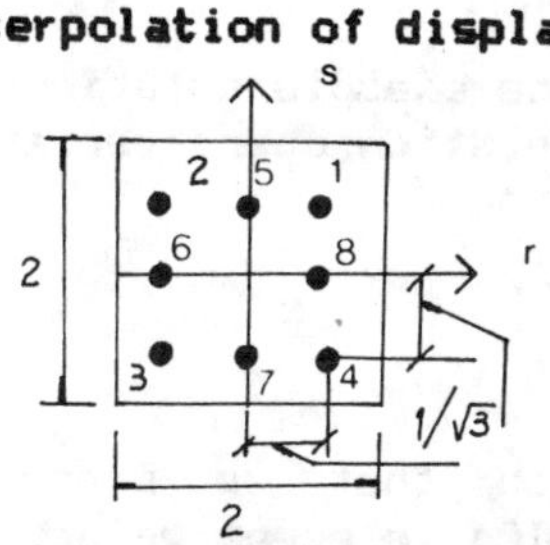

(c) Points used for interpolation of in-layer strains

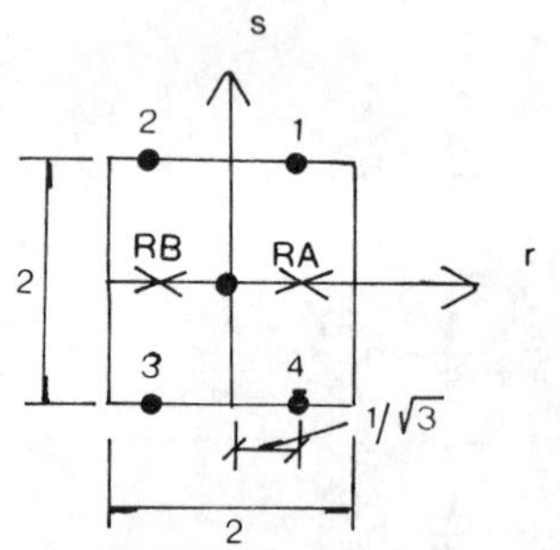

(d) Points used for interpolation of transverse shear strain $\tilde{\varepsilon}_{rt} \underline{g}^r \underline{g}^t$

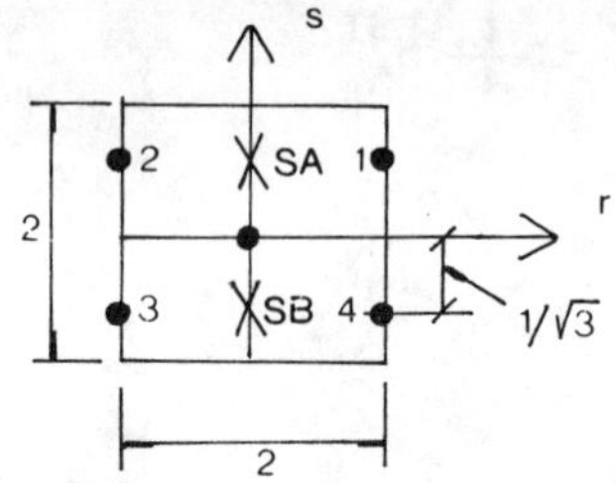

(e) Points used for interpolation of transverse shear strain $\tilde{\varepsilon}_{st} \underline{g}^s \underline{g}^t$

FIGURE 18 - MITC8 shell element

5.1.3 Some remarks regarding the 8-node element

(i) With the interpolation functions available, the implementation of the element for linear and nonlinear analysis follows the usual procedures. For geometric nonlinear analysis we have implemented the total Lagrangian formulation (allowing for very large displacements and rotations but restricted to small strain conditions).

(ii) The Patch Tests - using elements with straight sides - are all passed exactly. However, relatively small errors in displacements and stresses arise when the element sides are curved, or when the mid-side nodes are not placed at their mid-side physical location.

(iii) The results obtained in the example solutions given in the next Section demonstrate the solution characteristics of the MITC8 in linear and nonlinear analyses.

5.2 NUMERICAL EXPERIMENTATION

Following the same approach that we presented in the previous section for the case of the MITC4 element we are going to report in this section upon numerical experimentation with the MITC8.

The numerical solutions are all obtained using (3x3) Gauss integration in (r_1, r_2) and 2 and 4 Gauss points in r_3 for elastic and elasto-plastic analyses respectively.

5.2.1 Convergence

(i) Stability

The eigenvalues of the stiffness matrices of undistorted and distorted elements were calculated. In all cases, as expected, the element displayed the six rigid body modes and no spurious zero energy modes.

(ii) Consistency

The experiences with the Patch Test were reported in 5.1.3.

5.2.2 Linear analyses

In Fig.19 we show the MITC8 results for a plane stress analysis, compared against the results we obtained using a standard 8-node isoparametric A-I-Z or plane stress element. Although for this problem the MITC8 results are slightly less accurate than the standard isoparametric element results, it has to be remembered that the main purpose of the in-layer strain interpolation in Eqns. (21)-(28) is to avoid membrane locking while not deteriorating too much the plane stress results, as it can be seen in this example.

$L = 56$
$b = 20$
$d = 10$
$E = 7 \times 10^4$
$\nu = 0.25$
$p = 25$
h (thickness) = 1

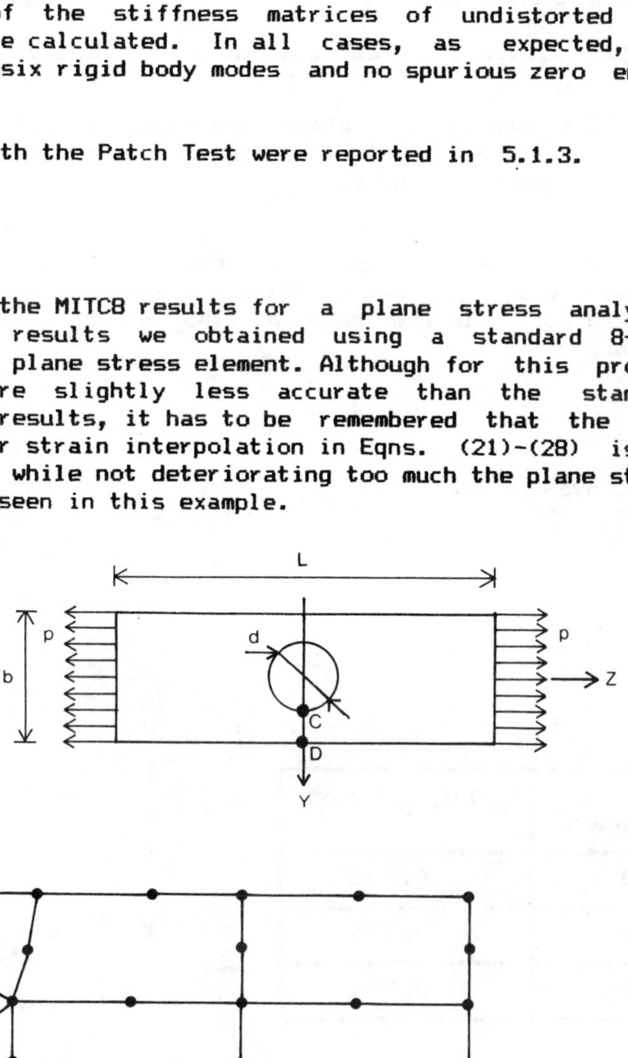

Mesh used

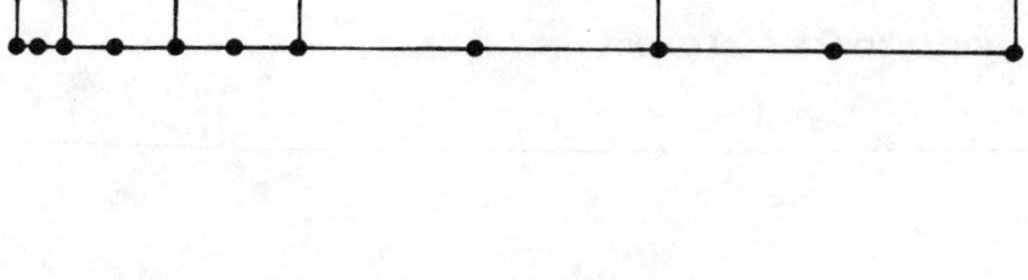

Calculated at point	$\tau_{zz}^{FEM} / \tau_{zz}^{ANALYT.}$	
	Isop. element 3x3x2 integrn.	MITC8
C	0.953	0.943
D	1.013	1.038

FIGURE 19 – Analysis of a plate with a hole in plane stress using the MITC8 element. The stresses are directly calculated at points C and D.

In Fig.20 we present the results corresponding to a curved cantilever under constant bending. Please compare with the results in Fig.4 and notice in the case of the MITC8 element the non-locking behavior.

R = 20
h (thickness) = 0.2
$E = 2.1 \times 10^6$
$\nu = 0.3$

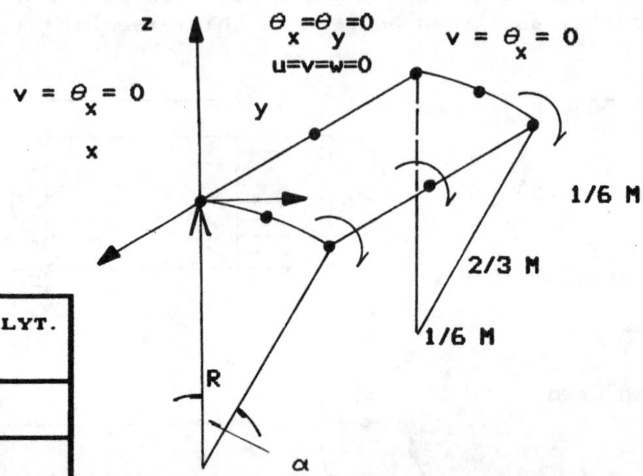

α [degrees]	$\theta^{FEM} / \theta^{ANALYT.}$
30	0.999
60	0.994
90	0.981

(a) Results using undistorted element

γ [degrees]	$\theta^{FEM} / \theta^{ANALYT.}$
0	0.999
3	1.000
5	1.001

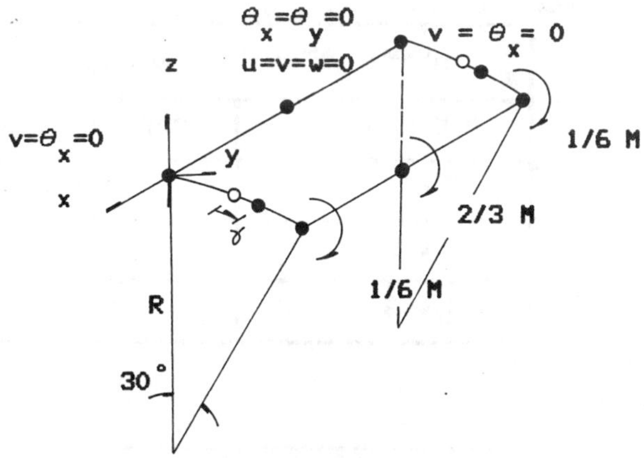

γ [degrees]	$\theta^{FEM} / \theta^{ANALYT.}$
0	0.999
3	0.974
5	0.928

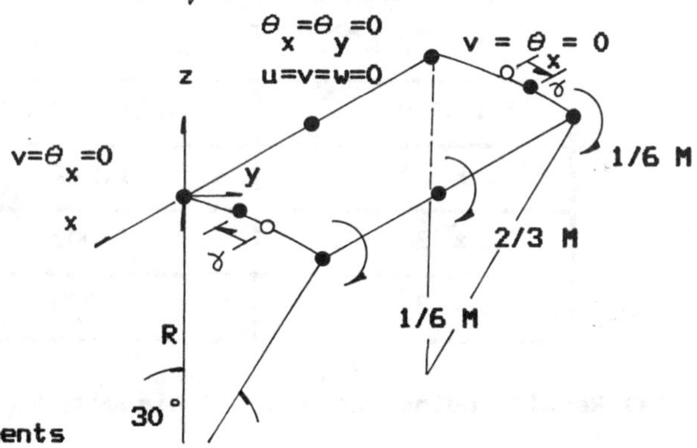

(b) Results using distorted elements

FIGURE 20 - Analysis of a curved cantilever subjected to a constant bending moment using one MITC8 element

In Figs. 21 to 23 we present some other MITC8 results for linear benchmarks.

L = 20
thickness = 0.02
$E = 2.1 \times 10^6$
$\nu = 0.3$

Boundary conditions

- Simply supported edge $w = 0$
- Clamped edge $w = 0$
  $\theta_t = 0$

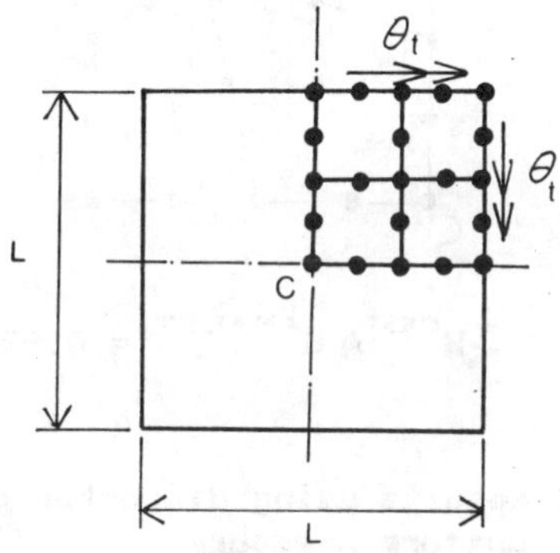

2x2 mesh

$W^{FEM}/W^{ANALYT.}$ FOR SIMPLY SUPPORTED PLATE AT C		
M E S H	UNIFORM PRESSURE	CONCENTRATED LOAD
1 x 1	0.993	1.003
2 x 2	1.000	0.998
3 x 3	1.000	0.999

$W^{FEM}/W^{ANALYT.}$ FOR CLAMPED PLATE AT C		
M E S H	UNIFORM PRESSURE	CONCENTRATED LOAD
1 x 1	1.240	1.118
2 x 2	1.005	1.000
3 x 3	1.005	1.001

(a) Results using undistorted elements

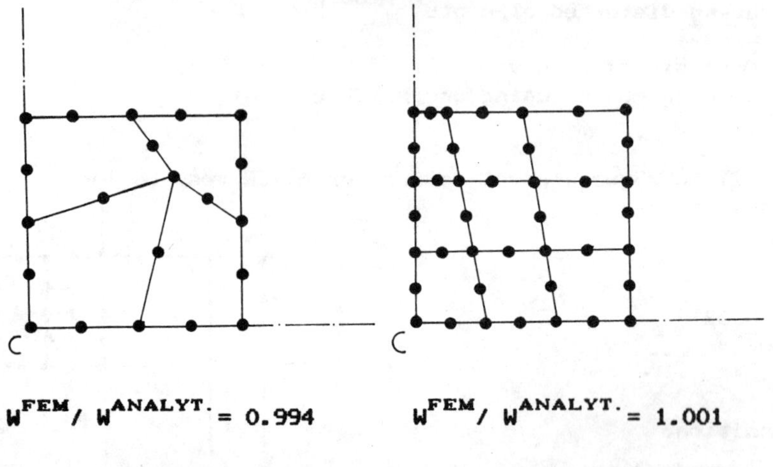

(b) Results using distorted elements, case of simply-supported plate and uniform pressure

FIGURE 21 - Analysis of a square plate using the MITC8 element

R/h = 100
L/R = 2
$E = 3 \times 10^7$
$\nu = 0.3$

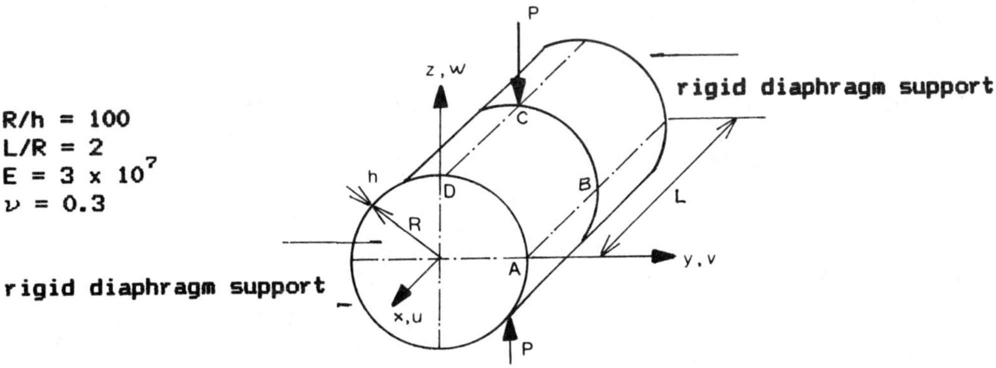

MESH FOR ABCD	$W_C^{FEM} / W_C^{ANALYT.}$
3 x 3	0.833
5 x 5	0.952
8 x 8	0.990
10 x 10	0.999

(a) Results using undistorted elements

α [degrees]	$W_C^{FEM} / W_C^{ANALYT.}$
0	0.952
1	0.951
1 1/2	0.950
2	0.949
2 1/2	0.949

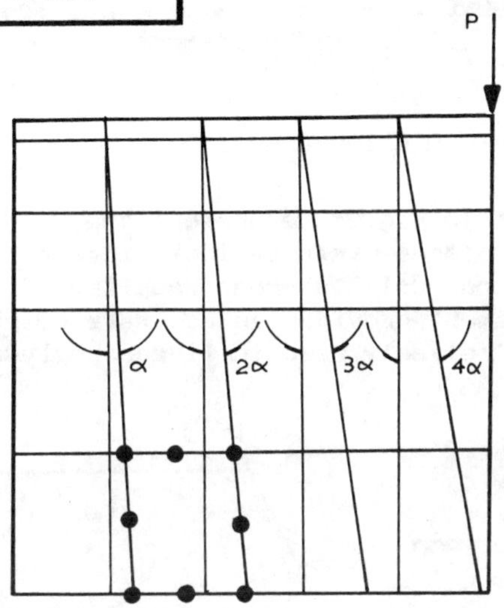

(b) Results using distorted elements (5 x 5 mesh)

FIGURE 22 - Analysis of a pinched cylinder using the MITC8 element

R = 300
L = 600
thickness = 3
E = 3×10^6
ν = 0
specific weight = 0.20833

rigid diaphragm

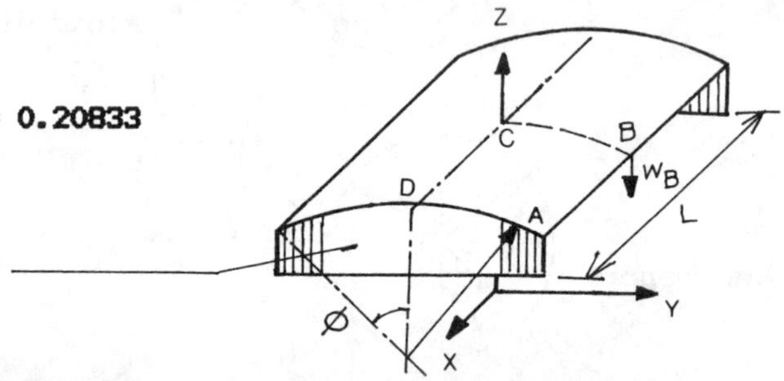

MESH FOR AREA ABCD	W_B^{FEM}
2 X 2	3.48
4 X 4	3.65
6 X 6	3.63

Analytical solution
W_B = 3.6

(DEEP SHELL THEORY)

FIGURE 23 - Analysis of the Scordelis-Lo shell using the MITC8 element

5.2.3 Nonlinear analyses

In Fig.24 we present the MITC8 results corresponding to a large displacements/rotations analysis of a cantilever compared against the 8-node A-I-Z element results. It is clear that while the MITC8 element provides an accurate solution, the 8-node A-I-Z element can only be reliably used in linear analyses.

L = 12
b = 1
h = 1
E = 1800
ν = 0

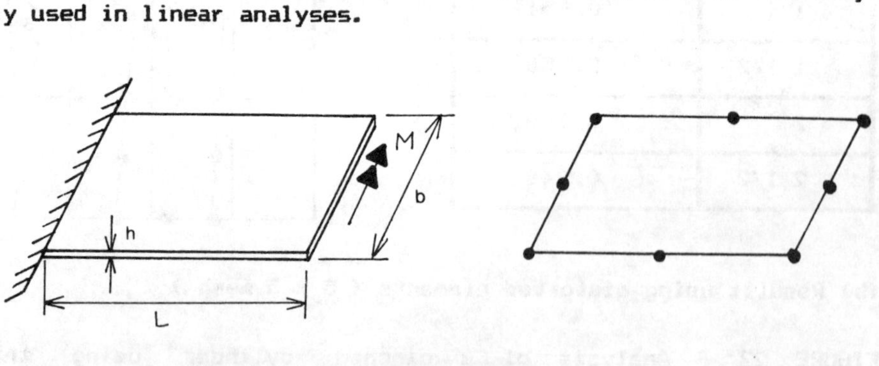

One 8-node element model

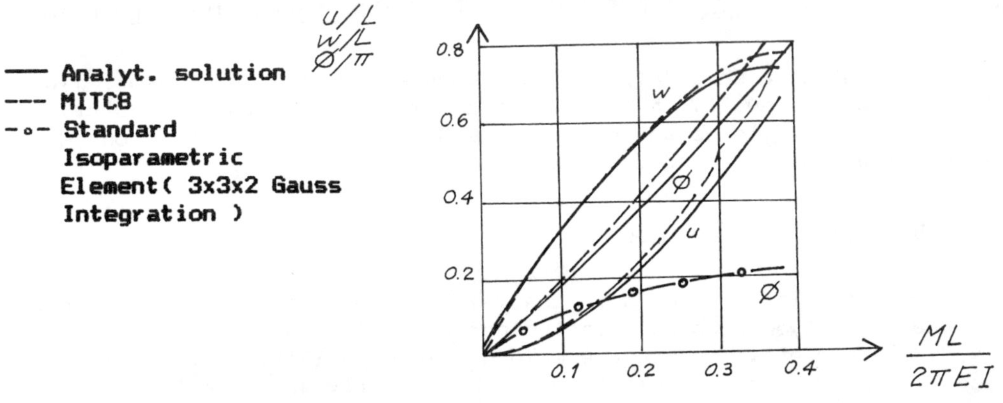

FIGURE 24 - Large displacement/rotation analysis of a cantilever using the MITC8 element

In Fig.25 the MITC8 results corresponding to the geometrically nonlinear analysis of an elastic initially flat plate are presented.

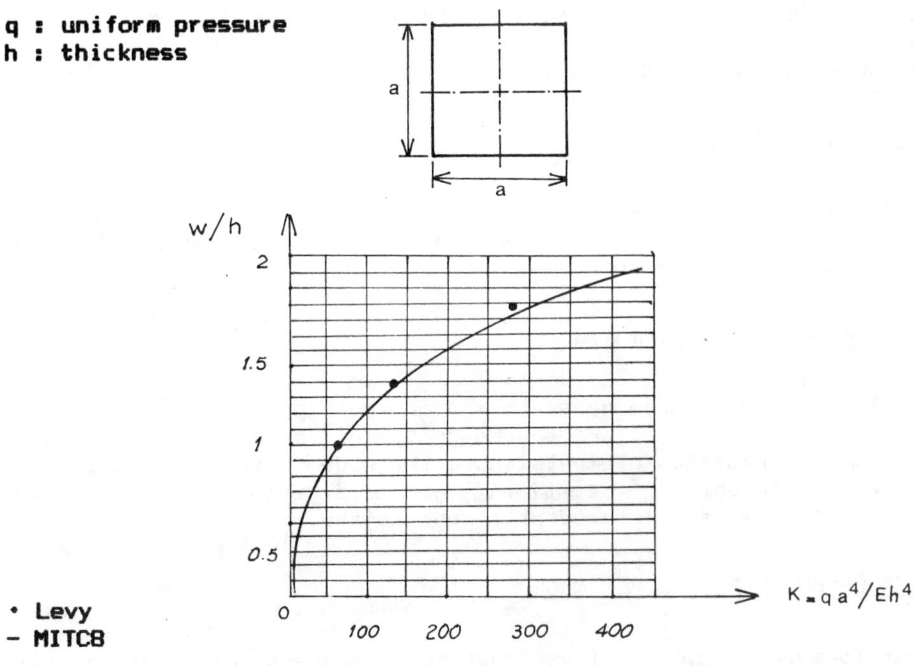

FIGURE - 25 Nonlinear analysis of a linear elastic simply supported plate under uniform pressure using the MITC8 element. Four elements were used to model one quarter of the plate

6. COMPUTATIONAL DETAILS FOR A NONLINEAR SHELL ELEMENT IMPLEMENTATION

Implementing a shell element in a general purpose nonlinear finite element code is usually an involved task because of the computational aspects that have to be taken into consideration in order to produce an efficient and user friendly software.

6.1 NONLINEAR STIFFNESS MATRIX

For geometrically nonlinear analysis the consistent stiffness matrix for the A-I-Z type elements (including the MITC4 and MITC8 elements) [14,21] is symmetric, hence there is no incompatibility with standard general purpose finite element codes that usually handle only symmetric matrices.

6.2 MID-SURFACE VECTOR SYSTEMS

As we have seen in previous sections the geometrically definition of A-I-Z type shell elements requires:

- mid surface nodes
- vectors that approximate the shell mid-surface normals (mid-surface vector systems)

There are two ways to define mid-surface vector systems:

6.2.1 Director vectors

Here the analyst (or the pre-processor) has to define at each mid-surface node a vector that it is the best approximation to the mid-surface normal at that point.

6.2.2 Program generated normals

At each node the program generates a normal to the interpolated mid-surface. At nodes that are shared by two or more elements there will be as many normals as elements share the node (remember that we are using elements with only C° continuity). The important point is that all the element normals generated at a node rotate together.

6.3 NUMBER OF SHELL D.O.F. PER NODE

In our formulation we use five 'natural' degrees of freedom per mid-surface shell node: three incremental displacements corresponding to the stationary global coordinate directions and two rotations α_k, β_k about the axes ${}^t\underline{V}_1^k$ and ${}^t\underline{V}_2^k$. The rotations define the change in the direction cosines of the director vector ${}^t\underline{V}_n^k$. We believe that the selected shell degrees of freedom are 'natural candidates' to describe the kinematics of the shell element because they directly derive from the usual degrees of freedom of solid mechanics when the shell kinematic assumptions are used. Also, transition elements with, both, shell mid-surface nodes (five degrees of freedom) and top and bottom surface nodes (three translational degrees of freedom per node) can directly be constructed. [5]

Even though the kinematics of the A-I-Z type shell elements is fully described with five degrees of freedom per mid-surface node, when used in an assemblage of elements it appears more convenient to use six degrees of freedom at each nodal point. However -since there is no stiffness corresponding to the rotation about the director vector ${}^t\underline{V}_n^k$- to always use six nodal degrees of freedom, it would be necessary to follow what is common practice: a small fictitious rotational spring is introduced at each element nodal point corresponding to the rotation about the director vector, after which the element degrees of freedom are transformed to correspond to the global axes. It is argued that if the spring stiffness is small enough, the error in the solution due to the artificial spring is acceptably small, but naturally the spring stiffness must still be large enough to allow the solution of the governing finite element equations (with six degrees of freedom at the shell nodes).

Our experience is that the approach of introducing the fictitious spring corresponding to the rotation normal to the shell mid-surface is usually acceptable in linear analysis, but much more care is necessary in nonlinear analysis; in particular when higher-order shell elements are employed with element internal nodes. The introduction of the artificial spring stiffness and the use of six degrees of freedom may result in artificial buckling loads and may inhibit the effective use of an automatic load stepping algorithm.

For the above reasons,we believe that shell elements are formulated in the most appropriate manner without the artificial spring stiffness. This means that the computer program with the element should allow a shell node to have either five or six degrees of freedom: the five natural nodal degrees of freedom are used when the shell model contains only stiffness corresponding to these five degrees of freedom and no globally aligned rotational boundary conditions are imposed; otherwise the five degrees of freedom are transformed to the six global degrees of freedom, so that the element can connect to other types of elements, appropriate boundary conditions can be imposed and so on.

6.4 MODELLING CONSIDERATIONS

If a proper software (including the options described above) is available, the analyst still has to use adequate models.

Let us consider the following modelling examples.

6.4.1 Shell intersections

In the example shown in Fig.26:

DV5: director vectors with 5 d.o.f./node
DV6: director vectors with 6 d.o.f./node
GN5: generated normals with 5 d.o.f./node
GN6: generated normal with 6 d.o.f./node
in this example the analyst should use either DV5 or GN6.

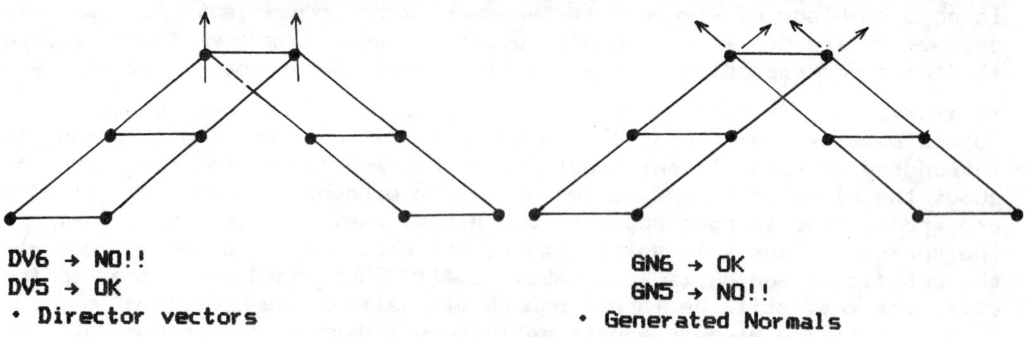

FIGURE 26 - Modelling considerations. Shell intersections

6.4.2 Stiffened shells

When a shell structure is stiffened we can model it using the A-I-Z type shell elements with A-I-Z beam elements (Timoshenko beam elements [5]), as shown in Fig.27.

In this cases, at the nodes that are shared by the shell and beam elements, 6 d.o.f. have to be used.

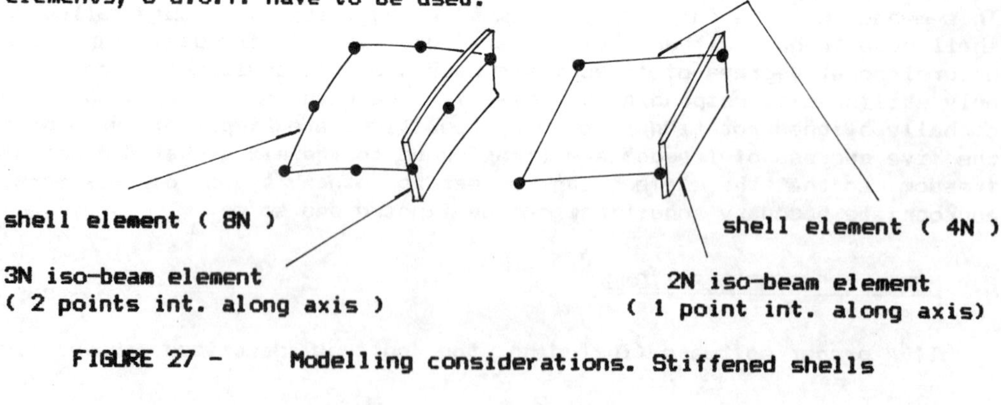

FIGURE 27 - Modelling considerations. Stiffened shells

6.4.3 Rotational boundary conditions

Using 5 d.o.f. with local rotations, the rotational boundary conditions keep changing during deformation (Fig.28).

Using 6 d.o.f. with global rotations it is the case of fixed rotational boundary conditions (Fig.28).

The analyst has to be aware of these facts and decide upon the type of model that best represents his physical problem.

• 5 d.o.f.

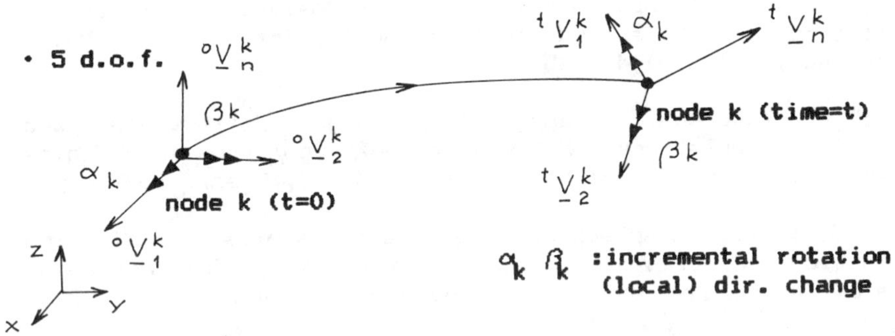

• 6 d.o.f.

global always

FIGURE 28 - Rotational boundary conditions in nonlinear analyses

7. CONCLUSIONS

Since the seventies the degenerated shell elements with C°continuity (Ahmad-Irons-Zienkiewicz) have been the most used tool for linear and nonlinear finite element analysis of shells.

Although reduced/selective integration has been for quite a long time the only available solution for overcoming the locking problem in A-I-Z type elements, it has been recognized that in order to provide engineers with more reliable analysis capabilities it is necessary to develop finite element formulations that overcome the locking problem without introducing zero energy modes.

The elements based on the method of mixed interpolation of tensorial components (MITC) fulfil the above criteria. In the present Chapter we commented on two elements based on the MITC method: the MITC4 and the MITC8 shell elements.

Using also the MITC method a 2D quadrilateral element was developed: the QMITC [22,23]. This element does not present spurious zero energy modes, satisfies Irons' Patch Test, does not lock in incompressible situations and for non-distorted geometries can exactly model a constant bending situation. At the present time we are working in the development of an improved MITC4 element that incorporates the QMITC formulation for the in-layer strain interpolation.

REFERENCES

1. Ahmad, S.; Irons, B.M. and Zienkiewicz, O.C.: Analysis of thick and thin shell structures by curved finite elements, Int. J. Numer. Methods in Eng., 2 (1970),419-451.

2. Ramm, E.: A plate/shell element for large deflections and rotations, in: Formulations and Computational Algorithms in Finite Element Analysis (Eds. K.J.Bathe et. al.), M.I.T. Press, 1977.

3. Kråkeland, B. : Large displacement analysis of shells considering elasto-plastic and elasto-viscoplastic materials, The Norwegian Institute of Technology, Univ. of Trondheim, Norway, Report 77-6 (1977).

4. Bathe, K.J. and Bolourchi, S.: A geometric and material nonlinear plate and shell element, Comp. & Structures, 11 (1979), 23-48.

5. Bathe, K.J.: Finite Element Procedures in Engineering Analysis, Prentice-Hall, Englewood Cliffs, New Jersey, 1982.

6. Bathe, K.J.; Dvorkin, E.N. and Ho, L.W.: Our discrete Kirchhoff and isoparametric shell elements for nonlinear analysis -an assessment, Comp. & Structures, 16 (1983), 89-98.

7. Belytschko, T.; Stolarski, H.; Liu, W.K.; Carpenter, N. and Ong, J.: Stress projection for membrane and shear locking in shell finite elements, Comp. Meth. Appl. Mechs. and Eng., 51 (1985), 221-258.

8. Zienkiewicz, O.C.; Taylor, R.L. and Too, J.M.: Reduced integration techniques in general analysis of plates and shells, Int. J. Numer. Methods in Eng., 3(1971), 275-290.

9. Dvorkin, E.N.: On nonlinear finite element analysis of shell structures, Ph.D.Thesis, Dept. of Mech. Eng., M.I.T., Cambridge, Mass., U.S.A.(1984)

10. Dvorkin, E.N. and Bathe, K.J.: A continuum mechanics based four node shell element for general nonlinear analysis, Eng. Computations, 1 (1984), 77-88.

11. Green, A.E. and Zerna W.: Theoretical Elasticity, Oxford University Press, 1968.

12. Irons, B.M. and Razzaque, A.:Experience with the patch test for convergence of finite elements, in: The Mathematical Foundations of the Finite Element Method with Applications to Partial Differential Equations (Ed. A.K. Aziz), Academic Press, 1972.

13. Zienkiewicz, O.C. and Taylor, R.L.: The Finite Element Method (Fourth Edition), McGraw Hill, 1989.

14. Dvorkin, E.N.; Oñate, E. and Oliver, J.:On a non-linear formulation for curved Timoshenko beam elements considering large displacement/rotation increments, Int. J. Numer. Methods in Eng., 26 (1988), 1597-1613.

15. Bathe, K.J. and Cimento, A.P. Some practical procedures for the solution of nonlinear finite element equations, Comp. Meth. Appl. Mechs. and Eng., 22 (1980), 59-85.

16. Bathe, K.J.and Dvorkin, E.N.:On the automatic solution of nonlinear finite element equations, Comp.& Structures, 17(1983), 871-879.

17. Strang, G. and Fix, G.J.: An Analysis of the Finite Element Method, Prentice Hall, 1973.

18. Bathe, K.J. and Dvorkin, E.N.: A four-node plate bending element based on Mindlin/Reissner plate theory and a mixed interpolation, Int. J. Numer. Methods in Eng., 21 (1985), 367-383.

19. Bathe, K.J. and Dvorkin, E.N.: A formulation of general shell elements -the use of mixed interpolation of tensorial components, Int. J. Numer. Methods in Eng., 22 (1986), 697-722.

20. Batoz, J.L.; Bathe, K.J. and Ho, L.W.: A study of three-node triangular plate bending elements, Int. J. Numer. Methods in Eng., 15 (1980), 1771-1812.

21. Oñate, E; Dvorkin, E.N.; Canga, M. and Oliver, J.: On the obtention of the tangent matrix for geometrically nonlinear analysis using continuum based beam/shell finite elements , in: Computational Mechanics of Nonlinear Response of Shells (Eds. W.B. Krätzig and E. Oñate), Springer-Verlag, 1990.

22. Dvorkin, E.N. and Vassolo, S.I.: A quadrilateral 2D finite element based on mixed interpolation of tensorial components, Eng. Computations, 6 (1989), 217-224.

23. Dvorkin, E.N. and Assanelli, A.P.: Elasto-plastic analysis using a quadrilateral 2D element based on mixed interpolation of tensorial components, in: Computational Plasticity (Eds. D.R.J. Owen et al.), Pineridge Press, 1989.

NONLINEAR STABILITY ANALYSIS OF SHELLS WITH THE FINITE ELEMENT METHOD

W. Wagner
University of Hannover, Hannover, Germany

Abstract

The investigation of the nonlinear response of shell structures requires besides knowledge about geometrical and material nonlinear behaviour the insight in the stability response. Here three main aspects arise. These are associated with the detection of singular points (e.g. limit or bifurcation points), the path–following in the pre– and postcritical range and a branch–switching between different paths . These problems are treated in this paper using the finite element method. For this purpose we summarize the necessary finite element formulations, where we emphasize the higher order derivatives. In a next section we present a brief overview on path–following methods. A main aspect of stability analysis is the detection of singular points. Thus, we introduce a definition of singular points, derive methods to detect the type of singular point and report possibilities to treat the stability considerations in an accompanying way. Furthermore we discuss modern concepts to calculate singular points directly using so called extended systems. Remarks on branch–switching procedures terminate the theoretical considerations. At the end of the paper some numerical examples are given to illustrate the derived methods and algorithms.

1. Introduction

The investigation of the nonlinear response of shell structures requires besides knowledge about the geometrical and material nonlinear behaviour the insight in the stability reaction.s

Here three main aspects arise. These are associated with the detection of singular points (e.g. limit or bifurcation points) and the path–following in the pre– and postcritical range and a branch–switching between different paths .

In the finite element literature mostly the so called arc–length procedures are applied to follow nonlinear solutions paths, see e.g. Riks [1,2], Crisfield [3], Ramm [4]. These methods work well even in difficult situations if a consistent formulation is used. No difference has to be made between primary and secondary branch.

Different possibilities exist for the computation of singular points on the load deflection curve. Simple methods for this purpose are given by inspection of the determinant of the tangent stiffness matrix $\mathbf{K}_T$, the number of negative diagonal elements of $\mathbf{K}_T$ or the calculation of the current stiffness parameter, Bergan et. al. [5]. More elaborate calculations are associated with the solution of eigenvalue problems based on different theoretical approaches.

These methods do not provide a tool to calculate stability points accurately since the basis of path–following is an incremental procedure.

If a stability point has to be computed within a certain accuracy a bi–section method for the detèrminant of the Hessian is a simple but sometimes slow approach, see e.g. Decker, Keller [6], Wagner, Wriggers [7]). A more efficient framework is provided by a Newton–type method for the direct calculation of stability points. The basis for this procedure is an extension of the nonlinear set of equations by a constraint condition which introduces additional information about the stability points. Here different possibilities exist, see e.g. the overview article of Mittelmann, Weber [8]. Such so called extended systems open the opportunity to compute limit or bifurcation points directly. These techniques lead with a consistent linearization of the extended system to a Newton scheme with quadratical convergence behaviour, see Moore, Spence [9]. Furthermore, the extended system provides enough information to distinguish between limit and bifurcation points. For a finite element implementation of this method, see Wriggers, Wagner, Miehe [10] and Wriggers, Simo [11].

Once a stability point has been detected one has to decide whether the primary or secondary branch of the global solution curve should be followed. In case of a limit point there exists only a primary branch which can be calculated using a path–following method. If, in a bifurcation point, the investigation of the secondary branch is necessary a branch–switching procedure has to be employed. In many engineering applications buckling mode injection which is equivalent to the superposition of buckling modes onto the displacement field at the bifurcation point is used to arrive at the secondary path. For this purpose an eigenvalue problem has to be solved which is expensive. However a sound mathematical treatment starts from the so called bifurcation equation, see Decker, Keller [6], Wagner [18] which need further information about the curvature of the parametrized solution curve.

Thus the paper is organized as follows:

In section 2 some basic finite element formulations are given to introduce the later on used stability considerations. Path–following methods are discussed briefly in section 3. The discussion of the stability behaviour is given in the following sections. In Section 4 a definition of singular points is given. Section 5 describes incremental methods to find singular points whereas section 6 is associated with the direct calculation of singular points. Remarks on branch–switching procedures are given in section 7. At the end of the paper some numerical examples to illustrate the derived methods and algorithms.

2. Basic Nonlinear Finite Element Formulation

This section describes briefly the basic equations for the finite element formulation used in this paper. All equations are given for three–dimensional elasticity. For beam or shell structures the corresponding equations have to be used. The restriction to elastic deformations is sufficient for most purposes concerning stability problems.

For elastic bodies the principle of virtual work can be stated with respect to the reference configuration in the following form

$$D\Pi(\mathbf{u})\cdot\boldsymbol{\eta}_1 = \int_B \mathbf{S}\cdot\mathbf{E}'(\mathbf{u},\boldsymbol{\eta}_1)\,dv - \int_B \rho\hat{\mathbf{b}}\cdot\boldsymbol{\eta}_1\,dv - \int_{\partial B_\sigma} \hat{\mathbf{t}}\cdot\boldsymbol{\eta}_1\,da = 0 \tag{2.1}$$

with the Green–Lagrangian strain tensor $\mathbf{E} = 1/2(\mathbf{F}^T\mathbf{F} - \mathbf{1})$, the variation of $\mathbf{E}$ given by $\mathbf{E}'(\mathbf{u},\boldsymbol{\eta}_1) = \frac{1}{2}(\mathbf{F}^T\mathrm{Grad}\,\boldsymbol{\eta}_1 + \mathrm{Grad}^T\boldsymbol{\eta}_1\,\mathbf{F})$, and the 2nd Piola–Kirchhoff stress tensor $\mathbf{S}$. $\rho\hat{\mathbf{b}}$ are the body forces and $\hat{\mathbf{t}}$ describes the surface tractions acting on ∂B_σ whereas $\boldsymbol{\eta}_1$ are the virtual displacements or test functions with $\boldsymbol{\eta}_1 = \mathbf{0}$ on ∂B_u.

For the further developments within this paper we need depart from the weak form also the tangent operator and its derivative with respect to the displacement field. The tangent operator follows from the linearization of the principle of virtual work, eq. (1) for St. Venant material

$$D[\,D\Pi(\mathbf{u})\cdot\boldsymbol{\eta}_1\,]\cdot\boldsymbol{\eta}_2 = \int_B \{\mathbf{E}'(\mathbf{u},\boldsymbol{\eta}_1)\cdot\mathbb{D}\,[\mathbf{E}'(\mathbf{u},\boldsymbol{\eta}_2\,] + \mathbf{S}\cdot\mathbf{E}''(\boldsymbol{\eta}_1,\boldsymbol{\eta}_2)\}dv\,, \tag{2.2}$$

with

$$
\begin{aligned}
\mathbb{D} &= \frac{\partial \mathbf{S}}{\partial \mathbf{E}}\,, \\
\mathbf{E}'(\mathbf{u},\boldsymbol{\eta}_2) &= \frac{1}{2}\,(\,\mathbf{F}^T\,\mathrm{Grad}\,\boldsymbol{\eta}_2 + \mathrm{Grad}^T\,\boldsymbol{\eta}_2\,\mathbf{F}\,)\,, \\
\mathbf{E}''(\boldsymbol{\eta}_1,\boldsymbol{\eta}_2) &= \frac{1}{2}\,(\,\mathrm{Grad}^T\,\boldsymbol{\eta}_1\,\mathrm{Grad}\,\boldsymbol{\eta}_2 + \mathrm{Grad}^T\,\boldsymbol{\eta}_2\,\mathrm{Grad}\,\boldsymbol{\eta}_1\,)\,.
\end{aligned} \tag{2.3}
$$

Eqs. (2.1)–(2.3) are standard within nonlinear finite element methods. For the later introduced extended system we define also the next derivative by repeating the linearization procedure in a formal way. We obtain

$$
\begin{aligned}
D\{\,D[\,D\Pi(\mathbf{u})\cdot\boldsymbol{\eta}_1\,]\cdot\boldsymbol{\eta}_2\,\}\cdot\boldsymbol{\eta}_3 = \int_B \{\,&\mathbf{E}''(\boldsymbol{\eta}_1,\boldsymbol{\eta}_3)\cdot\mathbb{D}\,[\,\mathbf{E}'(\mathbf{u},\boldsymbol{\eta}_2)] \\
&+ \mathbf{E}'(\mathbf{u},\boldsymbol{\eta}_1)\cdot\mathbb{D}\,[\,\mathbf{E}''(\boldsymbol{\eta}_2,\boldsymbol{\eta}_3)\,] + \mathbf{E}''(\boldsymbol{\eta}_1,\boldsymbol{\eta}_2)\cdot\mathbb{D}\,[\mathbf{E}'(\mathbf{u},\boldsymbol{\eta}_3)\,]\,\}dv
\end{aligned} \tag{2.4}
$$

with

$$
\begin{aligned}
\mathbf{E}'(\mathbf{u},\boldsymbol{\eta}_3) &= \frac{1}{2}\,(\,\mathbf{F}^T\,\mathrm{Grad}\,\boldsymbol{\eta}_3 + \mathrm{Grad}^T\,\boldsymbol{\eta}_3\,\mathbf{F}\,)\,, \\
\mathbf{E}''(\boldsymbol{\eta}_1,\boldsymbol{\eta}_3) &= \frac{1}{2}\,(\,\mathrm{Grad}^T\,\boldsymbol{\eta}_1\,\mathrm{Grad}\,\boldsymbol{\eta}_3 + \mathrm{Grad}^T\,\boldsymbol{\eta}_3\,\mathrm{Grad}\,\boldsymbol{\eta}_1\,)\,, \\
\mathbf{E}''(\boldsymbol{\eta}_1,\boldsymbol{\eta}_1) &= \frac{1}{2}\,(\,\mathrm{Grad}^T\,\boldsymbol{\eta}_1\,\mathrm{Grad}\,\boldsymbol{\eta}_2 + \mathrm{Grad}^T\,\boldsymbol{\eta}_2\,\mathrm{Grad}\,\boldsymbol{\eta}_1\,)\,, \\
\mathbf{E}''(\boldsymbol{\eta}_2,\boldsymbol{\eta}_3) &= \frac{1}{2}\,(\,\mathrm{Grad}^T\,\boldsymbol{\eta}_2\,\mathrm{Grad}\,\boldsymbol{\eta}_3 + \mathrm{Grad}^T\,\boldsymbol{\eta}_3\,\mathrm{Grad}\,\boldsymbol{\eta}_2\,)\,.
\end{aligned} \tag{2.5}
$$

The finite element discretization of B is given by

$$
B^h = \bigcup_{e=1}^{n_e} \Omega_e
$$

with n_e elements. Within a single element Ω_e the displacement field $\mathbf{u}$, the test functions $\boldsymbol{\eta}_1$ and the higher order displacement fields $\boldsymbol{\eta}_2$ and $\boldsymbol{\eta}_3$ are approximated by

$$
\mathbf{u}^h = \mathbf{N}\,\mathbf{v}_e, \quad \boldsymbol{\eta}_1^h = \mathbf{N}\,\delta\mathbf{v}_e, \quad \boldsymbol{\eta}_2^h = \mathbf{N}\,\Delta\mathbf{v}_e, \quad \boldsymbol{\eta}_3^h = \mathbf{N}\,d\mathbf{v}_e. \tag{2.6}
$$

Here the matrix $\mathbf{N}$ contains the shape functions whereas $\mathbf{v}_e, \delta\mathbf{v}_e, \Delta\mathbf{v}_e$ and $d\mathbf{v}_e$ are the nodal values. Within standard finite element formulations we introduce the matrices $\mathbf{B}^l$ und $\mathbf{B}^{nl}$ containing derivations of the shape functions, which have to be specified for the chosen theory. The components of the Green–Lagrangian strains $\mathbf{E} = 1/2(\mathbf{F}^T\mathbf{F} - \mathbf{1})$ are approximated by the vectors

$$
\begin{aligned}
\mathbf{E}_e^h &= [\mathbf{B}^l + \frac{1}{2}\mathbf{B}^{nl}(\mathbf{v}_e)]\,\mathbf{v}_e = \bar{\mathbf{B}}(\mathbf{v}_e)\,\mathbf{v}_e\,, \\
\mathbf{E}_e{}'^h &= [\mathbf{B}^l + \quad \mathbf{B}^{nl}(\mathbf{v}_e)]\,\delta\mathbf{v}_e = \mathbf{B}(\mathbf{v}_e)\,\delta\mathbf{v}_e\,.
\end{aligned} \tag{2.7}
$$

With (2.7) the discrete form of the virtual work is given for an element by

$$G_e^h\,(\mathbf{v}_e,\delta\mathbf{v}_e) = \delta\mathbf{v}_e^T\{\int\limits_{\Omega_e}\mathbf{B}^T(\mathbf{v}_e)\mathbf{S}(\mathbf{v}_e)\,d\Omega - \int\limits_{\Omega_e}\mathbf{N}^T\rho_R\hat{\mathbf{b}}\,d\Omega - \int\limits_{\Gamma_e}\mathbf{N}^T\hat{\mathbf{t}}\,d\Gamma\} \qquad (2.8)$$

where the vector $\mathbf{S}$ contains the components of the 2nd Piola-Kirchhoff stress tensor.

The assembling procedure to the global system leads to an algebraic system of equations. The operator $\bigcup$ is used to describe the assembling process which yields the global residual $\mathbf{G}$

$$\delta\mathbf{v}^T\,\mathbf{G}\,(\,\mathbf{v}\,) = \bigcup_{e=1}^{n_e} G_e^h\,. \qquad (2.9)$$

Based on (2.9) we obtain for arbitrary test functions $\delta\mathbf{v}$ the algebraic equilibrium equations which are nonlinear in $\mathbf{v}$

$$\mathbf{G}\,(\mathbf{v},\lambda) = \mathbf{R}\,(\mathbf{v}) - \lambda\,\mathbf{P} = \mathbf{0}. \qquad (2.10)$$

The vector $\mathbf{P}$ contains all loading terms, $\mathbf{R}$ denotes the stress divergence and the load parameter λ is introduced to describe the process of loading.

The linearization of (2.10) at $\bar{\mathbf{v}}$ leads to the definition of the tangent stiffness matrix of a finite element. With eq. (2.3) and (2.7) it remains

$$D\,G_e^h\,(\bar{\mathbf{v}}_e,\eta_{1e})\,\eta_{2e} = \delta\mathbf{v}_e^T\int\limits_{\Omega_e}[\,\mathbf{B}^T(\bar{\mathbf{v}}_e)\,\mathbf{D}\,\mathbf{B}(\bar{\mathbf{v}}_e) + \hat{\mathbf{G}}^T\hat{\mathbf{S}}(\bar{\mathbf{v}}_e)\hat{\mathbf{G}}\,]\;d\Omega\,\Delta\mathbf{v}_e = \delta\mathbf{v}_e\mathbf{K}_{Te}\Delta\mathbf{v}_e\,. \qquad (2.11)$$

The matrix $\hat{\mathbf{S}}$ contains values of the stress vector $\mathbf{S} = \mathbf{D}\,\tilde{\mathbf{B}}(\bar{\mathbf{v}}_e)\,\bar{\mathbf{v}}_e$. Standard assembly procedures denoted by the operator $\bigcup$ for the element stiffness matrices (2.11) lead to the global tangent stiffness matrix $\mathbf{K}_T$

$$\delta\mathbf{v}^T\,\mathbf{K}_T\,\Delta\mathbf{v} = \bigcup_{e=1}^{n_e}\delta\mathbf{v}_e\,\mathbf{K}_{T_e}\,\Delta\mathbf{v}_e\,. \qquad (2.12)$$

The discrete form of the derivation of the tangent stiffness matrix is based on eq. (2.5). Using finite element approximations (2.7) of strains and virtual strains we arrive on the element level at

$$\begin{aligned} D\,[\,D\,G_e\,\eta_{2e}\,]\,\eta_{3e} &= \delta\mathbf{v}_e^T[\mathbf{K}_{Te}(\bar{\mathbf{v}}_e,\bar{\mathbf{v}}_e)\Delta\mathbf{v}_e]_{,v}\,d\mathbf{v}_e \\ &= \delta\mathbf{v}_e^T\mathbf{h}_e(\bar{\mathbf{v}}_e,d\mathbf{v}_e,\Delta\mathbf{v}_e)\,, \end{aligned} \qquad (2.13)$$

with the vector

$$\mathbf{h}_e = \int\limits_{\Omega_e} [\mathbf{B}^T(\bar{\mathbf{v}}_e)\mathbf{D}\mathbf{B}^{nl}(d\mathbf{v}_e) + \mathbf{B}^{nlT}(d\mathbf{v}_e)\mathbf{D}\mathbf{B}(\bar{\mathbf{v}}_e) + \hat{\mathbf{G}}^T \Delta\hat{\mathbf{S}}(\bar{\mathbf{v}}_e, d\mathbf{v}_e)\hat{\mathbf{G}}]\, d\Omega\, \Delta\mathbf{v}_e\,. \tag{2.14}$$

Abbreviation $(...),v$ indicates the directional derivative with respect to the displacements. The matrix $\Delta\hat{\mathbf{S}}$ contains components of the stress vector given by $\Delta\mathbf{S}(\bar{\mathbf{v}}_e, d\mathbf{v}_e) = \mathbf{D}\mathbf{B}(\bar{\mathbf{v}}_e)d\mathbf{v}_e$.

3. Path–following methods

Standard finite element approximations lead to a system of algebraic equations, see (2.10) in section 2,

$$\mathbf{G}(\mathbf{v},\lambda) = \mathbf{R}(\mathbf{v}) - \lambda\,\mathbf{P} = \mathbf{0}, \qquad \mathbf{v} \in \mathbb{R}^N \tag{3.1}$$

Algorithms for the solution of (3.1) are standard. Algorithms which can be used in all situations are so called arc–length–, continuation– or path–following methods These methods are well established, see e.g. Riks [1,2], Crisfield [3], Ramm [4], Wagner, Wriggers [7], Schweizerhof, Wriggers [12]. Since arc-length methods are essential to follow postbuckling branches, we will briefly discuss a general form of these procedures.

The main idea of continuation methods is to add a single constraint equation to the nonlinear set of equations (3.1). Thus a special form of an extended system is used which can be stated as

$$\tilde{\mathbf{G}}(\mathbf{v},\lambda) = \left\{ \begin{array}{c} \mathbf{G}(\mathbf{v},\lambda) \\ f(\mathbf{v},\lambda) \end{array} \right\} = \mathbf{0}\,, \tag{3.2}$$

with a general form of the constraint equation $f(\mathbf{v},\lambda) = 0$. Based on this formulation special methods like load control, displacement control or arc–length procedures can be deduced. An important aspect for the iteration behaviour within the nonlinear calculations is the consistent linearization of the above given extended system at a known state $\mathbf{v}_i, \lambda_i$, which leads to

$$\left(\begin{array}{cc} \mathbf{G}_v & -\mathbf{G}_\lambda \\ \mathbf{f}_v^T & f_\lambda \end{array} \right)_i \left\{ \begin{array}{c} \Delta\mathbf{v} \\ \Delta\lambda \end{array} \right\}_i = - \left\{ \begin{array}{c} \mathbf{G} \\ f \end{array} \right\}_i \tag{3.3}$$

Here we have introduced the short hand notation for the directional derivatives

$$\begin{aligned} D_v \mathbf{G}\,\Delta\mathbf{v} = \mathbf{G}_v\,\Delta\mathbf{v} &= \frac{d}{d\epsilon}[\mathbf{G}(\mathbf{v}_i + \epsilon\Delta\mathbf{v}, \lambda_i)]\Big|_{\epsilon=0} \\ D_\lambda \mathbf{G}\Delta\lambda = \mathbf{G}_\lambda\,\Delta\lambda &= \frac{d}{d\epsilon}[\mathbf{G}(\mathbf{v}_i, \lambda_i + \epsilon\Delta\lambda)]\Big|_{\epsilon=0} \\ D_v\, f\,\Delta\mathbf{v} = \mathbf{f}_v^T\,\Delta\mathbf{v} &= \frac{d}{d\epsilon}[f(\mathbf{v}_i + \epsilon\Delta\mathbf{v}, \lambda_i)]\Big|_{\epsilon=0} \\ D_\lambda\, f\,\Delta\lambda = f_\lambda\,\Delta\lambda &= \frac{d}{d\epsilon}[f(\mathbf{v}_i, \lambda_i + \epsilon\Delta\lambda)]\Big|_{\epsilon=0} \end{aligned} \tag{3.4}$$

Here, $\mathbf{G}_v$ is the tangent stiffness matrix $\mathbf{K}_T$ whereas $\mathbf{G}_\lambda$ is a given load pattern defined by $\mathbf{P} = -\mathbf{G}_\lambda$, see also equation (3.1).

The, in general, non-symmetric system of equations (3.3) is commonly solved by a partitioning method also called bordering algorithm leading to two equations for $\Delta\mathbf{v}_i$ and $\Delta\lambda_i$, From $(3.3)_1$ the two partial solutions

$$\Delta\mathbf{v}_{Pi} = (\mathbf{K}_{Ti})^{-1}\,\mathbf{P}\,, \quad \Delta\mathbf{v}_{Gi} = -(\mathbf{K}_{Ti})^{-1}\,\mathbf{G}_i, \tag{3.5}$$

are available which can be combined to the total displacement increment

$$\Delta\mathbf{v}_i = \Delta\lambda_i \Delta\mathbf{v}_{Pi} + \Delta\mathbf{v}_{Gi}\,. \tag{3.6}$$

The unknown increment of the load factor λ has to be calculated from the second equation $(3.3)_2$

$$\Delta\lambda_i = -\frac{f_i + \mathbf{f}_{vi}^T \Delta\mathbf{v}_{Gi}}{f_{\lambda i} + \mathbf{f}_{vi}^T \Delta\mathbf{v}_{Pi}}\,. \tag{3.7}$$

Such methods are called arc–length procedures or in a more general context path–following or continuation methods. A large number of variants exists which only differ in the formulation of the constraint equation, see e.g. Riks [1], Keller [13], Rheinboldt [14], Crisfield [3], Ramm [4]. An overview may be found in e.g. Riks [15]. Necessary for quadratical convergence behaviour is the consistent linearization of the constraint equation., see e.g. Schweizerhof, Wriggers [12].

The associated algorithm for continuation methods is given for a specified arc–length ds in BOX 3.1.

The algorithm differs from an incremental Newton procedure only by the fact that within the continuation procedure the constraint equation $f(\mathbf{v}, \lambda) = 0$ yields the load parameter, a second backsubstitution for $\mathbf{P}$ and the predictor calculation for $\Delta\mathbf{v}_0$ and $\Delta\lambda_0$.

1.	calculate predictor	$\mathbf{K}_T\, \Delta\, \mathbf{v}_{P0} = \mathbf{P}$
2.	calculate load increment	$\Delta\lambda_0 = \dfrac{ds}{\sqrt{(\Delta \mathbf{v}_{P0})^T \Delta \mathbf{v}_{P0}}}$
3.	loop over all iterations i	$\mathbf{K}_T\, \Delta\, \mathbf{v}_{Pi} = \mathbf{P}$ $\mathbf{K}_T\, \Delta\, \mathbf{v}_{Gi} = -\mathbf{G}(\mathbf{v}_i, \lambda_i)$
4.	calculate increments	$\Delta\lambda_i = -\dfrac{f_i + \mathbf{f}_{vi}^T \Delta \mathbf{v}_{Gi}}{f_{\lambda i} + \mathbf{f}_{vi}^T \Delta \mathbf{v}_{Pi}}$ $\Delta \mathbf{v}_i = \Delta\lambda_i \Delta \mathbf{v}_{Pi} + \Delta \mathbf{v}_{Gi}$
5.	Update	$\lambda_{i+1} = \lambda_i + \Delta\lambda_i\,, \quad \mathbf{v}_{i+1} = \mathbf{v}_i + \Delta \mathbf{v}_i$
6.	convergence criterion	if $\|\mathbf{G}(\mathbf{v}_{i+1}, \lambda_{i+1})\| \leq$ TOL Stop else go to 3

BOX 3.1 : General algorithm for arc–length procedures

In the following BOX 3.2 some selected constraint equations are depicted.

name	Fig.	constraint equation
load control	3.1a	$f = \lambda - c$
displacement control Batoz, Dhatt [16]	3.1b	$f = v_a - c$
arc–length method Riks [1]	3.2	$f = (\mathbf{v}_m - \bar{\mathbf{v}})^T(\mathbf{v} - \mathbf{v}_m) + (\lambda_m - \bar{\lambda})(\lambda - \lambda_m)$
arc–length method Fried[17], Wagner[18]	3.3	$f = \Delta \mathbf{v}_P^T (\mathbf{v} - \mathbf{v}_m) + (\lambda - \lambda_m)$
arc–length method Crisfield [3]	3.4	$f = \sqrt{(\mathbf{v} - \bar{\mathbf{v}})^T(\mathbf{v} - \bar{\mathbf{v}}) + (\lambda - \bar{\lambda})^2} - ds$

BOX 3.2 : Examples for constraint equations

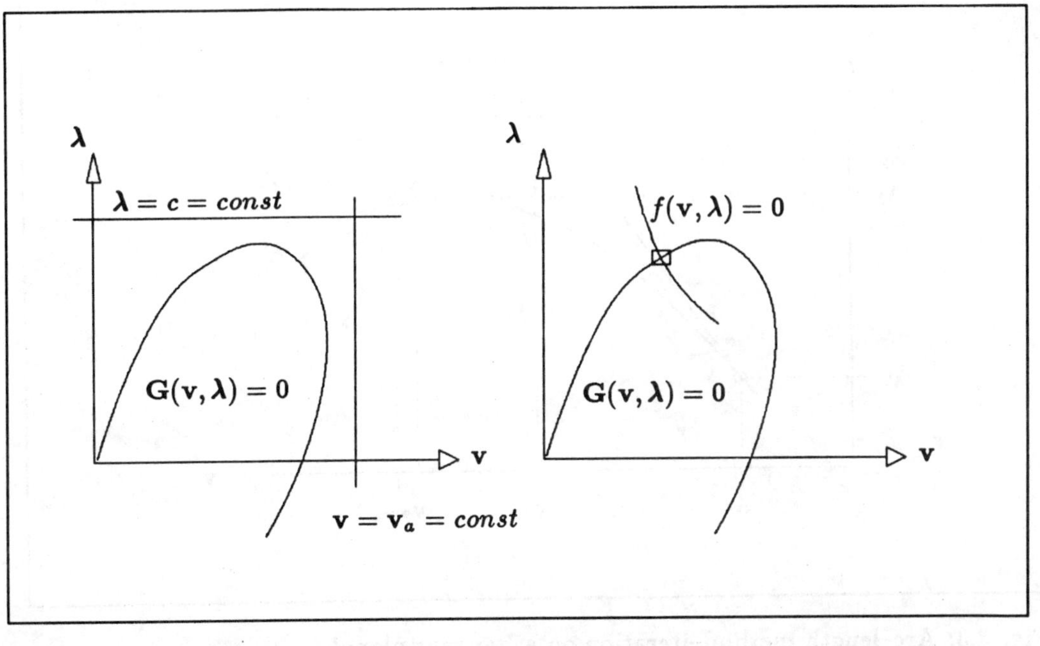

Fig. 3.1: Load-deflection diagrams

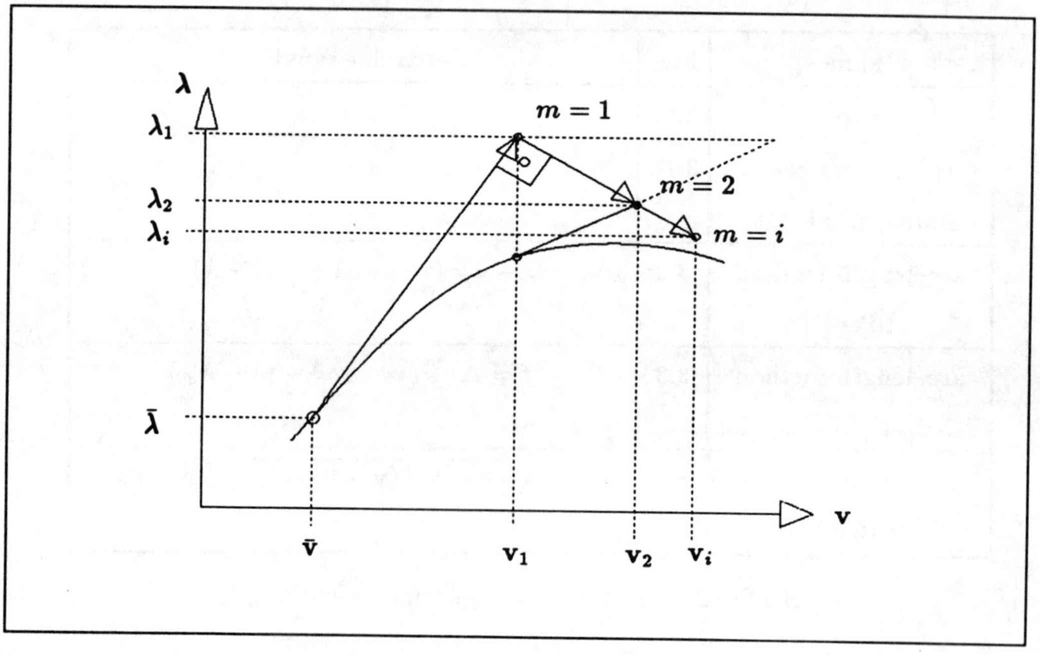

Fig. 3.2: Arc–length method–iteration on a 'normal plane'

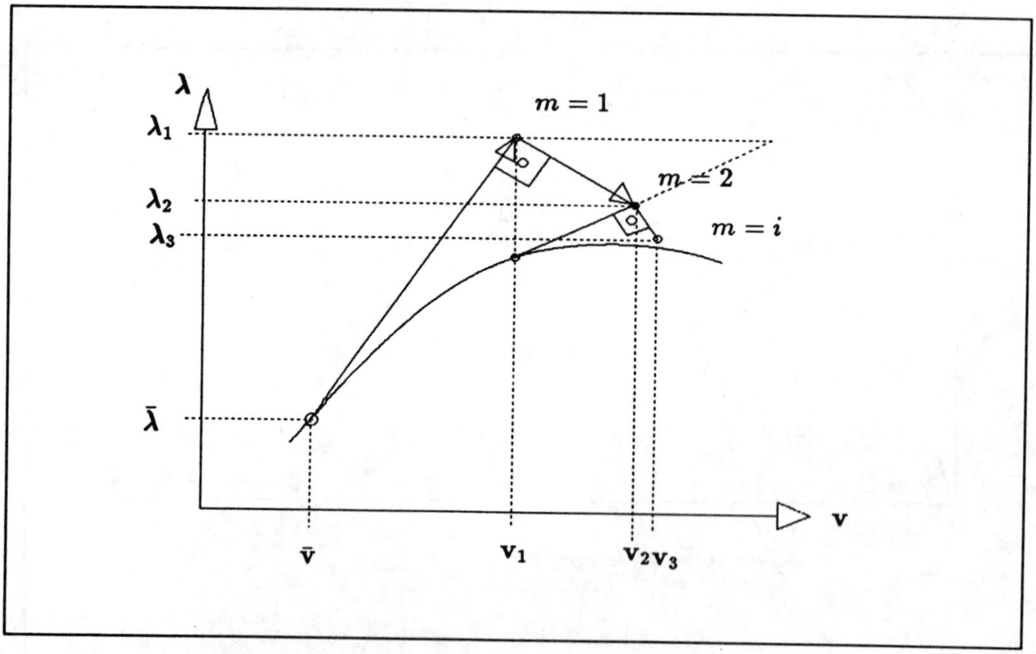

Fig. 3.3: Arc–length method–iteration on a 'tangent plane'

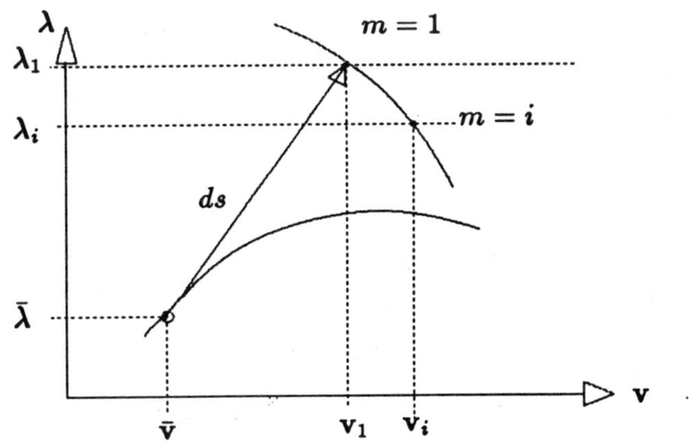

Fig. 3.4: Arc–length method–iteration on a 'sphere'

A detailed description of these constraint equations may be found in the given references or in e.g. Wagner, Wriggers [7].

Example

The following example shows a comparison of the iteration behaviour due to the chosen constraint equation. The example is a deep clamped–hinged arch and is described in Zienkiewicz/Taylor [19],volume 2, page 306–307.

Fig. 3.5: Clamped–hinged deep arch

In Fig 3.5 the complete load–deflection diagram is given. The finite element calculation has been carried out with the finite element program FEAP of R.L. Taylor, described in Zienkiewicz, Taylor [19], volume 2, chapter 16. Without differences two element types have been used, the element of Simo, Wriggers, Schweizerhof, Taylor [20] and the element of Wagner [21], which is similar to Wagner [22].

To compare the constraint equations we introduce as an example at $P = 412.9\,kN, v = 25.8\,cm$ a very large predictor step to the point $P = 918.3\,kN, v = 75.25\,cm$. The load increment has been chosen to $\Delta\lambda P = 300\,kN$. In the following corrector iteration different types of constraint equations has been tested. Due to the large predictor step we have a higher number of iteration steps.

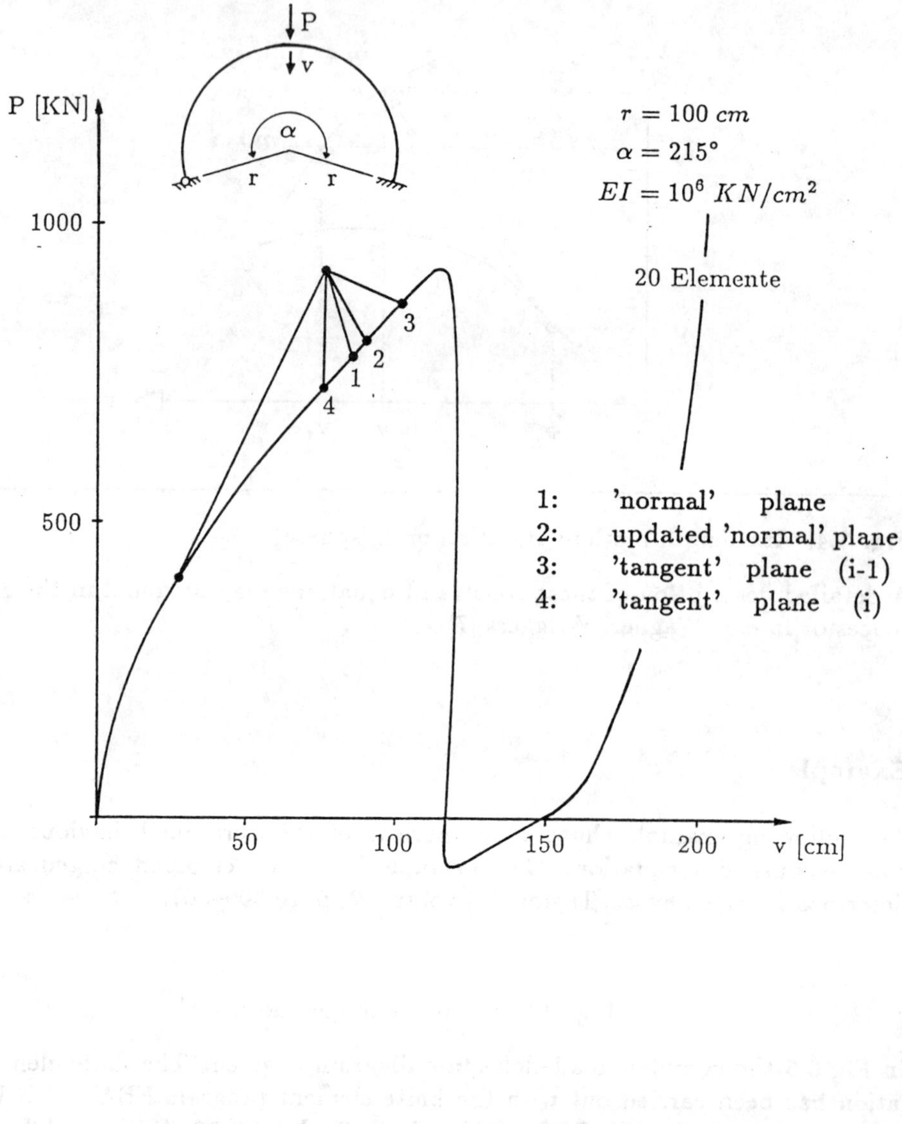

Fig. 3.5: Clamped–hinged deep arch

Iteration-number	'normal plane' $\|\|G\|\|$	λ	v	'Secant plane' $\|\|G\|\|$	λ	v
1	$6.6 \cdot 10^6$	34.86	98.34	$6.6 \cdot 10^6$	34.86	98.34
2	$3.2 \cdot 10^5$	3.28	99.52	$3.2 \cdot 10^5$	4.35	102.05
3	$2.5 \cdot 10^4$	2.67	94.57	$8.5 \cdot 10^4$	3.12	101.17
4	$2.8 \cdot 10^4$	2.65	92.45	$1.5 \cdot 10^4$	2.74	93.85
5	$4.3 \cdot 10^3$	2.55	87.48	$4.9 \cdot 10^3$	2.64	89.89
6	$8.6 \cdot 10^2$	2.56	85.42	$5.4 \cdot 10^3$	2.64	89.70
7	$5.6 \cdot 10^2$	2.56	85.42	$5.0 \cdot 10^2$	2.63	89.10
8	$3.1 \cdot 10^0$	2.56	85.42	$1.2 \cdot 10^2$	2.63	89.09
9	$2.3 \cdot 10^{-4}$	2.56	85.42	$4.0 \cdot 10^{-1}$	2.63	89.09
10				$3.6 \cdot 10^{-5}$	2.63	89.09

Table 3.1: Convergence behaviour in the corrector iteration

Iteration-number	'old tangent plane' $\|\|G\|\|$	λ	v	'new tangente plane' $\|\|G\|\|$	λ	v
1	$6.6 \cdot 10^6$	34.77	98.31	$6.6 \cdot 10^6$	-51.12	67.55
2	$3.2 \cdot 10^5$	2.56	97.76	$4.9 \cdot 10^5$	3.12	69.44
3	$2.4 \cdot 10^4$	3.58	106.75	$3.9 \cdot 10^4$	2.28	72.27
4	$5.6 \cdot 10^4$	3.13	106.56	$1.9 \cdot 10^4$	2.34	73.04
5	$4.6 \cdot 10^3$	2.98	103.91	$2.9 \cdot 10^3$	2.37	75.26
6	$1.3 \cdot 10^3$	2.91	102.10	$2.2 \cdot 10^2$	2.37	75.74
7	$3.8 \cdot 10^3$	2.90	102.09	$2.5 \cdot 10^1$	2.37	75.74
8	$3.8 \cdot 10^1$	2.90	102.12	$6.8 \cdot 10^{-2}$	2.37	75.74
9	$5.2 \cdot 10^{-1}$	2.90	102.12	$2.9 \cdot 10^{-6}$	2.37	75.74
10	$3.2 \cdot 10^{-5}$	2.90	102.12			

Table 3.2: Convergence behaviour in the corrector iteration

The iteration behaviour can be seen in tables 3.1 and 3.2 whereas in Fig. 3.5 the starting point of the iteration and the final equilibrium point on the solution curve are connected by a straight line.

The example shows that no significant differences occur for the number of iterations. On the other hand the resulting solution points differ. This effect vanishes if a 'normal' arc–length step is chosen.

4. Definition of Singular Points

The nonlinear structural response of a shell is governed by its stability behaviour. Thus, besides the calculation of a nonlinear solution path using arc–length procedures additional task has to be done.

This means one has to introduce algorithms to find the stability points. In engineering literature these points are called stability points, sometimes also instability points, which may be more correct, whereas in the mathematical literature the name singular points is used.

To find these points we have to introduce some definitions.

A classical linear engineering definition for singular points is given by the following idea, see e.g. Brendel, Ramm [23]: "An equilibrium point $\mathbf{G}^G(\bar{\mathbf{v}}, \bar{\lambda})$ is a singular point if there exists for the same load level $\bar{\lambda}$ another equilibrium point $\mathbf{G}^N(\tilde{\mathbf{v}}, \bar{\lambda})$ in the infinitesimal vicinity of $(\bar{\mathbf{v}}, \bar{\lambda})$".

The result is given by a Taylor series for $\mathbf{G}^G$ at $(\bar{\mathbf{v}}, \bar{\lambda})$ with respect to $\mathbf{v}$.

$$\mathbf{G}^N(\tilde{\mathbf{v}}, \bar{\lambda}) = \mathbf{G}^N(\bar{\mathbf{v}} + \Delta\mathbf{v}, \bar{\lambda}) = \mathbf{G}^G(\bar{\mathbf{v}}, \bar{\lambda}) + \mathbf{K}_T(\bar{\mathbf{v}}, \bar{\lambda})\,\Delta\mathbf{v} + \ldots \quad . \tag{4.1}$$

With the assumptions $\mathbf{G}^G(\bar{\mathbf{v}}, \bar{\lambda}) = \mathbf{0}$ and $\mathbf{G}^N(\bar{\mathbf{v}} + \Delta\mathbf{v}, \bar{\lambda}) = \mathbf{0}$ the final result for a singular point $\bar{\mathbf{v}}, \bar{\lambda})$ is

$$\mathbf{K}_T(\bar{\mathbf{v}}, \bar{\lambda})\Delta\mathbf{v} = \mathbf{0} \tag{4.2}$$

This result can be achieved in a more mathematically way, if we discuss the behaviour of the solution path $\mathbf{G}(\mathbf{v}, \lambda) = \mathbf{0}$.

Assume a solution point $\mathbf{G}(\bar{\mathbf{v}}, \bar{\lambda}) = \mathbf{0}$, assume furthermore that the solution curve is smooth in the vicinity of $(\bar{\mathbf{v}}, \bar{\lambda})$. If the derivative $\mathbf{G}_v$ is nonsingular which is equivalent to the statement that $\mathbf{G}_v^{-1}$ exists, then the implicit function theorem, see Hildebrandt, Graves [24] assures

1.) there exists a solution curve through $(\bar{\mathbf{v}}, \bar{\lambda})$
2.) there exists a displacement increment which solves the equation

$$\dot{\mathbf{G}} = \mathbf{G}_{,s} = \mathbf{G}_v \Delta \mathbf{v} + \mathbf{G}_\lambda \Delta \lambda = \mathbf{0}$$

$\mathbf{G}_{,s}$ is the derivative of $\mathbf{G}$ with respect to a path–paramter s which means the tangent on the solution curve in $(\bar{\mathbf{v}}, \bar{\lambda})$. Such a point is called a regular point. If the implicit function theorem does not hold, which means $\mathbf{G}_v^{-1}$ does not exist, then the point is called a singular point of the solution curve. Thus one has to look for the zero eigenvectors φ of $\mathbf{G}_v$. In our notation we end up with the problem

$$[\mathbf{K}_T - \omega \mathbf{1}]\varphi = \mathbf{0} \tag{4.3}$$

which is for $\omega = 0$ equal to $\mathbf{K}_T\varphi = \mathbf{0}$, see eq. 4.2.

The different types of singular points can then be examined by analyzing the path derivatives $\dot{\mathbf{G}}$ and $\ddot{\mathbf{G}}$, where $(\dot{...})$ denotes the derivative with respect to the path parameter s. In this case $\dot{\mathbf{G}}$ is the tangent and $\ddot{\mathbf{G}}$ the curvature of the equilibrium curve $\mathbf{G} = \mathbf{0}$. For simplification we assume within the following analysis that only simple bifurcation points occur. The tangent to $\mathbf{G}$ is given by

$$\dot{\mathbf{G}} = \mathbf{G}_v \Delta \mathbf{v} + \mathbf{G}_\lambda \Delta \lambda = \mathbf{0}\,. \tag{4.4}$$

To discuss singular points we introduce the eigenvalue problem

$$(\mathbf{G}_v - \omega \mathbf{1})\varphi = \mathbf{0}\,, \tag{4.5}$$

Left and right eigenvector are identical due to the symmetry of $\mathbf{G}_v$. At singular points the eigenvalue ω is zero and a determination of such points is based on the product of the eigenvector and the tangent, eq. (4.4)

$$\varphi^T[\mathbf{G}_v \Delta \mathbf{v} + \mathbf{G}_\lambda \Delta \lambda] = 0\,, \tag{4.6}$$

which leads with (4.4) to the equation

$$\varphi^T \mathbf{G}_\lambda \Delta \lambda = 0\,. \tag{4.7}$$

The discussion of (4.7) furnishes the well known condition for limit and bifurcation points

$$\varphi^T \mathbf{G}_\lambda = \begin{cases} = 0 \text{..... simple bifurcation point} \\ \neq 0 \text{..... limit point} \quad . \end{cases} \tag{4.8}$$

To solve the nonlinear system of equations (4.4) defined by the tangent, $\mathbf{G}_v \, \Delta \mathbf{v} = -\Delta\lambda \, \mathbf{G}_\lambda$ we introduce an orthogonal projector, see e.g. Chan [25],

$$\mathbf{P}_\perp = \mathbf{1} - \varphi \varphi^T\,. \tag{4.9}$$

With the condition $\boldsymbol{\varphi}^T \mathbf{G}_v = 0$, see eq. (4.5), the following system of equations has to be solved

$$\mathbf{G}_v \, \Delta \mathbf{v} = -\mathbf{P}_\perp \, \mathbf{G}_\lambda \, \Delta\lambda = -\mathbf{G}_\lambda \, \Delta\lambda + (\boldsymbol{\varphi}^T \mathbf{G}_\lambda) \, \Delta\lambda \, \boldsymbol{\varphi} . \tag{4.10}$$

Based on the assumption

$$\Delta \mathbf{v} = \Delta\lambda \, \mathbf{v}_0 + \alpha \boldsymbol{\varphi} \tag{4.11}$$

two solutions are possible: $\boldsymbol{\varphi}^T \mathbf{G}_\lambda = 0$ and $\Delta\lambda = 0$, see eq. (4.7). In case of $\boldsymbol{\varphi}^T \mathbf{G}_\lambda = 0$ (bifurcation point) the result is

$$\left.\begin{aligned} \mathbf{G}_v \, \Delta\lambda \, \mathbf{v}_0 &= -\Delta\lambda \, \mathbf{G}_\lambda \\ \mathbf{G}_v \, \alpha \boldsymbol{\varphi} &= \ \ \alpha \mathbf{0} \end{aligned}\right\} \Longrightarrow \quad \Delta \mathbf{v} = \Delta\lambda \, \mathbf{v}_0 + \alpha \boldsymbol{\varphi} , \tag{4.12}$$

whereas $\Delta\lambda = 0$ (limit point) leads to

$$\mathbf{G}_v \, \alpha \boldsymbol{\varphi} = \ \ \alpha \mathbf{0} \qquad \} \quad \Longrightarrow \quad \Delta \mathbf{v} = \alpha \boldsymbol{\varphi} . \tag{4.13}$$

Due to the singularity (rank deficiency of 1) of the system $\mathbf{G}_v \mathbf{v}_0 = -\mathbf{G}_\lambda$ we have to add one constraint equation. Normally the orthogonality condition $\boldsymbol{\varphi}^T \mathbf{v}_0 = 0$ is used.

More information for example on the type of the bifurcation point is given by the second derivative of $\mathbf{G}$ which is stated above

$$\ddot{\mathbf{G}} = (\mathbf{G}_v \Delta \mathbf{v})_{,v} \, \Delta \mathbf{v} + 2 \mathbf{G}_{v\lambda} \Delta\lambda \Delta \mathbf{v} + \mathbf{G}_v \Delta\Delta \mathbf{v} + \mathbf{G}_\lambda \lambda \Delta\lambda^2 + \mathbf{G}_\lambda \Delta\Delta\lambda = \mathbf{0} . \tag{4.14}$$

Multiplication of $\ddot{\mathbf{G}}$ by the eigenvector $\boldsymbol{\varphi}$ yields under the conditions for a bifurcation point $\boldsymbol{\varphi}^T \mathbf{G}_\lambda = 0$ and $\boldsymbol{\varphi}^T \mathbf{G}_v = 0$

$$\boldsymbol{\varphi}^T [\mathbf{G}_v \Delta \mathbf{v}]_{,v} \, \Delta \mathbf{v} + 2 \boldsymbol{\varphi}^T [\mathbf{G}_v \mathbf{v}]_{,\lambda} \, \Delta\lambda + \boldsymbol{\varphi}^T \mathbf{G}_{\lambda\lambda} \Delta\lambda^2 = 0. \tag{4.15}$$

With eq. (4.11) we obtain from (4.15) the so called bifurcation equation

$$a\alpha^2 + 2b\alpha\Delta\lambda + c\Delta\lambda^2 = 0 , \tag{4.16}$$

with the constants

$$\begin{aligned} a &= \boldsymbol{\varphi}^T [\mathbf{G}_v \boldsymbol{\varphi}]_{,v} \, \boldsymbol{\varphi} \\ b &= \boldsymbol{\varphi}^T \{ [(\mathbf{G}_v \boldsymbol{\varphi}]_{,v} \, \mathbf{v}_0 + [\mathbf{G}_v \boldsymbol{\varphi}]_{,\lambda} \} \\ c &= \boldsymbol{\varphi}^T \{ [\mathbf{G}_v \mathbf{v}_0]_{,v} \, \mathbf{v}_0 + 2 \, [\mathbf{G}_v \mathbf{v}_0]_{,\lambda} + \mathbf{G}_{\lambda\lambda} \} \\ D &= b^2 - a\,c . \end{aligned} \tag{4.17}$$

and the solution for the load direction

$$\Delta\lambda_{1,2} = -\frac{b\alpha}{c} \pm \frac{\alpha}{c}\sqrt{b^2 - a\,c} \quad . \tag{4.18}$$

For $D > 0$ a real square root exists which leads to two distinct solution curves in the bifurcation point. The singular point associated with $a \neq 0$ is called in the engineering literature asymmetric bifurcation point, see e.g. Koiter [26], whereas in the mathematical literature it is entitled simple transcritical bifurcation point, see e.g. Jepson, Spence [27].

For the special case $a = \varphi^T[\mathbf{G}_v\varphi]_{,v}\,\varphi = 0$ the solution of (4.16) yields

$$\Delta\lambda_1 = -\frac{2\,b}{c}\,\alpha \quad \Delta\lambda_2 = 0\,. \tag{4.19}$$

This point is called a symmetric or simple pitchfork bifurcation point. The evaluation of $\Delta\Delta\lambda_2$ indicates whether the symmetric bifurcation point is stable or unstable.

In addition we have for $\varphi^T\mathbf{G}_\lambda = 0, D = 0$ a cusp point and for $\varphi^T\mathbf{G}_\lambda = 0, D < 0$ a simple isola formation point, see e.g. Jepson, Spence [27], Wagner [18].

Furthermore we can discuss singular points obeying the condition $\Delta\lambda = 0$. Since a denotes the curvature of the solution path $\mathbf{G} = \mathbf{0}$, the condition $a \neq 0$ is associated with a limit point or simple quadratic turning point. Thus the type of limit point is determined by the sign of a. A special turning point can be found for $a = 0$ and is called a simple cubic turning point. For a further discussion see e.g. Wagner [18].

With respect to the engineering problem the variables a,b and c are given by the quantities

$$\begin{aligned} a =& \varphi^T\,[\mathbf{K}_T\,\varphi]_{,v}\;\varphi, \\ b =& \varphi^T\,[\mathbf{K}_T\,\varphi]_{,v}\;\mathbf{v}_0 + \varphi^T\,[\mathbf{K}_T\,\varphi]_{,\lambda}\,, \\ c =& \varphi^T\,[\mathbf{K}_T\,\mathbf{v}_0]_{,v}\;\mathbf{v}_0 + 2\varphi^T\,[\mathbf{K}_T\,\mathbf{v}_0]_{,\lambda} - \varphi^T\,\mathbf{P}_\lambda\,. \end{aligned} \tag{4.20}$$

Summarizing the above stated results singular points can be classified according to the criteria

Limit points:

$$\mathbf{L} = \begin{cases} \boldsymbol{\varphi}^T \mathbf{P} \neq 0 \; a \neq 0 : & \text{simple quadratic turning point,} \\ \boldsymbol{\varphi}^T \mathbf{P} \neq 0 \; a = 0 : & \text{simple cubic turning point.} \end{cases} \tag{4.21$_1$}$$

Bifurcation points:

$$\mathbf{B} = \begin{cases} \boldsymbol{\varphi}^T \mathbf{P} = 0 \; a \neq 0 \,\text{and}\, D > 0 : & \text{simple trans-critical bifurcation point,} \\ \boldsymbol{\varphi}^T \mathbf{P} = 0 \; a = 0 \,\text{and}\, b \neq 0 : & \text{simple pitch-fork bifurcation point,} \\ \boldsymbol{\varphi}^T \mathbf{P} = 0 \; D < 0 : & \text{simple isola formation point.} \end{cases} \tag{4.21$_2$}$$

These criteria can be readily implemented in a finite element code if information about the directional derivative of $\mathbf{K}_T$ is available. In section 2 this derivative is presented in equation (2.13). For the vectors $\boldsymbol{\varphi}$ and $\mathbf{v}_0$ it holds

$$\mathbf{h}(\bar{\mathbf{v}}, \boldsymbol{\varphi}, \mathbf{v}_0) = (\mathbf{K}_T(\bar{\mathbf{v}}, \bar{\mathbf{v}})\boldsymbol{\varphi})_{,v}\, \mathbf{v}_0 \tag{4.22}$$

with the vector

$$\mathbf{h}_e = \int_{\Omega_e} [\mathbf{B}^T(\bar{\mathbf{v}}_e)\mathbf{D}\mathbf{B}^{nl}(\mathbf{v}_{0e}) + \mathbf{B}^{nlT}(\mathbf{v}_{0e})\mathbf{D}\mathbf{B}(\bar{\mathbf{v}}_e) + \hat{\mathbf{G}}^T \Delta\hat{\mathbf{S}}(\bar{\mathbf{v}}_e, \mathbf{v}_{0e})\hat{\mathbf{G}}]\, d\Omega\, \boldsymbol{\varphi}_e \,. \tag{4.23}$$

Thus all necessary derivations in the parameters a, b and c can be calculated. A disadvantage is that the calculation of $\mathbf{h}_e$ has to be implemented in a finite element code for each element. On the other hand if the computation of the $\mathbf{B}$–matrices $\mathbf{B}(\delta\mathbf{v}, \Delta\mathbf{v})$, $\mathbf{B}^{nl}(\delta\mathbf{v}, \Delta\mathbf{v})$ and the stress vector $\mathbf{S}(\delta\mathbf{v}, \Delta\mathbf{v})$ is available there is no difficulty in formulating $\mathbf{h}_e$. These matrices are usable if the tangent stiffness matrix $\mathbf{K}_T$ is implemented.

As an example mor information is given for a beam element in Wriggers, Wagner, Miehe [10].

Alternatively it is practicable to introduce the derivative of $\mathbf{K}_T$ by a numerical differentiation procedure, see e.g. Keller [1977], Rheinboldt [1976] or Wriggers, Simo [1989]. Thus the algorithm is element independent.

To create a numerical approximation that preserves the quadratic rate of convergence of Newton's method, we first recall that the vector $\mathbf{K}_T \boldsymbol{\varphi}$ can be represented as the directional derivative of the residual $\mathbf{G}$ in the direction of $\boldsymbol{\varphi}$, according to the expression

$$\mathbf{K}_T\,\boldsymbol{\varphi} = D_v\,\mathbf{G}(\mathbf{v}\,,\lambda)\,\boldsymbol{\varphi} = \frac{d}{d\epsilon}\,\mathbf{G}(\mathbf{v}+\epsilon\,\boldsymbol{\varphi}\,,\lambda)\Big|_{\epsilon=0}\,. \tag{4.24}$$

By exploiting the symmetry of the second derivative of $\mathbf{G}$, we can express the directional derivative of $\mathbf{K}_T\,\boldsymbol{\varphi}$ in the direction of $\Delta\mathbf{v}$ in the following equivalent form

$$\begin{aligned} D_v\,[\,\mathbf{K}_T\,\boldsymbol{\varphi}\,]\,\Delta\mathbf{v} &= D_v\,[\,D_v\,\mathbf{G}(\mathbf{v}\,,\lambda)\,\boldsymbol{\varphi}\,]\,\Delta\mathbf{v} \\ &= D_v\,[\,D_v\,\mathbf{G}(\mathbf{v}\,,\lambda)\,\Delta\mathbf{v}\,]\,\boldsymbol{\varphi}\,. \end{aligned} \tag{4.25}$$

With this expression in hand, we show next that the vectors $\mathbf{h}$ in the algorithm described above can be computed with only one additional evaluation of the tangent stiffness. To this end, we use the definition of directional derivative to obtain the following expression:

$$D_v\,[\,\mathbf{K}_T\,\boldsymbol{\varphi}\,]\,\Delta\mathbf{v} = \frac{d}{d\epsilon}\,[\,\mathbf{K}_T(\mathbf{v}+\epsilon\,\boldsymbol{\varphi})\,]\,\Delta\mathbf{v}\,\Big|_{\epsilon=0}\,. \tag{4.26}$$

This equation is now recast in an alternative format amenable to numerical approximation. We set

$$D_v\,[\,\mathbf{K}_T\,\boldsymbol{\varphi}\,]\,\Delta\mathbf{v} = \lim_{\epsilon=0}\,\frac{1}{\epsilon}\,[\,\mathbf{K}_T(\mathbf{v}+\epsilon\,\boldsymbol{\varphi})\,\Delta\mathbf{v} - \mathbf{K}_T(\mathbf{v})\,\Delta\mathbf{v}\,]\,. \tag{4.27}$$

By suitable selecting the parameter ϵ we obtain

$$D_v\,[\,\mathbf{K}_T\,\boldsymbol{\varphi}\,]\,\Delta\mathbf{v} \approx \frac{1}{\epsilon}\,[\,\mathbf{K}_T(\mathbf{v}+\epsilon\,\boldsymbol{\varphi})\,\Delta\mathbf{v} - \mathbf{K}_T(\mathbf{v})\,\Delta\mathbf{v}\,]\,. \tag{4.28}$$

The application of this approximation to the directional derivative of ..... leads to the following algorithmic expressions for the vectors $\mathbf{h}$

$$\begin{aligned} \mathbf{h}_a &\approx \frac{1}{\epsilon}\,[\,\mathbf{K}_T(\bar{\mathbf{v}}+\epsilon\,\boldsymbol{\varphi})\,\boldsymbol{\varphi}\,]\,, \\ \mathbf{h}_b &\approx \frac{1}{\epsilon}\,[\,\mathbf{K}_T(\bar{\mathbf{v}}+\epsilon\,\boldsymbol{\varphi})\,\mathbf{v}_0 - \mathbf{P}\,] \\ \mathbf{h}_c &\approx \frac{1}{\epsilon}\,[\,\mathbf{K}_T(\bar{\mathbf{v}}+\epsilon\,\mathbf{v}_0)\,\mathbf{v}_0 - \mathbf{P}\,] \end{aligned} \tag{4.29}$$

which can be introduced in eq. (4.20).

REMARK:

i. Observe that at this stage $\mathbf{P}$ is already computed.

ii. The matrix multiplications in (4.29) can be carried out at the element level, and the matrix $\mathbf{K}_T(\mathbf{v}+\epsilon\,\boldsymbol{\varphi})$ need not be assembled at the global level. The standard assembly procedure has to be performed only on the vectors $\mathbf{h}_i$, $\quad i = a, b, c$.

iii. A suitable selection of the parameter ϵ in (4.29) is crucial to the success of the method. This choice depends on the vectors $\mathbf{v}$ and $\boldsymbol{\varphi}$ and on the computer precision. An estimate for ϵ may be found in Dennis, Schnabel [28]. Based on numerical experience we choose $10^{-8} < \epsilon < 10^{-5}$.

5. Incremental stability analysis

For highly nonlinear shell problems incremental–iterative strategies using path–following methods are necessary. When following nonlinear solution paths the stability behaviour of the structure has to be investigated in an accompanying way which may be very time consuming. Thus different methods have been introduced to detect singular points based on $\mathbf{K}_T\boldsymbol{\varphi} = \mathbf{0}$, see eq. (4.2).

A simple method for this purpose is the inspection of the determinant of the tangent stiffness matrix $\mathbf{K}_T$. The homogeneous system of equations $\mathbf{K}_T\boldsymbol{\varphi} = \mathbf{0}$ has non–trivial solutions only for $det\mathbf{K}_T = 0$. The calculation of $det\mathbf{K}_T$ is possible without additional effort within the process of triangularization of equations (3.5). It holds

$$det\,\mathbf{K}_T = \prod_{i=1}^{ndof} D_{ii}\,, \tag{5.1}$$

with

$$\mathbf{K}_T = \mathbf{LDL}^T\,.$$

To avoid extremely large values for $det\,\mathbf{K}_T$ within the numerical calculations it is advantageous to scale the determinant

$$\det \bar{\mathbf{K}}_T = \frac{\det {}_k\mathbf{K}_{T\,s}}{\det {}_1\mathbf{K}_{T\,s}} \qquad \text{with} \quad \det {}_k\mathbf{K}_{T\,s} = \prod_{i=1}^{ndof} \frac{{}_kD_{ii}}{\|{}_1\mathbf{D}\|}\,. \tag{5.2}$$

In this equation the subscripts k and 1 denote the actual equilibrium state and the first linear step of the incrementation respectively.

In a $det\,\bar{\mathbf{K}}_T - v$ - diagram stability points can be seen as intersections of the v-axis. No distinction can be made between limit and bifurcation points so that further methods of inspection are needed.

Clear information about the equilibrium state is given by the sign of diagonal elements D_{ii} . One or more negative diagonal elements indicate unstable equilibrium states.

Another simple method which gives information about the stiffness behaviour of a structure is associated with the current stiffness parameter Sp , Bergan et.al. [5]. Sp is defined by

$$Sp = \frac{\Delta_1\mathbf{v}_P^T\,\mathbf{P}}{\Delta_k\mathbf{v}_P^T\,\mathbf{P}} \quad = \frac{\text{const.}}{\Delta_k\mathbf{v}_P^T\,\mathbf{P}}\,. \tag{5.3}$$

Sp tends to zero if limit points are reached but gives no information about bifurcation points. A simple test can be made by the discussion of $\boldsymbol{\varphi}^T\mathbf{P}$, see eq. (4.8).

A more elaborate technique is given by introducing an inverse iteration. This method is valid for the lowest eigenvalue. Consequently this technique can be used for the eigenvalue problem $[\mathbf{K}_T - \omega\,\mathbf{1}]\,\boldsymbol{\varphi} = \mathbf{0}$. With an inverse iteration the associated eigenvector is calculated by

$$\boldsymbol{\varphi}_{i+1} = \mathbf{K}_T^{-1}\boldsymbol{\varphi}_i\,. \tag{5.4}$$

The eigenvalue ω ican be found by the Rayleigh – quotient

$$\omega_{i+1} = \frac{\boldsymbol{\varphi}_{i+1}^T\boldsymbol{\varphi}_i}{\boldsymbol{\varphi}_{i+1}^T\boldsymbol{\varphi}_{i+1}}\,. \tag{5.5}$$

Hence, the following procedure can be introduced within an incremental solution process based on a given equilibrium point $\mathbf{G} = \mathbf{0}$.

i. Make some steps (5.4). Start with $\boldsymbol{\varphi}_1 = \mathbf{1}$. Each step is one backsolve. $\mathbf{K}_T$ has not to be calculated because this matrix is available from the last iteration step. Own numerical experience show that only 1–3 backsolves have to be done. It is not necessary to calculate $\boldsymbol{\varphi}$ exactly because we are not at a singular point.

ii. Calculate an approximation for ω using eq. (5.5).

iii. $\omega \begin{cases} > 0 & \text{stable point} \\ = 0 & \text{singular point} \\ < 0 & \text{unstable point} \end{cases}$

An advantage of this procedure that an approximation of the eigenvector is calculated too.

The clearest but most expensive results can be achieved by solving the eigenvalue problem $[\mathbf{K}_T - \omega\,\mathbf{1}]\,\boldsymbol{\varphi} = \mathbf{0}$ itself. Hence, this should be done only in the vicinity of a singular points. The decision where this vicinity is, can be made by the above mentioned simple methods.

In the engineering literature often other eigenvalue problems are solved. These are extensions of the classical and linear eigenvalue analysis. Here the actual tangent stiffness matrix will be splitted into the linear stiffness matrix $\mathbf{K}_L$, the initial displacement matrix $\mathbf{K}_U$ and the initial stress matrix $\mathbf{K}_\sigma$

$$\mathbf{K}_T(\bar{\mathbf{v}},\bar{\lambda}) = \mathbf{K}_L + \mathbf{K}_U(\bar{\mathbf{v}},\bar{\lambda}) + \mathbf{K}_\sigma(\bar{\mathbf{v}},\bar{\lambda}) \tag{5.6}$$

A definition of these matrices can be found in Zienkiewicz, Taylor [19]. The idea is to increase the displacement and load dependent terms $\mathbf{K}_U$ and $\mathbf{K}_\sigma$ by a load factor Λ and construct a linear eigenvalue problem for Λ. Thus two eigenvalue problems are possible.

$$\begin{aligned} [\mathbf{K}_L + \mathbf{K}_U + \Lambda \mathbf{K}_\sigma]\,\boldsymbol{\varphi} &= \mathbf{0} \\ [\mathbf{K}_L + \Lambda(\mathbf{K}_U + \mathbf{K}_\sigma)]\,\boldsymbol{\varphi} &= \mathbf{0} \end{aligned} \tag{5.7}$$

The result for the stability analysis can be stated as

$$\Lambda \quad \begin{cases} > 1 & \text{stable point} \\ = 1 & \text{singular point} \\ < 1 & \text{unstable point} \end{cases} \tag{5.8}$$

Note that for the original eigenvalue problem $[\mathbf{K}_T - \omega \mathbf{1}]\,\boldsymbol{\varphi} = \mathbf{0}$, eq.(4.3), the condition

$$\omega \quad \begin{cases} > 0 & \text{stable point} \\ = 0 & \text{singular point} \\ < 0 & \text{unstable point} \end{cases} \tag{5.9}$$

holds.

It is obvious that for $\Lambda = 1$ and $\omega = 0$ the same eigenvalue problems will be solved. In a load–deflection diagram the $\Lambda\,\lambda\mathbf{P} - v$ – curve intersects the $\lambda\mathbf{P} - v$ – curve at singular points. An example will be given in section 8. A comparison of the eigenvalue problems (4.3) and $(5.7_1), (5.7_2)$ can be found in Wagner, Wriggers [7].

All stated methods base on an incremental strategy. Thus only approximations for the singular point and the associated eigenvectors are available. In the next section we discuss methods how to find these points within a given accuracy.

6. Direct calculation of singular points

The addition of the constraint $det\,\mathbf{K}_T$ to the nonlinear equilibrium equation (2.10) leads to the following extended system

$$\bar{\mathbf{G}}(\mathbf{v}, \lambda) = \left\{ \begin{array}{c} \mathbf{G}(\mathbf{v}, \lambda) \\ det\,\mathbf{K}_T(\mathbf{v}, \lambda) \end{array} \right\} = 0\,. \tag{6.1}$$

This system was considered by Abott [29] for the indirect computation of nontrivial bifurcation points. Strategies based on equation (6.1) have been widely used in engineering applications, see e.g. Brendel, Ramm [23]. However no method for a direct computation of singular points has been developed so far in the engineering literature using equation (6.1). In most applications the constraint $det\,\mathbf{K}_T = 0$ is basis for an accompanying investigation to detect stability points or for extrapolation techniques, see section 5. Often (6.1) is combined with arc–length methods which yields the system

$$\tilde{\mathbf{G}}(\mathbf{v},\lambda) = \left\{ \begin{array}{c} \mathbf{G}(\mathbf{v},\lambda) \\ f(\mathbf{v},\lambda) \\ det\,\mathbf{K}_T(\mathbf{v},\lambda) \end{array} \right\} = 0 \tag{6.2}$$

where $f(\mathbf{v},\lambda)$ is a constraint equation associated with path–following procedures. Here $det\,\mathbf{K}_T$ checks whether a singular point has been passed during path following. With this knowledge we can construct a bi–section method in which the direction and arc–length of the continuation method is changed according to the sign of $det\,\mathbf{K}_T$. The associated algorithm, see BOX 6.1, is very simple to implement, however, it only converges linearly to the singular point.

* Follow nonlinear solution path until $det\,\mathbf{K}_T$ changes sign:

1. Do $i = 1, 2, ...$ *until convergence*
2. Change load direction and arc–length: $ds_i = ds_{i-1}/2$
3. Compute new solution with algorithm in BOX 2
   - 3.1 Convergence check: IF $det\,\mathbf{K}_T < EPS \Longrightarrow STOP$
   - 3.2 IF sign of $det\,\mathbf{K}_T$ changes again go to 2.)

     ELSE change arc–length: $ds_{i+1} = ds_i/2$ and go to 3.)

BOX 6.1: Bi–section algorithm to compute singular points

The algorithm has been applied successfully to elasticity problems.

In the following we like to construct an algorithm which exhibits faster convergence than the bi–section scheme. Thus a Newton–scheme is necessary.

An alternative approach to construct an extended system for the computation of stability points bases on the eigenvalue problem

$$(\mathbf{K}_T - \omega\,\mathbf{1})\,\boldsymbol{\varphi} = \mathbf{0} \tag{6.3}$$

which is equivalent to the constraint $det\,\mathbf{K}_T = 0$. Since at singular points the eigenvalue ω of $\mathbf{K}_T$ is zero we obtain $\mathbf{K}_T\boldsymbol{\varphi} = \mathbf{0}$. Thus an extended system associated with this constraint may be stated as follows

$$\widehat{\mathbf{G}}(\mathbf{v},\lambda,\boldsymbol{\varphi}) = \left\{ \begin{array}{c} \mathbf{G}(\mathbf{v},\lambda) \\ \mathbf{K}_T(\mathbf{v},\lambda)\,\boldsymbol{\varphi} \\ l\,(\boldsymbol{\varphi}) \end{array} \right\} = \mathbf{0}\,. \tag{6.4}$$

This extended system has been used e.g. in Werner, Spence [30] and Weinitschke [31] for the calculation of limit and simple bifurcation points, for a combination with finite element formulations, see Wriggers, Wagner, Miehe [10].

In equation (6.4) $l(\boldsymbol{\varphi})$ denotes some normalizing functional which has to be included to ensure a non zero eigenvector $\boldsymbol{\varphi} \in \mathbb{R}^N$ at the singular point. The following expressions

$$\begin{aligned} l(\boldsymbol{\varphi}) &= \|\boldsymbol{\varphi}\| - 1 = 0\,, \\ l(\boldsymbol{\varphi}) &= \mathbf{e}_i^T \boldsymbol{\varphi} - \hat{\varphi}_0 = 0\,, \\ l(\boldsymbol{\varphi}) &= \mathbf{P}^T \boldsymbol{\varphi} - 1 = 0\,, \end{aligned} \tag{6.5}$$

where $\mathbf{e}_i$ is a vector containing zeros and a value 1 at the chosen place associated with i. The constant scalar φ_0 is computed prior to the first iteration from the starting vector $\boldsymbol{\varphi}_0$ for the eigenvector: $\hat{\varphi}_0 = \frac{\mathbf{e}_i^T \boldsymbol{\varphi}_0}{\|\boldsymbol{\varphi}_0\|}$. Note, that constraint $(6.5)_3$ leads only to limit points which comes from (4.8). In our computations we have used the first equation which seems to be more robust. Extended systems like (6.4) are associated with $2N+1$ unknowns. Thus the numerical effort to solve such systems is considerably increased. However this apparent disadvantage can be circumvented by using a bordering algorithm for the solution of the extended set of equations.

It should be noted that equation (6.4) automatically yields the eigenvector $\boldsymbol{\varphi}$ associated with the limit or bifurcation point which may be applied for branch-switching.

The incremental solution scheme using Newtons method leads to the formulation

$$\begin{aligned} \widehat{\mathbf{K}}_{T\,i}\, \Delta\mathbf{w}_i &= -\widehat{\mathbf{G}}(\mathbf{w}_i) \\ \mathbf{w}_{i+1} &= \mathbf{w}_i + \Delta\mathbf{w}_i \end{aligned} \tag{6.6}$$

where

$$\widehat{\mathbf{K}}_{T\,i}\Delta\mathbf{w} = \frac{d}{d\epsilon}[\widehat{\mathbf{G}}(\mathbf{w}_i + \epsilon\Delta\mathbf{w})]\bigg|_{\epsilon=0} \quad \text{with} \quad \mathbf{w} = \begin{Bmatrix} \mathbf{v} \\ \lambda \\ \boldsymbol{\varphi} \end{Bmatrix}.$$

Here the vector $\mathbf{w}$ contains $2N+1$ unknown variables as defined above.

In detail equation (6.6) looks as follows

$$\begin{bmatrix} \mathbf{K}_T & \mathbf{0} & -\mathbf{P} \\ D_v(\mathbf{K}_T\boldsymbol{\varphi}) & \mathbf{K}_T & D_\lambda(\mathbf{K}_T\boldsymbol{\varphi}) \\ \mathbf{0}^T & \frac{\boldsymbol{\varphi}^T}{\|\boldsymbol{\varphi}\|} & 0 \end{bmatrix} \begin{Bmatrix} \Delta\mathbf{v} \\ \Delta\boldsymbol{\varphi} \\ \Delta\lambda \end{Bmatrix} = - \begin{Bmatrix} \mathbf{G}(\mathbf{v},\lambda) \\ \mathbf{K}_T(\mathbf{v},\lambda)\boldsymbol{\varphi} \\ \|\boldsymbol{\varphi}\| - 1 \end{Bmatrix} \tag{6.7}$$

where $D_v(\mathbf{K}_T\boldsymbol{\varphi})$ and $D_\lambda(\mathbf{K}_T\boldsymbol{\varphi})$ have already be defined in (2.13) and (4.22). The computation of these derivatives within the finite element setting has been discussed in section 4.

1. Solve

$$\mathbf{K}_T\,\Delta\,\mathbf{v}_P = \mathbf{P}\,, \quad \mathbf{K}_T\,\Delta\,\mathbf{v}_G = -\mathbf{G}\,.$$

2. Compute within a loop over all finite elements

$$\mathbf{h}_P = D_v\,(\mathbf{K}_T\,\boldsymbol{\varphi})\,\Delta\mathbf{v}_P + D_\lambda\,(\mathbf{K}_T\,\boldsymbol{\varphi})\,,$$
$$\mathbf{h}_G = D_v\,(\mathbf{K}_T\,\boldsymbol{\varphi})\,\Delta\mathbf{v}_G\,.$$

3. Solve

$$\mathbf{K}_T\,\Delta\boldsymbol{\varphi}_P = -\mathbf{h}_P\,, \quad \mathbf{K}_T\,\Delta\boldsymbol{\varphi}_G = -\mathbf{h}_G\,.$$

4. Compute new increments

$$\Delta\lambda = \frac{-\boldsymbol{\varphi}^T\,\Delta\boldsymbol{\varphi}_G + \|\boldsymbol{\varphi}\|}{\boldsymbol{\varphi}^T\,\Delta\boldsymbol{\varphi}_P}\,,$$
$$\Delta\mathbf{v} = \Delta\lambda\,\Delta\mathbf{v}_P + \Delta\mathbf{v}_G\,.$$

5. Update displacements, eigenvector and load parameter

$$\lambda = \lambda + \Delta\lambda\,, \quad \mathbf{v} = \mathbf{v} + \Delta\mathbf{v} \quad \text{and} \quad \boldsymbol{\varphi} = \Delta\lambda\,\Delta\boldsymbol{\varphi}_P + \Delta\boldsymbol{\varphi}_G\,.$$

BOX 6.2: Algorithm for the direct computation of singular points

Using a block elimination technique a solution scheme of (6.7) can be developed which only needs one factorization of the tangent stiffness $\mathbf{K}_T$. BOX 6.2 below shows the algorithm for one iteration step.

The derivation of the tangent matrix can be calculated using standard linearization procedures. We refer to section 4 for details.

Before computing the directional derivative of $\mathbf{K}_T\,\boldsymbol{\varphi}$ we note that the resulting matrix is never needed in the algorithm described in BOX 6.2. Thus we compute only the vector

$$\begin{aligned}\mathbf{h}_j &= [\,(\mathbf{K}_T\boldsymbol{\varphi})],_v\,\Delta\mathbf{v}_j \\ &= \bigcup_{e=1}^{n_e}\int_{\Omega_e}\{\mathbf{B}^{nlT}(\Delta\mathbf{v}_{je})\mathbf{D}\mathbf{B}(\mathbf{v}_e) + \mathbf{B}^T(\mathbf{v}_e)\mathbf{D}\mathbf{B}^{nl}(\Delta\mathbf{v}_{je}) + \hat{\mathbf{G}}^T\Delta\hat{\mathbf{S}}\hat{\mathbf{G}}\}\boldsymbol{\varphi}_e d\Omega.\end{aligned} \tag{6.8}$$

where $j = P$ or $j = G$. $\Delta\hat{\mathbf{S}}$ contains the incremental stresses given by $\Delta\mathbf{S} = \mathbf{D}\,\mathbf{B}(\mathbf{v}_e)\,\Delta\mathbf{v}_{je}$. Furthermore a numerical differentiation is possible, see section 4. This results in the following expressions for the vectors $\mathbf{h}_P$ and $\mathbf{h}_G$

$$\begin{aligned}\mathbf{h}_P &\approx \frac{1}{\epsilon}\,[\,\mathbf{K}_T(\mathbf{v}+\epsilon\,\boldsymbol{\varphi})\,\Delta\mathbf{v}_P - \mathbf{P}\,]\,,\\ \mathbf{h}_G &\approx \frac{1}{\epsilon}\,[\,\mathbf{K}_T(\mathbf{v}+\epsilon\,\boldsymbol{\varphi})\,\Delta\mathbf{v}_G + \mathbf{G}\,]\,.\end{aligned} \tag{6.9}$$

Observe that at this stage $\mathbf{P}$ and $\mathbf{G}$ are already computed. Consequently, the computation of $\mathbf{h}$ through the expressions in (6.9) involves only one additional evaluation of the stiffness matrix; i.e., $\mathbf{K}_T(\mathbf{v}+\epsilon\,\boldsymbol{\varphi})$.

A detailed description can be found in Wriggers, Wagner, Miehe [10].

7. Calculation of secondary branches

The treatment of bifurcation problems especially the calculation of secondary branches requires additional considerations. Near stability points the associated eigenvalue problem has to be solved in order to calculate the number of existing branches. The number of possible branches in a bifurcation point is determined by the number of zero eigenvalues ω_j. The zero eigenvectors $\boldsymbol{\varphi}_j$ indicate the direction of the solution to be followed for the calculation of a secondary branch associated with $\omega_j = 0$.

A simple but effective engineering approach is to perturb the equilibrium state in a singular point by these eigenvectors. This perturbation is performed by adding the scaled eigenvector to the deformed configuration in the following way

$$\mathbf{v}_j = \bar{\mathbf{v}} + \xi_j \frac{\boldsymbol{\varphi}_j}{\|\,\boldsymbol{\varphi}_j\,\|}\,. \tag{7.1}$$

$\mathbf{v}_j$ denotes the perturbed deformed configuration which is used as starting vector for branch switching. $\bar{\mathbf{v}}$ is the displacement vector of the last converged solution and $\boldsymbol{\varphi}_j$ is the *j–th* eigenvector. ξ_j denotes a scaling factor which value is critical for a successful branch–switching.

Since $\boldsymbol{\varphi}_j$ can be viewed as a pre–deformation at $\bar{\mathbf{v}}$ the magnitude of ξ_j can be estimated by

$$\frac{\xi_j = \pm\|\mathbf{v}\|}{\tau_j.} \tag{7.2}$$

Here, τ_j is a factor which appears also in the description of imperfections. Thus, in the engineering range the value of τ_j will be of the order of 100 approximately. At simple bifurcation points usually three directions can be followed. One direction is related to the primary branch and the other two are associated with the secondary branch. For $\xi_j = 0$ the solution remains on the primary branch. For $\xi_j \neq 0$ the sign of ξ_j controls which part of the secondary branch is calculated.

The above described method can be implemented in existing finite element codes in a simple way.

A mathematically sound branch–switching technique follows from the bifurcation equation, see section 4,

$$a\,\alpha^2 + 2\,b\alpha\,\Delta\lambda + c\,\Delta\lambda^2 = 0 \tag{7.3}$$

which is derived in e.g. Decker, Keller [6], Wagner [18]. The constants a, b, c are defined by (4.17). Thus with the knowledge of (4.22) or (4.29) these constants can be computed. Then the above quadratic equation can be solved for a given α which can be interpreted as the scaling factor ξ since $\Delta\mathbf{v} = \Delta\lambda\,\bar{\mathbf{v}} + \alpha\,\boldsymbol{\varphi}$. For a further discussion we refer to Decker, Keller [6] or Wagner [18]. However, it should be remarked that for the practically important case of symmetric bifurcation the above mentioned engineering approach coincides with the use of the bifurcation equation (7.3).

Once a secondary branch is reached a further path–following can be done using standard arc–length schemes, see section 3.

8. Examples

All in this paper discussed algorithms and finite element formulations are introduced in the finite element program **F E A P** of R.L. Taylor, described in Zienkiewicz, Taylor [19].

8.1 Cylindrical shell segment under point load

This example is a classical test example to discuss the behaviour of nonlinear shell formulations. Often it is used to test elements with finite rotations but due to the deformation behaviour shown above a shell theory with moderate rotations is sufficient for accurate results. The geometrical and material data are shown in Fig. 8.1. The boundary conditions are 'free' at the curved edges and 'hinged' at the straight edges.

Fig. 8.1: Cylindrical shell segment under point load

The load deflection behaviour is governed by a snap–through response of the shell segment. Thus arc–length methods are necessary to calculate the nonlinear solution path. Also – in this special case due to a single load – a displacement control method can be used successful.

The employed element is described in Stein, Wagner Wriggers [32]. The load deflection path for the center displacement is depicted in Fig. 8.2, based on a 4*4 mesh for one quarter of the shell segment. Due to the snap–through behaviour there exist two limit points on the load deflection path. In section 5 we have introduced different methods to describe the stability behaviour. Based on eq. (5.1) the $det\mathbf{K}_T - v$ curve intersects the v–axis at the singular points. Note that the determinant is scaled due to (5.2). Thus the curve starts with a value of 1 which bases on the results for the first load step. If we solve in each load step the eigenvalue problems (4.3) and (5.7_2 two other curves occur in Fig. 8.2. For the eigenvalue problem $[\mathbf{K}_T - \omega \mathbf{1}]\,\boldsymbol{\varphi} = \mathbf{0}$ (eq. 4.3) we have at singular points intersections of the curve with the v–axis whereas the eigenvalue problem $[\mathbf{K}_L + \Lambda(\mathbf{K}_U + \mathbf{K}_\sigma)]\,\boldsymbol{\varphi} = \mathbf{0}$ leads to intersections with the load deflection path at singular points. Note that these complete curves are in general not calculated.

Fig. 8.2: Stability analysis of a cylindrical shell segment

If we are not interested in the complete load deflection path the extended system (6.4) can be used to find the first limit point directly starting from $(\mathbf{v}, \lambda) = (\mathbf{0}, 0)$.

The results for the iteration behaviour are shown in Box 8.1. After 3 iterations the limit point has been found within a certain accuracy. The following 2 iterations demonstrate the Newton-like iteration behaviour in the vicinity of the solution point. The value of $\boldsymbol{\varphi}^T\mathbf{P}$ (see eq. (4.8)) is 1.0 mm * 0.25 kN = 0.25 kNmm, which indicates a limit point (1 mm is the scaled value of the eigenvector at the center and 0.25 kN is a quarter of the center load P = 1 kN). For comparison we calculate results with the classical and linear buckling analysis. These methods can be understand as linearized versions of the eigenvalue problems defined in eq. (5.7).

As in Box 8.1 shown these results, especially for the classical buckling analysis, deviate strongly from the nonlinear results. Thus a careful use of these methods in the nonlinear case is recommended.

	Extended System			class. buck. anal.		lin. buck. anal.	
No. It.	$\|\|\mathbf{G}\|\|$	P [kN]	v [mm]	P [kN]	v [mm]	P [kN]	v [mm]
1	$2.464 \cdot 10^{-02}$	4.018	10.2464	10.967	2.7308	3.283	0.817
2	$3.199 \cdot 10^{+00}$	2.376	10.9799				
3	$2.224 \cdot 10^{-01}$	2.235	10.7842				
4	$2.346 \cdot 10^{-02}$	2.232	10.7646				
5	$1.420 \cdot 10^{-07}$	2.232	10.7662				
6	$5.835 \cdot 10^{-10}$	2.232	10.7663				

Box 8.1: Iteration behaviour within the calculation of the limit point

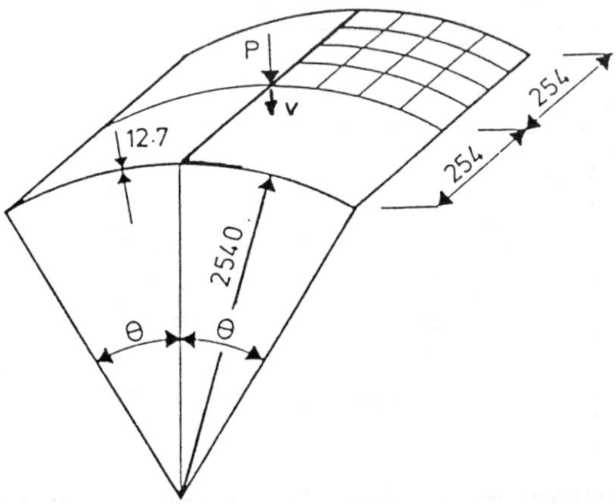

$$E = 3102.75\, N/mm^2$$
$$\nu = 0.3$$
$$\Theta = 0.1\, rad$$
$$R = 2540\, mm$$
$$L = 254\, mm$$
$$t = 12.7\, mm$$
$$\Delta P = 1\, kN$$

Fig. 8.1: Cylindrical shell segment under point load

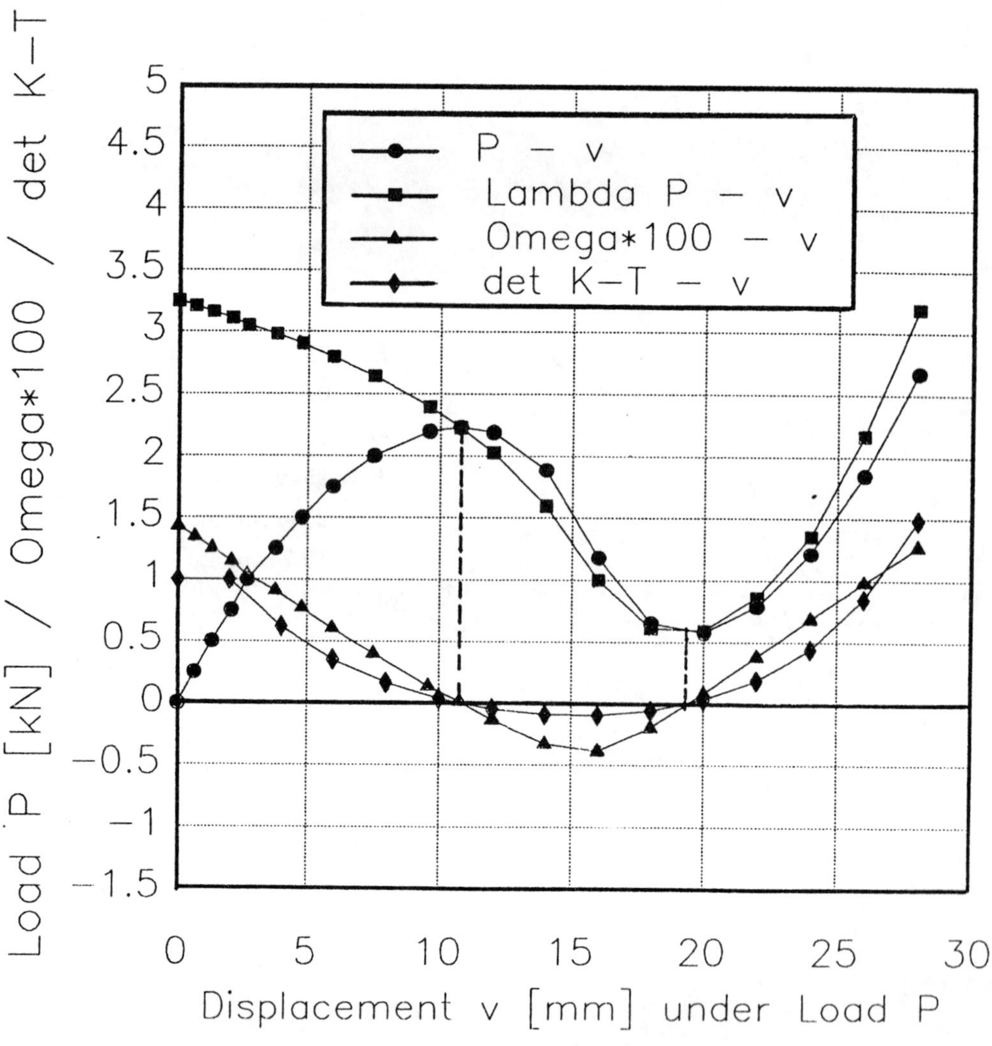

Fig. 8.2: Stability analysis of a cylindrical shell segment

8.2 Cylindrical shell segment under uniform load

In this example the occurrence of a symmetric bifurcation point is discussed.

The shell segment shown in Fig. 8.3 is similar to example 1. In contrast to 8.1 all edges are clamped, another thickness is chosen and the load is a uniform load.

All known results base on the introduction of a double symmetry, see e.g. Horrigmoe [33], Noor [34]. Thus only one quarter of the system is used within the finite element analysis. The associated load–deflection curve for the center deflection w_c is stated in Fig. 8.3 for a 4*4 (reduced) mesh.

Fig. 8.3: Symmetrical load deflection path of a cylindrical shell segment

Own calculations show that using the complete system a symmetry breaking bifurcation point occur at a load level of $p = 1.802\,kN/m^2$. In Fig. 8.4 the relevant part of the load–deflection path is shown. Between the symmetry breaking bifurcation point ($p = 1.802\,kN/cm^2\,, w_c = 5.490\,mm$) and the bifurcation point ($p = 3.052\,kN/cm^2\,, w_c = 11.033\,mm$) we have an symmetric and an antisymmetric solution branch, which can be seen in Fig. 8.4 in detail.

Fig. 8.4: Symmetric and antisymmetric load–deflection paths $p - w_c$

The associated deformation behaviour is characterized by the first two eigenvectors - calculated in the symmetry breaking bifurcation point, see Fig. 8.5. The Null–eigenvector φ_1 defines the antisymmetric deformation behaviour whereas the second eigenvector φ_2 belongs to the symmetric path.

Fig. 8.5: 1st and 2nd eigenvector at the symmetry breaking bifurcation point

The different behaviour on both paths can be seen clearly if we look at the load–deflection paths for the points L and R, see Fig. 8.6. Thus we have in Fig. 8.6 one symmetric solution ($w_s = w_L = w_R$) and two antisymmetric solutions ($w_L \neq w_R$). The singular point is a stable symmetric bifurcation point.

Fig. 8.6: Symmetric and antisymmetric load–deflection paths $p - w_s$, $p - w_L$, $p - w_R$

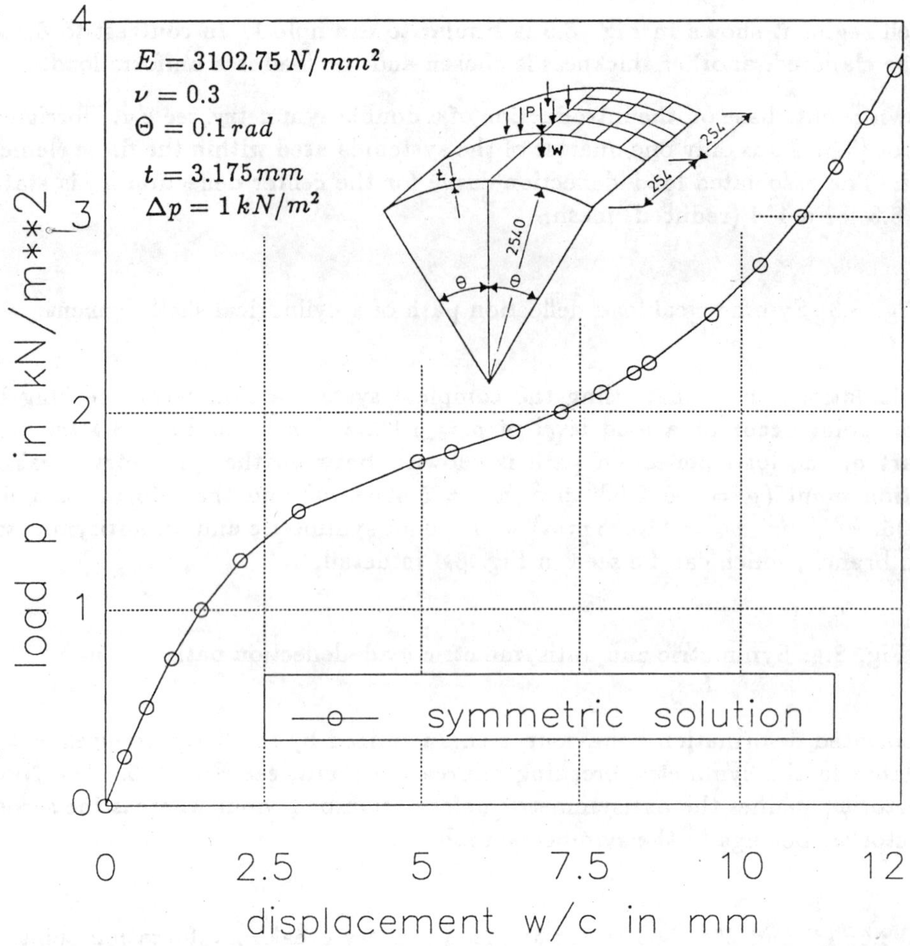

Fig. 8.3: Symmetrical load deflection path of a cylindrical shell segment

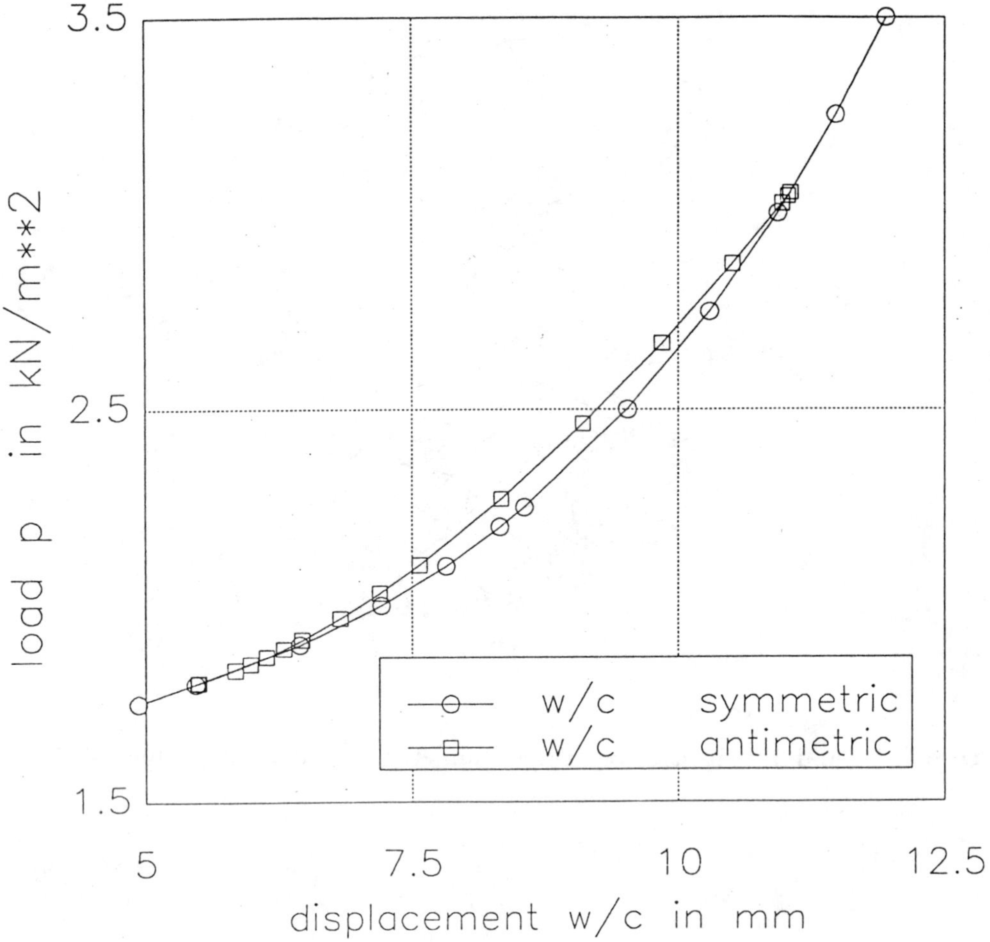

Fig. 8.4: Symmetric and antisymmetric load–deflection paths $p - w_c$

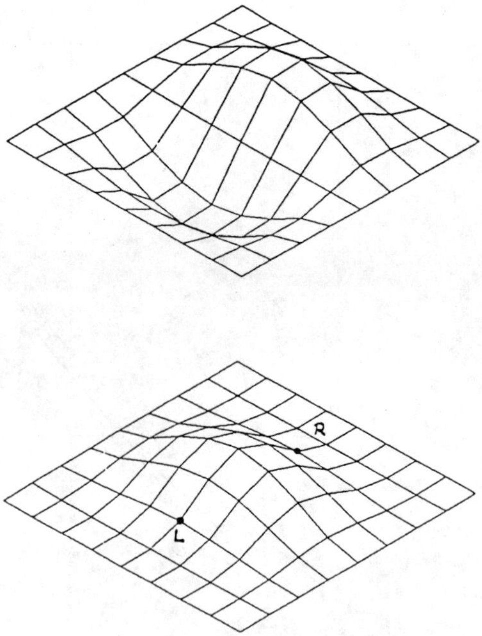

Fig. 8.5: 1st and 2nd eigenvector at the symmetry breaking bifurcation point

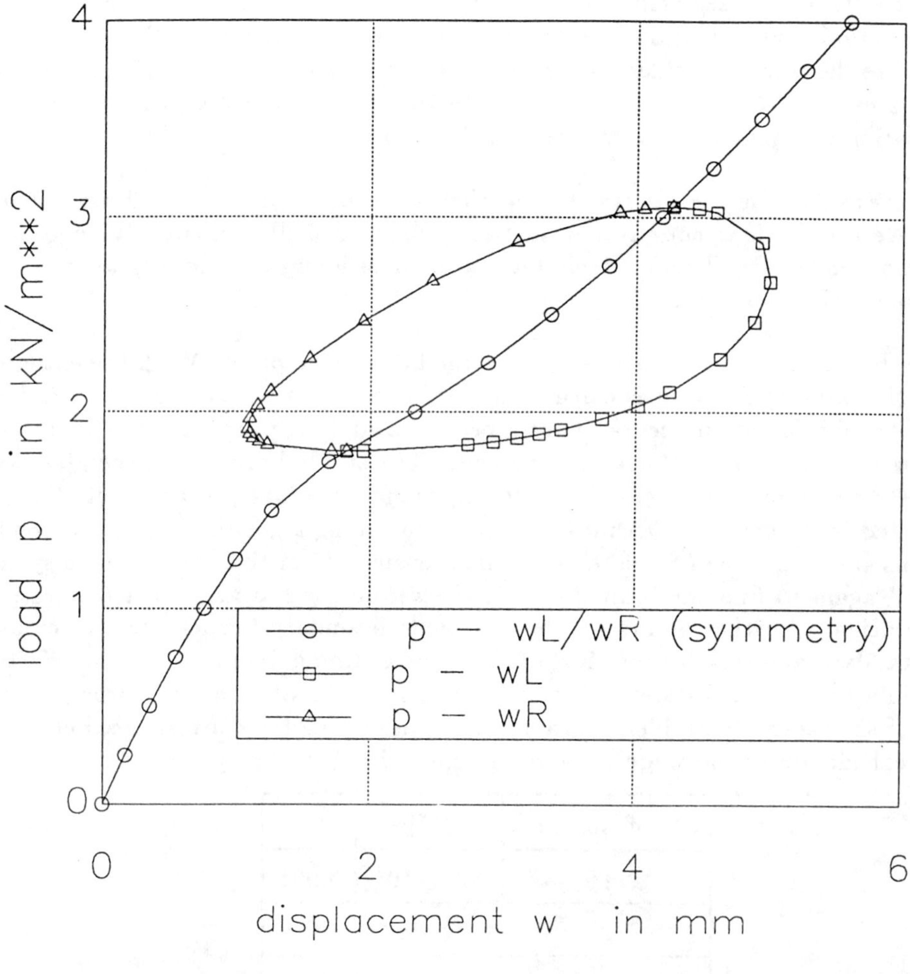

Fig. 8.6: Symmetric and antisymmetric load–deflection paths $p - w_s$, $p - w_L$, $p - w_R$

8.3 Cylindrical shell segment under axial load

In this example we investigate the buckling behaviour of an axially loaded simply supported cylindrical segment shown in Figure 8.7. We have treated this problem in earlier studies to test branch switching procedures, see Wagner, Wriggers [7] and to compare the results of different shell theories and associated finite element formulations, see Stein, Wagner, Wriggers [32]. In this paper we want to investigate the computation of singular points via extended systems.

The discretization of one quarter of the shell is realized with a 8*8 finite element mesh. We use the four node isoparametric cylindrical shell element with one point integration and a stabilization technique to control hourglass modes, see e.g. Belytschko and Tsay [35], Wagner [36].

The pre–buckling behaviour is linear up to the bifurcation point. With the extended system the bifurcation point is calculated within 7 Newton steps, see Fig. 8.7. Here asymmetrical bifurcation occurs. Thus a branch–switching procedure has to be used to follow the different postbuckling branches. The stable branch is associated with negative deflections w whereas the unstable branch exhibits positive deflections w as depicted in Figure 8.7. Within the branch–switching algorithm, only 4 corrector iterations are necessary to reach the secondary branch. After this the extended system was used again to find the limit point on the secondary branch. Here 6 iterations were sufficient. Further points on the nonlinear secondary branch were calculated using standard arc–length procedures. Thus the example demonstrate the effective applicability of extended systems, branch–switching and path–following procedures to nonlinear shell stability problems. Additionally, in Box 8.2 the convergence behaviour for the calculation of the singular point is depicted.

no. of iteration	$\|\mathbf{G}\|$	λ
1	$3.53 \cdot 10^{2}$	0.991
2	$6.22 \cdot 10^{-3}$	1.010
3	$4.85 \cdot 10^{0}$	1.002
4	$6.85 \cdot 10^{-1}$	1.001
5	$2.81 \cdot 10^{-2}$	1.000
6	$1.72 \cdot 10^{-4}$	1.000
7	$2.81 \cdot 10^{-8}$	1.000

Box 8.2: Convergence behaviour within the computation of the first singular point

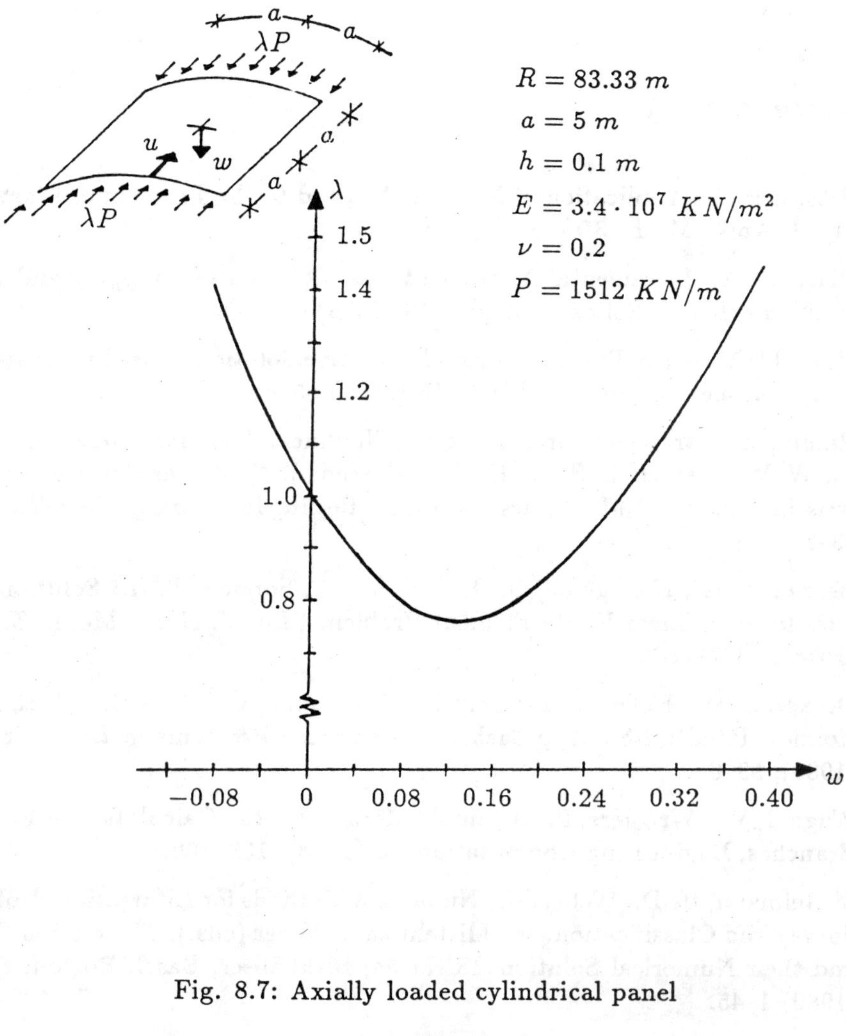

Fig. 8.7: Axially loaded cylindrical panel

Fig. 8.7: Axially loaded cylindrical panel

9. References

1. Riks, E.: The Application of Newtons Method to the Problem of Elastic Stability, J. Appl. Mech. **39** (1972) 1060–1066.

2. Riks, E.: An Incremental Approach to the Solution of Snapping and Buckling Problems, Int. J. Solids & Struct. **15** (1979) 529–551.

3. Crisfield, M. A.: A Fast Incremental/Iterative Solution Procedure that Handles Snap Through, Comp. & Struct. **13** (1981) 55–62.

4. Ramm, E.: Strategies for Tracing the Nonlinear Response Near Limit Points. In: W. Wunderlich, E. Stein, K.–J. Bathe (eds.): Nonlinear Finite Element Analysis in Structural Mechanics, Springer, Berlin, Heidelberg, New–York (1981) 63–89.

5. Bergan, P. G., Horrigmoe, G., Krakeland, B., Soreide, T. H.: Solution Techniques for Non-linear Finite Element Problems, Int. J. Num. Meth. Engng. **12** (1978), 1677–1696.

6. Decker, D. W., Keller, H. B.: Solution Branching–A Constructive Technique, in: Holmes, P.(ed): New Approaches to Nonlinear Problems in Dynamics, SIAM (1980), 53–69.

7. Wagner, W., Wriggers, P.: A Simple Method for the Calculation of Postcritical Branches, Engineering Computations, **5** (1988), 103–109.

8. Mittelmann, H.-D., Weber, H.: Numerical Methods for Bifurcation Problems - a Survey and Classification. In: Mittelmann, Weber (eds.): Bifurcation Problems and their Numerical Solution, ISNM 54, Birkhäuser, Basel, Boston, Stuttgart (1980) 1–45.

9. Moore, G., Spence, A.: The Calculation of Turning Points of Nonlinear Equations, SIAM, J. Numer. Anal. Comput. **17** (1980), 567–575.

10. Wriggers, P., Wagner, W., Miehe, C.: A Quadratically Convergent Procedure for the Calculation of Stability Points in Finite Element Analysis, Comp. Meth. Appl. Mech. Engng. **70** (1988), 329–347.

11. Wriggers, P., Simo, J.C.: A General Procedure for the Direct Computation of Turning and Bifurcation Points, Int. J. Num. Meth. Engng. **30** (1990), 155–176.

12. Schweizerhof, K. H., Wriggers, P.: Consistent Linearization for Path Following Methods in Nonlinear FE Analysis, Comp. Meth. Appl. Mech. Engng. **59** (1986), 261–279.

13. Keller, H. B.: Numerical Solution of Bifurcation and Nonlinear Eigenvalue Problems. In: Rabinowitz, P. (ed.): Application of Bifurcation Theory, Academic Press, New York (1977), 359–384.

14. Rheinboldt, W. C.: Numerical Analysis of Continuation Methods for Nonlinear Structural Problems, Comp. & Struct. **13** (1981) 103–113.

15. Riks, E.: Some Computational Aspects of Stability Analysis of Nonlinear Structures, Comp. Meth. Appl. Mech. Engng. **47** (1984), 219–260.

16. Batoz, J. L., Dhatt, G.: Incremental Displacement Algorithms for Non-Linear Problems, Int. J. Num. Meth. Engng., **14** (1979), 1262-1267.

17. Fried, I.: Orthogonal Trajectory Accession to the nonlinear Equilibrium Curve, Comp. Meth. Appl. Mech. Engng. **47** (1984) 283–297.

18. Wagner, W.: Zur Behandlung von Stabilitätsproblemen der Elastostatik mit der Methode der Finiten Elemente, Forschungs- und Seminarberichte aus dem Bereich der Mechanik der Universität Hannover, F91/1 (1991).

19. Zienkiewicz O. C., Taylor, R. L.: The Finite Element Method, Vol.1-2, 4. Edn., Mc Graw–Hill, London, 1989/1991.

20. Simo, J. C., Wriggers, P., Schweizerhof, K., Taylor, R. L.: Finite Deformation Postbuckling Analysis Involving Inelasticity and Contact Constraints, Int. J. Num. Meth. Engng., **23** (1986), 779-800.

21. Wagner, W.: A simple Finite Element Model for Beams with Finite Rotations, in preparation.

22. Wagner, W.: A Finite Element Model for Nonlinear Shells of Revolution with Finite Rotations, Int. J. Num. Meth. Engng. **29** (1990), 1455–1471.

23. Brendel, B., Ramm, E.: Nichtlineare Stabilitätsuntersuchungen mit der Methode der finiten Elemente, Ing. Archiv 51 (1982), 337–362.

24. Hildebrandt, T. H., Graves, L. M.: Implicit Functions and their Differentials in General Analysis. A.M.S. Transactions **29** (1927), 127–153.

25. Chan, T. F.: Deflation Techniques and Block–Elimination Algorithms for Solving Bordered Singular Systems, SIAM, J. Sci. Stat. Comput. **5** (1984), 121–134.

26. Koiter W. T.: On the Stability of Elastic Equilibrium, Translation of 'Over de Stabiliteit von het Elastisch Evenwicht', Polytechnic Institute Delft, H. J. Paris Publisher Amsterdam 1945, NASA TT F-10,833, 1967.

27. Jepson A. D., Spence, A.: Folds in Solutions of two Parameter Systems and their Calculation, SIAM, J. Numer. Anal. **22** (1985), 347–369.

28. Dennis, J. E., Schnabel, R. B.: Numerical Methods for Unconstrained Optimization and Nonlinear Equations, Prentice–Hall Inc., Englewood Cliffs, New Jersey (1983).

29. Abbott, J. P.: An Efficient Algorithm for the Determination of certain Bifurcation Points, J. Comp. Appl. Math. **4**(1978), 19–27.

30. Werner, B., Spence, A.: The Computation of Symmetry-Breaking Bifurcation Points, SIAM J. Num. Anal. **21** (1984), 388–399.

31. Weinitschke, H. J.: On the Calculation of Limit and Bifurcation Points in stability Problems of Elastic Shells, Int. J. Solids Struct. **21** (1985), 79–95.

32. Stein, E., Wagner, W., Wriggers, P.: Concepts of Modeling and Discretization of Elastic Shells for Nonlinear Finite Element Analysis, in: Proceedings of the Mathematics of Finite Elements and Applications VI MAFELAP 1987 Conference, ed. J. R. Whiteman, Academic Press, London (1988), 205–232.

33. Horrigmoe, G.: Finite Element Analysis of Free-Form Shells, Rep. No. 77-2, Inst. for Statikk, Division of Structural Mechanics, The Norwegian Instiute of Technology, University of Trondheim, Norway.

34. Noor, A. K.: Recent Advances in Reduction Methods for Nonlinear Problems, Comp. & Struct. **13** (1981) 31–44.

35. Belytschko, T., Tsay, C. S.: A Stabilization Procedure for the Quadrilateral Plate Element with One–Point Quadrature. Int. J. Num. Meth. Engng., **19** (1983), 405–419.

36. Wagner, W.: Zur Formulierung eines Zylinderschalenelementes mit vollständig reduzierter Integration, ZAMM, **68** (1988),T430–433.

COMPOSITE AND SANDWICH SHELLS

F. G. Rammerstorfer
Vienna Technical University, Vienna, Austria

K. Dorninger
Saturn Corp., Troy, MI, USA

A. Starlinger
NASA Lewis Research Center, Cleveland, OH, USA

ABSTRACT

Finite shell elements for geometrically and materially nonlinear stress and stability analyses of layered fibre reinforced composite shells as well as sandwich shells with orthotropic core material are presented. The element formulations are based on the degeneration principle including large displacements with an efficient analytical thickness integration. Some micromechanical models are discussed which allow the estimation of over-all material laws used in the shell analyses. Examining suitable failure criteria and taking into account post-cracking, post-wrinkling and post-buckling stiffnesses allow for the investigation of progressive damage. Furthermore, anisotropic plasticity effects are taken into account for consideration of metal matrix composite (MMC) shells. Some computed examples and comparisons with experimental results illustrate that an extended range of composite specific problems can be covered by these laminated and sandwich elements.

1. INTRODUCTION

Composite or sandwich shell structures are often used in advanced lightweight constructions. The prediction of their limit loads is crucial for fully utilizing the material capabilities. Either buckling - local or global - of the structure or the strength of the material (or both) limit the sustainable load of the shell. Special emphasis will be put on investigating these two limiting factors, especially for layered fibre reinforced polymer matrix (FRP) composite and sandwich shells. In case of metal matrix composite (MMC) shells the onset of plastification of the matrix is generally not a failure criterion. Hence, the anisotropic plastic behavior must be taken into account in the analysis of the load carrying capacity of MMC shells.

By using special shell finite elements developed for laminated fibre-reinforced composites (LFC-element [1], LCSLFC-element [2] or D3MMC-element [3] and for sandwich shells (LM-D3SW-element [4]), all implemented in CARINA [5], these limit load analyses including the post failure and post buckling regime can be conducted for arbitrary shell structures. A survey of some of these elements is presented in [6].

The onset of failure, particularly matrix or fibre cracking in fibre reinforced plastic (FRP) shells as well as wrinkling or intracell buckling of the face layers in sandwich shells, and the effective safety margin to overall collapse can be computed, which is of considerable practical relevance.

The complexity of the internal stress state typically found in layered composite materials and, in addition, the number of failure modes possible necessitate the introduction of some simplifications and assumptions in order to make post-failure analyses feasible.

2. LFC-ELEMENT FOR FRP-SHELLS

As a result of a critical examination of different theories used to derive shell finite elements, the degeneration principle (see e.g. [7] and corresponding contributions to this book [8,9]) was selected to form the basis of the LFC-element because all necessary extensions due to the layered material could be included in a straightforward way.

A detailed description of the theoretical background of the LFC-element along with a number of illustrative examples can be found in [10]. In this chapter only the outline of the element formulation is presented with special emphasis on the failure analysis modelling for FRP shells.

2.1 Basics

The element formulation is based on the degeneration principle. By using objective strain and stress measures, geometrically nonlinear behavior in terms of large deformations is included. The incremental finite element equation has the well known form:

$$[{}^m\underset{\approx}{\mathbf{K}}_e + {}^m\underset{\approx}{\mathbf{K}}_g](\Delta\mathbf{u})^1 = {}^{m+1}\underset{\sim}{\mathrm{r}} - ({}^m\underset{\sim}{\mathrm{f}} - \Delta\underset{\sim}{\mathrm{f}}_{th}) \qquad (2.1).$$

The usual iterative application of eqn. (2.1) in each increment improves the result up to a given accuracy.

For the updated Lagrange formulation the stiffness matrices and the nodal force vectors for one element (e) are given by the following integrals over the element volume (^{m}V):

$${}^m\underset{\approx}{\mathbf{K}}_e^{(e)} = \int_{{}^m\mathrm{V}} {}^m\underset{\approx}{\mathbf{B}}_l^T\, {}^m\underset{\approx}{\mathbf{C}}\, {}^m\underset{\approx}{\mathbf{B}}_l \, d^m\mathrm{V} \qquad (2.2),$$

$${}^m\underset{\approx}{\mathbf{K}}_g^{(e)} = \int_{{}^m\mathrm{V}} {}^m\underset{\approx}{\mathbf{B}}_{nl}^T\, {}^m\underset{\approx}{\mathcal{T}}\, {}^m\underset{\approx}{\mathbf{B}}_{nl} \, d^m\mathrm{V} \qquad (2.3),$$

$${}^m\underset{\sim}{\mathrm{f}}^{(e)} = \int_{{}^m\mathrm{V}} {}^m\underset{\approx}{\mathbf{B}}_l^T\, {}^m\underset{\sim}{\tau} \, d^m\mathrm{V} \qquad (2.4),$$

$$\Delta\underset{\sim}{\mathrm{f}}_{th}^{(e)} = \int_{{}^m\mathrm{V}} {}^m\underset{\approx}{\mathbf{B}}_l^T\, {}^m\underset{\approx}{\mathbf{C}}\, \underset{\sim}{\alpha}\Delta\vartheta \, d^m\mathrm{V} \qquad (2.5).$$

In eqn. (2.1) the sum in square brackets is often referred to as the current global tangent stiffness matrix ${}^m\underset{\approx}{\mathbf{K}}$ at state m, comprised of ${}^m\underset{\approx}{\mathbf{K}}_e$ (material stiffness matrix, depending on the current material matrix ${}^m\underset{\approx}{\mathbf{C}}$ which, in turn, depends on the local state of damage) and ${}^m\underset{\approx}{\mathbf{K}}_g$ (initial stress or geometrical stiffness matrix, depending explicitly on the Cauchy stress tensor, ${}^m\underset{\approx}{\mathcal{T}}$). The external load vector at the current state $m+1$, ${}^{m+1}\underset{\sim}{\mathrm{r}}$, is given by the surface and body forces, concentrated or distributed. ${}^m\underset{\sim}{\mathrm{f}}$ is the vector of internal forces corresponding to the stresses at state m (vector ${}^m\underset{\sim}{\tau}$) and $\Delta\underset{\sim}{\mathrm{f}}_{th}$ is the vector of internal nodal forces equivalent to stress increments resulting from a temperature increment $\Delta\vartheta$ and computed by using the direction dependent coefficients of linear thermal expansion (vector $\underset{\sim}{\alpha}$).

The concept used to describe the degenerated shell element's geometry and its deformations as well as the assumptions with respect to modified plane stress conditions are in close analogy to [7] and are described in detail in [1]. For example, the interpolation of the geometry is performed by

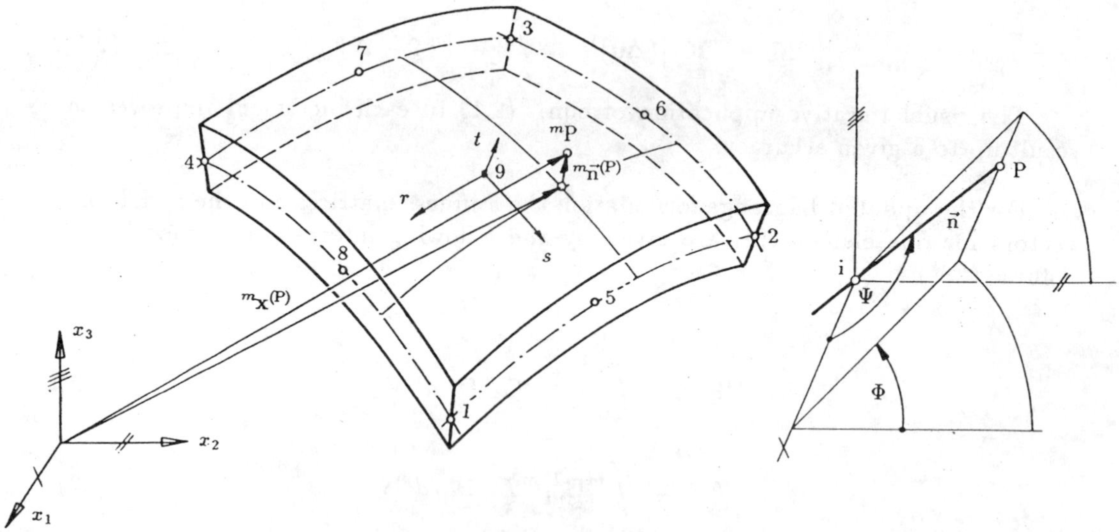

Fig. 2.1 Geometry of the degenerated shell element

$$
\begin{aligned}
{}^m x_1(r,s,t) &= \sum_{k=1}^{M} \phi^{(k)}(r,s)[{}^m x_1^{(k)} + t\frac{\mathrm{h}^{(k)}}{2} \cos {}^m\Psi^{(k)}] \\
{}^m x_2(r,s,t) &= \sum_{k=1}^{M} \phi^{(k)}(r,s)[{}^m x_2^{(k)} + t\frac{\mathrm{h}^{(k)}}{2} \sin {}^m\Psi^{(k)} \cos {}^m\Phi^{(k)}] \\
{}^m x_3(r,s,t) &= \sum_{k=1}^{M} \phi^{(k)}(r,s)[{}^m x_3^{(k)} + t\frac{\mathrm{h}^{(k)}}{2} \sin {}^m\Psi^{(k)} \sin {}^m\Phi^{(k)}]
\end{aligned}
\qquad (2.6),
$$

with $\phi^{(k)}(r,s)$ being standard 2/D shape functions (i.e. Lagrangian polynomials); M is the number of nodes forming the element, $\mathrm{h}^{(k)}$ is the thickness of the shell at node k and $\Psi^{(k)}$ and $\Phi^{(k)}$ are used to determine the position of the shell's normal at node k, see Fig.2.1. A more advanced description of the kinematics and essential improvements of this concept with regard to locking can be found in other contributions to this book, see [8,9].

2.2 Material Description of the Multilayer FRP-Composite

Due to the anisotropic and layered setup of composite shells the overall material matrix ${}^m\underset{\approx}{\mathbf{C}}$ in eqns. (2.2,2.5) and the vector of coefficients of thermal expansion $\underset{\sim}{\alpha}$ in eqn. (2.5) become position and orientation dependent. Each layer is assumed to have orthotropic material behavior with respect to its individual fibre-fixed local coordinate system l, q. The definition of the local system and the nomenclature for the material setup of the LFC-element are shown in Fig. 2.2.

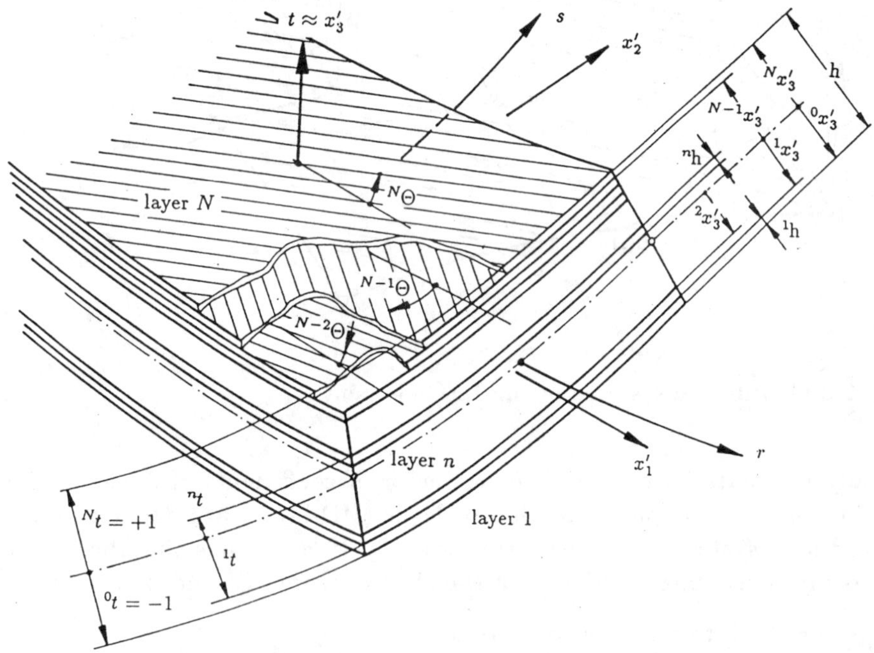

Fig. 2.2 Geometry and material setup of the LFC-element

Assuming a linear elastic material (in terms of Cauchy stresses and Almansi strains) the elasticity matrix and the vectors of stresses and strains can be defined corresponding to the modified plane stress conditions ($\tau'_{33} = 0$) resulting from the degeneration principle:

$$ {}^m\underset{\sim}{\tau}' = {}^m\underset{\approx}{\mathbf{C}}'({}^m\underset{\sim}{\varepsilon}' - \underset{\sim}{\alpha}'\,{}^m\vartheta) \qquad (2.7), $$

with ${}^m\vartheta$ being the temperature difference with respect to a stress-free reference temperature, and

$$ \underset{\sim}{\tau}'^T = (\tau'_{11}, \tau'_{22}, 0, \tau'_{12}, \tau'_{13}, \tau'_{23}), \qquad \underset{\sim}{\varepsilon}'^T = (\varepsilon'_{11}, \varepsilon'_{22}, 0, \gamma'_{12}, \gamma'_{13}, \gamma'_{23}) \qquad (2.8). $$

Equations (2.7,2.8) are referred to the shell's local coordinate system $\underset{\approx}{x}'$ (x_1' being tangential to the natural coordinate r and x_3' being normal to the shell's midsurface), see Fig. 2.3.

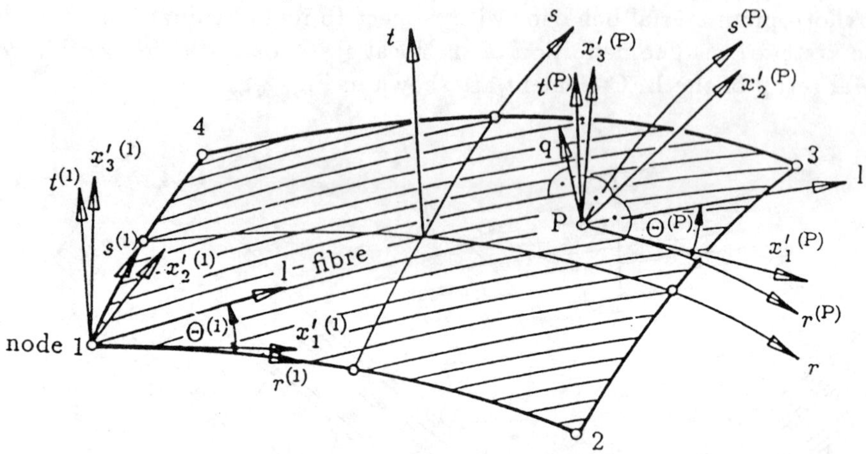

Fig. 2.3 Coordinate systems of the LFC-element

Usually the material matrix and the vector of coefficients of linear thermal expansion of the unidirectionally reinforced layer (UD-layer) are defined in the local layer coordinate system denoted by subscript L. If Θ denotes the fibre angle with respect to the coordinate system $\underset{\approx}{x}'$ of the shell element, $\underset{\approx}{C}'$ and $\underset{\approx}{\alpha}'$ are found from $\underset{\approx}{C}_L$ and $\underset{\approx}{\alpha}_L$ by the rotation transformation

$$\underset{\approx}{C}' = \underset{\approx}{T}^T(\Theta)\,\underset{\approx}{C}_L\,\underset{\approx}{T}(\Theta) \qquad (2.9),$$

$$\underset{\approx}{\alpha}' = \underset{\approx}{T}^{-1}(\Theta)\underset{\approx}{\alpha}_L \qquad (2.10),$$

with

$$\underset{\approx}{C}_L = \begin{pmatrix} E_{11} & E_{12} & 0 & 0 & 0 & 0 \\ & E_{22} & 0 & 0 & 0 & 0 \\ & & 0 & 0 & 0 & 0 \\ & sym. & & E_{44} & 0 & 0 \\ & & & & E_{55} & 0 \\ & & & & & E_{66} \end{pmatrix}, \quad \underset{\approx}{\alpha}_L = \begin{pmatrix} \alpha_l \\ \alpha_q \\ 0 \\ 0 \\ 0 \\ 0 \end{pmatrix} \qquad (2.11, 2.12),$$

$$\underset{\approx}{\mathbf{T}}(\Theta) = \begin{pmatrix} c^2 & s^2 & 0 & sc & 0 & 0 \\ s^2 & c^2 & 0 & -sc & 0 & 0 \\ 0 & 0 & 0 & 0 & 0 & 0 \\ -2sc & 2sc & 0 & c^2 - s^2 & 0 & 0 \\ 0 & 0 & 0 & 0 & c & s \\ 0 & 0 & 0 & 0 & -s & c \end{pmatrix} \qquad (2.13).$$

$$s := sin\Theta \qquad c := cos\Theta$$

In Figure 2.3 the corresponding denotations are shown.

Since element geometry and fibre-direction are independent of each other the fibre angle Θ is not necessarily constant within the element (see Fig. 2.3). Two ways for introducing such data are implemented in the LFC-element:

1) By assuming the fibres parallel within the shell's midsurface one can find a geometric relationship for the change of Θ (see [1]).

2) By specifying the Θ-angles at each nodal point of the element and using the 2/D shape functions $\phi^{(k)}(r,s)$ (from the interpolation of the geometry, eqn. (2.6)) one can interpolate the Θ-angle at any point.

The E_{ij} in $\underset{\approx}{C}_L$ are either calculated from material data given for the layer, or can be computed from the material data of fibre and matrix along with the fibre volume fraction.

If layer data E_l, E_q, ν_{ql}, G_{lq}, G_{lt}, G_{qt} are given (with l denoting the fibre direction; q and t transverse and thickness directions, respectively, and $\nu_{ql} = \frac{-\varepsilon_{qq}}{\varepsilon_{ll}}$ for uniaxial loading in l-direction), the E_{ij} values in $\underset{\approx}{C}_L$ are given by

$$\begin{aligned} E_{11} &= \frac{E_l}{1 - \nu_{ql}^2 \frac{E_q}{E_l}} \\ E_{12} &= \nu_{ql} E_{22} \\ E_{22} &= \frac{E_q}{1 - \nu_{ql}^2 \frac{E_q}{E_l}} \\ E_{44} &= G_{lq} \\ E_{55} &\approx E_{44} = G_{lq} \\ E_{66} &\approx G^{(M)} < G_{lq} \end{aligned} \qquad (2.14).$$

The superscript (M) denotes matrix material, and in the following relations (F)

indicates fibre properties.

Many theories for the determination of E_{ij} and α_{ij} from fibre and matrix data have been developed. In [11] a comprehensive overview of this topic is presented. The formulation implemented in the LFC-element was taken from [12] and is valid for a wide variety of fibre-reinforced plastics; the fibres are assumed to behave orthotropically (transversely isotropic) and the matrix isotropically. With ζ denoting the fibre-volume fraction, given by

$$\zeta = \frac{V^{(F)}}{V^{(F)} + V^{(M)}} \qquad (2.15),$$

the elements of the local compliance matrix follow to:

$$C_{11} = \frac{1}{E_l^{(F)}\zeta + E^{(M)}(1-\zeta)} \qquad (2.16),$$

$$C_{12} = -\frac{\nu_{ql}^{(F)}\zeta + \nu^{(M)}(1-\zeta)}{E_l^{(F)}\zeta + E^{(M)}(1-\zeta)} \qquad (2.17),$$

$$C_{22} = C_{22}^{(s)} - (0.2 + 0.4\zeta)(C_{22}^{(s)} - \widehat{C}_{22}^{(s)}) \qquad (2.18),$$

$$\text{with} \quad C_{22}^{(s)} = \frac{\zeta}{E_q^{(F)}} + \frac{1-\zeta}{E^{(M)}}$$

$$\widehat{C}_{22}^{(s)} = \frac{1}{E_l^{(F)}\zeta + E^{(M)}(1-\zeta)}$$

$$C_{44} = C_{44}^{(s)} - (0.4 + 0.4\zeta)(C_{44}^{(s)} - \widehat{C}_{44}^{(s)}) \qquad (2.19).$$

$$\text{with} \quad C_{44}^{(s)} = \frac{\zeta}{G_{lq}^{(F)}} + \frac{1-\zeta}{G^{(M)}}$$

$$\widehat{C}_{44}^{(s)} = \frac{1}{G_{lq}^{(F)}\zeta + G^{(M)}(1-\zeta)}$$

By inverting the compliance matrix the E_{ij} can be computed:

$$
\begin{aligned}
E_{11} &= \frac{C_{22}}{C_{11}C_{22} - C_{12}^2} \\
E_{12} &= \frac{-C_{12}}{C_{11}C_{22} - C_{12}^2} \\
E_{22} &= \frac{C_{11}}{C_{11}C_{22} - C_{12}^2} \\
E_{44} &= \frac{1}{C_{44}}
\end{aligned}
\qquad (2.20).
$$

The coefficients of linear thermal expansion for the UD-layer referred to the local layer coordinate system can be obtained from the constituents' data [12]:

$$
\begin{aligned}
\alpha_l &= \frac{E_l^{(F)}\alpha_l^{(F)}\zeta + E^{(M)}\alpha^{(M)}(1-\zeta)}{E_l^{(F)}\zeta + E^{(M)}(1-\zeta)} \\
\alpha_q &= \Big(\big(\alpha_q^{(F)} - \nu_{ql}^{(F)}(\alpha_l - \alpha_l^{(F)})\big)\zeta + \big(\alpha^{(M)} - \nu^{(M)}(\alpha_l - \alpha^{(M)})\big)(1-\zeta)\Big)
\end{aligned}
\qquad (2.21).
$$

There are some alternative approaches and one can derive benefit from using available software packages which perform the homogenization based on given material properties of the constituents of the composite and their geometrical configuration. For example in [13] the Mori-Tanaka averaging scheme is used. The fibres are assumed to behave orthotropically (transversely isotropic) and the matrix isotropically.

After transforming the locally defined material laws of all UD-layers of the laminate into the x_1', x_2', x_3' coordinate system and using the isoparametric concept one can express the material matrix and the vector of coefficients of linear thermal expansion of layer n as a function of the natural coordinates r, s and obtain

$$
\underset{\approx}{C}'(r,s,t) = \begin{cases} {}^1\underset{\approx}{C}'(r,s) \\ {}^2\underset{\approx}{C}'(r,s) \\ \vdots \\ {}^N\underset{\approx}{C}'(r,s) \end{cases}
\qquad
\underset{\sim}{\alpha}'(r,s,t) = \begin{cases} {}^1\underset{\sim}{\alpha}'(r,s) & -1 \le t \le {}^1t \\ {}^2\underset{\sim}{\alpha}'(r,s) & {}^1t < t \le {}^2t \\ \vdots & \vdots \\ {}^N\underset{\sim}{\alpha}'(r,s) & {}^{N-1}t < t \le +1 \end{cases}
\qquad (2.22),
$$

with

$$
{}^nt = -1 + \frac{2}{h}\sum_{j=1}^{n} {}^jh \qquad (2.23),
$$

$$ {}^{n}\underset{\approx}{\mathbf{C}}' = \underset{\approx}{\mathbf{T}}^{T}({}^{n}\Theta(r,s))\; {}^{n}\underset{\approx}{\mathbf{C}}_{L}\; \underset{\approx}{\mathbf{T}}({}^{n}\Theta(r,s)) \qquad (2.24), $$

$$ {}^{n}\underset{\approx}{\alpha}' = \underset{\approx}{\mathbf{T}}^{-1}({}^{n}\Theta(r,s))\; {}^{n}\underset{\approx}{\alpha}_{L} \qquad (2.25) $$

for the multi-layer compound with layer number $n = 1, \ldots, N$.

Regarding the definition of ${}^{j}\mathrm{h}$ and ${}^{n}\Theta$ see Fig. 2.2.

The usual degeneration principle includes the assumption that normal vectors remain straight during the deformation. In order to reduce the error resulting from this kinematic restriction shear-correction factors can be introduced. *Noor and Peters* [14] proposed a predictor-corrector approach which not only improves the overall shear response but also the computation of interlaminar stress quantities. *Dorninger* [2] introduced additional degrees of freedom for each layer; see Chapter 3. As long as rather thin shells are considered the use of such corrections can be omitted.

2.3 Stiffness Expressions

Computation of the element stiffness matrix requires a three-dimensional integration (see eqns. (2.2–2.5)) which usually is performed by some sort of numerical integration technique. Since we are dealing with multi-layer shells (with a very large number of layers allowed), where each layer requires at least one (even better two or more) integration points over its thickness, the numerical effort increases rapidly with the number of layers. An effective way to overcome this difficulty is the use of a quasi-analytical thickness integration as described for homogeneous shells e.g. in [15,16]. This, however, requires an assumption on the $t-$dependence of the Jacobian matrix (see e.g. [17]). In the present work the $t-$dependence is neglected. This approximation is acceptable as long as the shells are thin, their curvature is moderate and their thickness does not vary too much.

The displacements follow from the description of the geometry (eqn. (2.6)) and, therefore, the $t-$dependence of the displacement derivatives can be formulated explicitely:

$$ {}^{m}\mathrm{u}_{i,j}(r,s,t) = {}^{m}x_{i,j} - {}^{0}x_{i,j} = {}^{m}\overline{\mathrm{u}}_{i,j}(r,s) + t\, {}^{m}\widehat{\mathrm{u}}_{i,j}(r,s), \qquad i,j = 1,2,3 \qquad (2.26), $$

with

$$
\begin{aligned}
{}^{m}\overline{\mathrm{U}}_{1,j} &= \sum_{k=1}^{M}[\phi_{,j}^{(k)}\,{}^{m}\mathrm{u}_1^{(k)} + \phi^{(k)}\frac{\partial t}{\partial^m x_j}\frac{\mathrm{h}^{(k)}}{2}(\cos{}^{m}\Psi^{(k)} - \cos{}^{0}\Psi^{(k)})] \\
{}^{m}\overline{\mathrm{U}}_{2,j} &= \sum_{k=1}^{M}[\phi_{,j}^{(k)}\,{}^{m}\mathrm{u}_2^{(k)} + \phi^{(k)}\frac{\partial t}{\partial^m x_j}\frac{\mathrm{h}^{(k)}}{2}(\sin{}^{m}\Psi^{(k)}\cos{}^{m}\Phi^{(k)} - \sin{}^{0}\Psi^{(k)}\cos{}^{0}\Phi^{(k)})] \qquad (2.27) \\
{}^{m}\overline{\mathrm{U}}_{3,j} &= \sum_{k=1}^{M}[\phi_{,j}^{(k)}\,{}^{m}\mathrm{u}_3^{(k)} + \phi^{(k)}\frac{\partial t}{\partial^m x_j}\frac{\mathrm{h}^{(k)}}{2}(\sin{}^{m}\Psi^{(k)}\sin{}^{m}\Phi^{(k)} - \sin{}^{0}\Psi^{(k)}\sin{}^{0}\Phi^{(k)})]
\end{aligned}
$$

and

$$
\begin{aligned}
{}^{m}\widehat{\mathrm{U}}_{1,j} &= \sum_{k=1}^{M}\phi_{,j}^{(k)}\frac{\mathrm{h}^{(k)}}{2}(\cos{}^{m}\Psi^{(k)} - \cos{}^{0}\Psi^{(k)}) \\
{}^{m}\widehat{\mathrm{U}}_{2,j} &= \sum_{k=1}^{M}\phi_{,j}^{(k)}\frac{\mathrm{h}^{(k)}}{2}(\sin{}^{m}\Psi^{(k)}\cos{}^{m}\Phi^{(k)} - \sin{}^{0}\Psi^{(k)}\cos{}^{0}\Phi^{(k)}) \qquad (2.28), \\
{}^{m}\widehat{\mathrm{U}}_{3,j} &= \sum_{k=1}^{M}\phi_{,j}^{(k)}\frac{\mathrm{h}^{(k)}}{2}(\sin{}^{m}\Psi^{(k)}\sin{}^{m}\Phi^{(k)} - \sin{}^{0}\Psi^{(k)}\sin{}^{0}\Phi^{(k)})
\end{aligned}
$$

where

$$
\phi_{,i}^{(k)} = \frac{\partial\phi^{(k)}}{\partial r}\frac{\partial r}{\partial^m x_i} + \frac{\partial\phi^{(k)}}{\partial s}\frac{\partial s}{\partial^m x_i} \qquad (2.29),
$$

and $\frac{\partial r}{\partial^m x_i}, \frac{\partial s}{\partial^m x_i}, \frac{\partial t}{\partial^m x_i}$ correspond to the elements of the inverse of the Jacobian matrix $\underset{\approx}{\mathbf{J}}$ at load state m.

With the definition of the Almansi strains and by invoking eqn. (2.26), one can extract the $t-$dependency of the strains:

$$
{}^{m}\varepsilon_{ij}(r,s,t) = {}^{m}\overline{\varepsilon}_{ij}(r,s) + t\,{}^{m}\widehat{\varepsilon}_{ij}(r,s) + t^2\,{}^{m}\widetilde{\varepsilon}_{ij}(r,s) \qquad i,j = 1,2,3 \qquad (2.30),
$$

with

$$
\begin{aligned}
{}^{m}\overline{\varepsilon}_{ij} &= \frac{1}{2}({}^{m}\overline{\mathrm{U}}_{i,j} + {}^{m}\overline{\mathrm{U}}_{j,i} - {}^{m}\overline{\mathrm{U}}_{k,i}\,{}^{m}\overline{\mathrm{U}}_{k,j}) \\
{}^{m}\widehat{\varepsilon}_{ij} &= \frac{1}{2}({}^{m}\widehat{\mathrm{U}}_{i,j} + {}^{m}\widehat{\mathrm{U}}_{j,i} - {}^{m}\overline{\mathrm{U}}_{k,i}\,{}^{m}\widehat{\mathrm{U}}_{k,j} - {}^{m}\widehat{\mathrm{U}}_{k,i}\,{}^{m}\overline{\mathrm{U}}_{k,j}) \qquad (2.31). \\
{}^{m}\widetilde{\varepsilon}_{ij} &= -\frac{1}{2}({}^{m}\widehat{\mathrm{U}}_{k,i}\,{}^{m}\widehat{\mathrm{U}}_{k,j})
\end{aligned}
$$

If the temperature field is assumed to be linearly distributed over the thickness of the shell, with $\bar{\vartheta}$ being the temperature load of the midsurface and $\hat{\vartheta}$ the temperature difference between opposite points on the two surfaces of the shell the vector of Cauchy stress components (eqn. (2.7)) becomes:

$$
\begin{aligned}
{}^m\underset{\sim}{\tau}(r,s,t) = {}^m\underset{\approx}{C}(r,s,t)\Big(\big[{}^m\underset{\sim}{\bar{\varepsilon}}(r,s) - {}^m\underset{\sim}{\alpha}(r,s,t)\,{}^m\bar{\vartheta}(r,s)\big] + \\
+ t\Big[{}^m\underset{\sim}{\hat{\varepsilon}}(r,s) - {}^m\underset{\sim}{\alpha}(r,s,t)\,\frac{{}^m\hat{\vartheta}(r,s)}{2}\Big] + t^2\,{}^m\underset{\sim}{\tilde{\varepsilon}}(r,s)\Big)
\end{aligned}
\qquad (2.32),
$$

where

$$
{}^m\underset{\approx}{C}(r,s,t) = {}^m\underset{\approx}{G}^T(r,s)\,\underset{\approx}{C}'(r,s,t)\,{}^m\underset{\approx}{G}(r,s) \qquad (2.33),
$$

$$
{}^m\underset{\sim}{\alpha}(r,s,t) = {}^m\underset{\approx}{G}^{-1}(r,s)\,\underset{\sim}{\alpha}'(r,s,t) \qquad (2.34).
$$

The matrix ${}^m\underset{\approx}{G}$ represents the transformation from the global $\underset{\sim}{x}$ system to the local $\underset{\sim}{x}'$ system. $\underset{\approx}{G}$ is composed of the elements of the Jacobian matrix and, therefore, if $\underset{\approx}{\mathbf{J}}$ is independent of t so is $\underset{\approx}{G}$. The $\underset{\approx}{\mathbf{B}}_l$ and $\underset{\approx}{\mathbf{B}}_{nl}$ matrices are constructed from derivatives of the shape functions and, therefore, they can be decomposed into a t–independent part ($\rightarrow \underset{\approx}{\overline{\mathbf{B}}}_l\,, \underset{\approx}{\overline{\mathbf{B}}}_{nl}$) and a part linear in t ($\rightarrow \underset{\approx}{\widehat{\mathbf{B}}}_l\,, \underset{\approx}{\widehat{\mathbf{B}}}_{nl}$). Now the stiffness matrices and nodal force vectors of eqns. (2.2–2.5) can be rewritten and the quasi-analytical thickness integration can be performed.

By introducing some abbreviations $\underset{\approx}{\mathbf{K}}_e^{(e)}, \underset{\approx}{\mathbf{K}}_g^{(e)}, \underset{\sim}{f}^{(e)}, \Delta \underset{\sim}{f}_{th}^{(e)}$ follow as:

$$
{}^m\underset{\approx}{\mathbf{K}}_e^{(e)} = \int_{-1}^{+1}\!\!\int_{-1}^{+1} \Big({}^m\underset{\approx}{\overline{\mathbf{B}}}_l^T\big({}^m\underset{\approx}{C}1\,{}^m\underset{\approx}{\overline{\mathbf{B}}}_l + {}^m\underset{\approx}{C}2\,{}^m\underset{\approx}{\widehat{\mathbf{B}}}_l\big) + {}^m\underset{\approx}{\widehat{\mathbf{B}}}_l^T\big({}^m\underset{\approx}{C}2\,{}^m\underset{\approx}{\overline{\mathbf{B}}}_l + {}^m\underset{\approx}{C}3\,{}^m\underset{\approx}{\widehat{\mathbf{B}}}_l\big)\Big)\det|{}^m\underset{\approx}{\mathbf{J}}|\;dr\,ds \qquad (2.35),
$$

$$
{}^m\underset{\approx}{\mathbf{K}}_g^{(e)} = \int_{-1}^{+1}\!\!\int_{-1}^{+1} \Big({}^m\underset{\approx}{\overline{\mathbf{B}}}_{nl}^T\big({}^m\underset{\approx}{S}1\,{}^m\underset{\approx}{\overline{\mathbf{B}}}_{nl} + {}^m\underset{\approx}{S}2\,{}^m\underset{\approx}{\widehat{\mathbf{B}}}_{nl}\big) + {}^m\underset{\approx}{\widehat{\mathbf{B}}}_{nl}^T\big({}^m\underset{\approx}{S}2\,{}^m\underset{\approx}{\overline{\mathbf{B}}}_{nl} + {}^m\underset{\approx}{S}3\,{}^m\underset{\approx}{\widehat{\mathbf{B}}}_{nl}\big)\Big)\det|{}^m\underset{\approx}{\mathbf{J}}|\;dr\,ds \qquad (2.36),
$$

$$
{}^m\underset{\sim}{f}^{(e)} = \int_{-1}^{+1}\!\!\int_{-1}^{+1} \Big({}^m\underset{\approx}{\overline{\mathbf{B}}}_l^T\,{}^m\underset{\sim}{S}1 + {}^m\underset{\approx}{\widehat{\mathbf{B}}}_l^T\,{}^m\underset{\sim}{S}2\Big)\det|{}^m\underset{\approx}{\mathbf{J}}|\;dr\,ds \qquad (2.37),
$$

$$ {}^m\Delta \underset{\sim}{f}^{(e)}_{th} = \int_{-1}^{+1}\int_{-1}^{+1} \left({}^m\underset{\approx}{\overline{B}}^T_l \left({}^m\underset{\sim}{C}1_{th}\, \Delta\overline{\vartheta} + {}^m\underset{\sim}{C}2_{th} \frac{\Delta\widehat{\vartheta}}{2} \right) + {}^m\underset{\approx}{\widehat{B}}^T_l \left({}^m\underset{\sim}{C}2_{th}\, \Delta\overline{\vartheta} + {}^m\underset{\sim}{C}3_{th} \frac{\Delta\widehat{\vartheta}}{2} \right) \right) \det|{}^m\underset{\approx}{J}|\, dr\, ds \tag{2.38}$$

where the following abbreviations are used:

$$ \begin{aligned} {}^m\underset{\sim}{S}1 &= {}^m\underset{\approx}{C}1\, {}^m\underset{\sim}{\overline{\varepsilon}} - {}^m\underset{\sim}{C}1_{th}\, {}^m\overline{\vartheta} + {}^m\underset{\approx}{C}2\, {}^m\underset{\sim}{\widehat{\varepsilon}} - {}^m\underset{\sim}{C}2_{th} \frac{{}^m\widehat{\vartheta}}{2} + {}^m\underset{\approx}{C}3\, {}^m\underset{\sim}{\widetilde{\varepsilon}} \\ {}^m\underset{\sim}{S}2 &= {}^m\underset{\approx}{C}2\, {}^m\underset{\sim}{\overline{\varepsilon}} - {}^m\underset{\sim}{C}2_{th}\, {}^m\overline{\vartheta} + {}^m\underset{\approx}{C}3\, {}^m\underset{\sim}{\widehat{\varepsilon}} - {}^m\underset{\sim}{C}3_{th} \frac{{}^m\widehat{\vartheta}}{2} + {}^m\underset{\approx}{C}4\, {}^m\underset{\sim}{\widetilde{\varepsilon}} \\ {}^m\underset{\sim}{S}3 &= {}^m\underset{\approx}{C}3\, {}^m\underset{\sim}{\overline{\varepsilon}} - {}^m\underset{\sim}{C}3_{th}\, {}^m\overline{\vartheta} + {}^m\underset{\approx}{C}4\, {}^m\underset{\sim}{\widehat{\varepsilon}} - {}^m\underset{\sim}{C}4_{th} \frac{{}^m\widehat{\vartheta}}{2} + {}^m\underset{\approx}{C}5\, {}^m\underset{\sim}{\widetilde{\varepsilon}} \end{aligned} \tag{2.39}$$

are "stress integrals" and

$$ {}^m\underset{\approx}{C}i(r,s) = {}^m\underset{\approx}{G}^T(r,s) \sum_{n=1}^{N} {}^n\underset{\approx}{C}'(r,s) \frac{{}^nt^i - {}^{n-1}t^i}{i}\, {}^m\underset{\approx}{G}(r,s), \qquad i = 1,\dots 5 \tag{2.40}$$

$$ {}^m\underset{\sim}{C}i_{th}(r,s) = {}^m\underset{\approx}{G}^T(r,s) \sum_{n=1}^{N} {}^n\underset{\approx}{C}'(r,s)\, {}^n\underset{\sim}{\alpha}'(r,s) \frac{{}^nt^i - {}^{n-1}t^i}{i}, \qquad i = 1,\dots 4 \tag{2.41}$$

are thickness integrals of the material matrix and the vector of linear thermal expansion. ${}^m\underset{\approx}{S}1$, ${}^m\underset{\approx}{S}2$, ${}^m\underset{\approx}{S}3$ are equivalent to ${}^m\underset{\sim}{S}1$, ${}^m\underset{\sim}{S}2$, ${}^m\underset{\sim}{S}3$, but the elements are rearranged in special matrix form, see [1].

The summations in eqns. (2.40,2.41) are independent of the state m (as long as local failure is not taken into account), thus they only have to be computed once (prior to the incremental analysis) for the complete prefailure analysis. This results in a decrease in the numerical effort in each increment due to the reduction of the 3/D numerical integration to a 2/D one. The more layers there are within the laminate the more efficient the analytical thickness integration becomes compared to the fully 3/D numerical integration.

2.4 Effects of Local Failure in FRP shells

Due to the nature of FRP composite material the failure behavior is very complex with numerous different failure modes possible. Therefore, an accurate prediction of failure with reasonable effort is almost impossible so that some restrictions have to be made. In this paper a simple way of dealing with this kind of behavior was chosen:

Failure is considered to occur within the layers only and here delamination is not taken into account. With respect to this and by assuming a linear elastic stress-strain relationship up to failure (which is valid for many FRPs) one can use proper strength criteria to determine onset of failure. Although the following procedures are based on "homogenized" material, the local damage of the composite, i.e. matrix or fibre cracking, can be estimated. In this paper FRP failure is indicated by a combination of two failure criteria, and two distinct failure modes are assumed:

A quadratic strength criterion, the Tsai-Wu-criterion [18] serves for predicting failure for stress states with relatively large transverse stress components which affect the matrix material rather than the fibres. Therefore, violation of this criterion can be related to matrix failure. Endurable stress states lie within a failure surface in the stress space

$$F_{01}\sigma_{ll} + F_{11}\sigma_{ll}^2 + F_{12}\sigma_{ll}\sigma_{qq} + F_{02}\sigma_{qq} + F_{22}\sigma_{qq}^2 + F_{44}\tau_{lq}^2 < 1 \qquad (2.42),$$

with

$$F_{01} = \frac{1}{\sigma_{lTu}} + \frac{1}{\sigma_{lCu}} \qquad F_{11} = \frac{-1}{\sigma_{lTu}\sigma_{lCu}} \qquad F_{44} = \frac{1}{\tau_{lqu}^2}$$
$$F_{02} = \frac{1}{\sigma_{qTu}} + \frac{1}{\sigma_{qCu}} \qquad F_{22} = \frac{-1}{\sigma_{qTu}\sigma_{qCu}} \qquad F_{12} = -\sqrt{F_{11}F_{22}} \qquad (2.43).$$

$\sigma_{ll}, \sigma_{qq}, \tau_{lq}$ are the in-plane engineering stresses. With respect to material damage we use the same laws for Cauchy stresses, too. The following notations are used:

σ_{lCu} maximum endurable, i.e. ultimate uniaxial compression stress in fibre direction,
σ_{lTu} ultimate tensile stress in fibre direction,
σ_{qCu} ultimate compression stress normal to fibre direction,
σ_{qTu} ultimate tensile stress normal to fibre direction,
τ_{lqu} ultimate shear stress.

In a certain sense it might be more convenient to express this criterion in the strain space:

$$G_{01}\varepsilon_{ll} + G_{11}\varepsilon_{ll}^2 + G_{12}\varepsilon_{ll}\varepsilon_{qq} + G_{02}\varepsilon_{qq} + G_{22}\varepsilon_{qq}^2 + G_{44}\gamma_{lq}^2 < 1 \qquad (2.44),$$

with

$$\begin{aligned} G_{01} &= F_{01}E_{11} + F_{02}E_{12} \\ G_{02} &= F_{02}E_{22} + F_{01}E_{12} \\ G_{11} &= F_{11}E_{11}^2 + F_{22}E_{12}^2 + F_{12}E_{11}E_{12} \\ G_{22} &= F_{22}E_{22}^2 + F_{11}E_{12}^2 + F_{12}E_{22}E_{12} \\ G_{12} &= 2E_{12}(F_{11}E_{11} + F_{22}E_{22}) + F_{12}(E_{12}^2 + E_{11}E_{22}) \\ G_{44} &= F_{44}E_{44}^2 \end{aligned} \qquad (2.45).$$

Since in many cases the Tsai-Wu-criterion overestimates the strength of the UD-layer in the case of stresses acting predominantly in fibre direction, a maximum stress limit in fibre direction is imposed:

$$\sigma_{lCu} < \sigma_{ll} < \sigma_{lTu} \qquad (2.46).$$

leading finally to a failure surface as shown in Fig. 2.4.

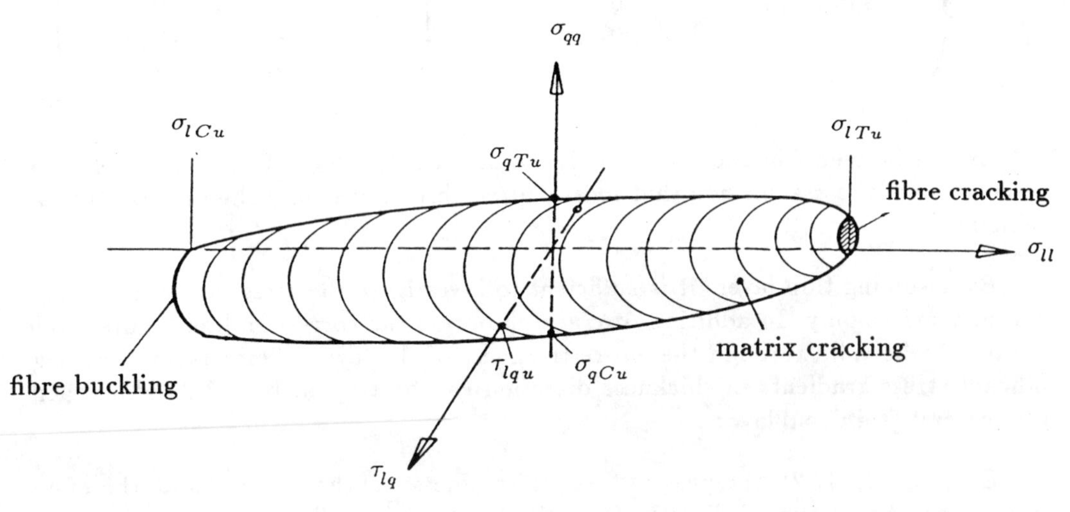

Fig. 2.4 Combined failure criterion

Post-cracking stiffnesses are introduced according to the two assumed failure modes. In the case of matrix cracking they take the form:

$$\underset{\approx}{C}_L = \begin{pmatrix} E_{11} & E_{12} & 0 & 0 & 0 & 0 \\ & E_{22} & 0 & 0 & 0 & 0 \\ & & 0 & 0 & 0 & 0 \\ & sym. & & E_{44} & 0 & 0 \\ & & & & E_{55} & 0 \\ & & & & & E_{66} \end{pmatrix} \xrightarrow{\text{matrix-cracking}} \begin{pmatrix} \beta_E E_{11} & 0 & 0 & 0 & 0 & 0 \\ & 0 & 0 & 0 & 0 & 0 \\ & & 0 & 0 & 0 & 0 \\ & sym. & & 0 & 0 & 0 \\ & & & & E_{55} & 0 \\ & & & & & E_{66} \end{pmatrix} \tag{2.47}$$

E_{12}, E_{22} and E_{44} are set to zero, which is a simple representation of the matrix stiffness being removed. The reason for introducing a correction factor β_E is twofold: First, a damaged matrix also reduces the stiffness in fibre direction and, second, for nonstraight (e.g. wavy) fibres the reduced support by the damaged matrix can lead to a further loss of stiffness. β_E depends on the composition of the layer and must be given as an additional material parameter.

In the case of fibre failure the reduced stiffnesses are modeled by:

$$\underset{\approx}{C}_L = \begin{pmatrix} E_{11} & E_{12} & 0 & 0 & 0 & 0 \\ & E_{22} & 0 & 0 & 0 & 0 \\ & & 0 & 0 & 0 & 0 \\ & sym. & & E_{44} & 0 & 0 \\ & & & & E_{55} & 0 \\ & & & & & E_{66} \end{pmatrix} \xrightarrow{\text{fibre-failure}} \begin{pmatrix} E_{22} & E_{12} & 0 & 0 & 0 & 0 \\ & E_{22} & 0 & 0 & 0 & 0 \\ & & 0 & 0 & 0 & 0 \\ & sym. & & E_{44} & 0 & 0 \\ & & & & E_{55} & 0 \\ & & & & & E_{66} \end{pmatrix} \tag{2.48}$$

As can be seen from eqn. (2.48), E_{11} is set equal to E_{22}. This is a simple representation of the assumption that locally after fibre failure only the matrix stiffness remains.

By assuming thin layers it is sufficient to investigate the stress state in the layers' midsurface only. In addition, if cracking occurs the corresponding failure mode is taken to be valid through the entire thickness of the layer. Thick layers with significant stress gradients in thickness direction can be treated by subdividing them into several "thin" sublayers.

Equations (2.47,2.48) represent "secant"-stiffnesses of the material and, therefore, stresses can be computed directly from the total strains. The following procedure, applied at each load step m at each stiffness sampling point r_i, s_j (= 2/D integration points) for all N layers of the laminate, accounts for stiffness changes due to material cracking:

- The local in-plane strains $\underset{\sim}{\varepsilon}_L$ at the midsurface of layer n with respect to the layer's local axis are derived from the global strain vector ${}^{m}\underset{\sim}{\varepsilon}$:

$$ {}^{n}\underset{\sim}{\varepsilon}_L(r_i,s_j) = \underset{\approx}{\mathbf{T}}({}^{n}\Theta(r_i,s_j))\; {}^{m}\underset{\approx}{\mathbf{G}}(r_i,s_j)\left({}^{m}\underset{\sim}{\overline{\varepsilon}}(r_i,s_j) + {}^{n}\bar{t}\;{}^{m}\underset{\sim}{\widehat{\varepsilon}}(r_i,s_j) + {}^{n}\bar{t}^{2}\;{}^{m}\underset{\sim}{\widetilde{\varepsilon}}(r_i,s_j)\right) \qquad (2.49), $$

with

$$ {}^{n}\bar{t} = \frac{{}^{n}t + {}^{n-1}t}{2} \qquad (2.50). $$

- With ${}^{n}\underset{\sim}{\varepsilon}_L$ and the corresponding local elasticity matrix ${}^{n}\underset{\approx}{\mathbf{C}}_L$ of the previous step an estimate of the stresses is computed:

$$ {}^{n}\underset{\sim}{\tau}_L(r_i,s_j) = {}^{n}\underset{\approx}{\mathbf{C}}_L\left({}^{n}\underset{\sim}{\varepsilon}_L(r_i,s_j) - {}^{n}\underset{\sim}{\alpha}_L(r_i,s_j)\left({}^{m}\bar{\vartheta}(r_i,s_j) + {}^{n}\bar{t}\,\frac{{}^{m}\widehat{\vartheta}(r_i,s_j)}{2}\right)\right) \qquad (2.51). $$

- In the next step the combined failure criterion is examined. If cracking is indicated by violation of eqn. (2.42) or eqn. (2.46) then the local stiffness matrix is changed according to eqn. (2.47) or eqn. (2.48). The elements of ${}^{n}\underset{\approx}{\mathbf{C}}_L$ to be reduced are properly subtracted from the $\underset{\approx}{\mathbf{C}}i'$ matrices and the $\underset{\sim}{\mathbf{C}}i'_{th}$ vectors:

$$ \underset{\approx}{\mathbf{C}}i'^{\,new} = \underset{\approx}{\mathbf{C}}i'^{\,old} - \underset{\approx}{\mathbf{T}}^{T}({}^{n}\Theta)\left({}^{n}\underset{\approx}{\mathbf{C}}_L^{old} - {}^{n}\underset{\approx}{\mathbf{C}}_L^{new}\right)\underset{\approx}{\mathbf{T}}({}^{n}\Theta)\frac{{}^{n}t^{i} - {}^{n-1}t^{i}}{i}, \quad i = 1,\dots 5 \qquad (2.52), $$

$$ \underset{\sim}{\mathbf{C}}i'^{\,new}_{th} = \underset{\sim}{\mathbf{C}}i'^{\,old}_{th} - \underset{\approx}{\mathbf{T}}^{-1}({}^{n}\Theta)\left({}^{n}\underset{\approx}{\mathbf{C}}_L^{old} - {}^{n}\underset{\approx}{\mathbf{C}}_L^{new}\right){}^{n}\underset{\sim}{\alpha}_L\frac{{}^{n}t^{i} - {}^{n-1}t^{i}}{i}, \quad i = 1,\dots 4 \qquad (2.53). $$

A flow chart of the complete algorithm can be found in [1]. Due to the simplifications made, the iteration scheme has to be chosen carefully and the influence of the mesh size on the accuracy has to be taken into account, see [1].

2.5 Algorithms for Stability Analysis

Equation (2.1) is formally the same relation as obtained for the isothermal case with the exception of the incremental thermal load vector $\Delta\underset{\sim}{\mathbf{f}}_{th}$. Hence, the algorithms developed for the detection of buckling loads of isothermally loaded structures, see e.g. [19] and further lectures of this Course, can be applied in the non-isothermal case, too. Since thermoelastic buckling is one of the most interesting instability phenomena in composite shell analysis, in the following descriptions of this Chapter particular emphasis lies on thermal loading. In many cases, especially if limit points appear, the observation of the "load displacement path" indicates critical mechanical or thermal load configurations. However, one should bear in mind that thermal loads correspond to deformation controlled processes rather than to load controlled ones, see [20].

According to the static stability criterion [21], bifurcation or limit points, i.e. bifurcation buckling or snap-through buckling, appear if at least one further equilibrium state exists adjacent to the original configuration without any load variation. Hence, in the critical state the equation

$$ {}^{*}\underset{\approx}{\mathbf{K}}\,\delta\underset{\sim}{u} = 0 \tag{2.54}$$

has a non-trivial solution $\delta\underset{\sim}{u} \neq 0$. If it is assumed that the material behaves linearly elastic and has temperature independent elastic properties and if, additionally, proportional mechanical and thermal loading

$$ {}^{m}\underset{\sim}{r} = {}^{m}\lambda\,\underset{\sim}{r}_{ref} \tag{2.55),}$$

$$ {}^{m}\vartheta = {}^{m}\lambda\,\vartheta_{ref} \tag{2.56}$$

is applied with $\underset{\sim}{r}_{ref}$ denoting the reference mechanical loading and ϑ_{ref} representing the reference thermal loading, then condition (2.54) can be rewritten as

$$ \left({}^{*}\underset{\approx}{\mathbf{K}}_e + \underset{\approx}{\mathbf{K}}_g\,({}^{*}\lambda)\right)\delta\underset{\sim}{u} = 0. \tag{2.57).}$$

The relation $\underset{\approx}{\mathbf{K}}_g\,(\lambda)$ is nonlinear. Having reached the state m we can ask for the multiplier ${}^{m}\eta$ which leads to

$$ {}^{*}\lambda = {}^{m}\eta\,{}^{m}\lambda \tag{2.58).}$$

After linearizing the nonlinear relation between $\underset{\approx}{\mathbf{K}}_g$ and λ and neglecting, for the time being, the difference between ${}^{m}\underset{\approx}{\mathbf{K}}_e$ and ${}^{*}\underset{\approx}{\mathbf{K}}_e$ we obtain a linear eigenvalue problem

$$ \left({}^{m}\underset{\approx}{\mathbf{K}}_e + {}^{m}\overline{\eta}\,{}^{m}\underset{\approx}{\mathbf{K}}_g\right)\delta\overline{\underset{\sim}{u}} = 0 \tag{2.59}$$

leading to an estimate for the critical load multiplier

$$^{*}\lambda \approx {}^{*}\overline{\lambda} = {}^{m}\overline{\eta}_1 \, {}^{m}\lambda \tag{2.60}$$

and an approximation for the fundamental buckling mode

$$\delta \underset{\sim}{u} \approx \delta \underset{\sim}{\overline{u}}_1 \tag{2.61}.$$

${}^{m}\overline{\eta}_1$ and $\delta \underset{\sim}{\overline{u}}_1$ are the smallest eigenvalue and the corresponding eigenvector of the eigenvalue problem eqn. (2.59), respectively.

The error caused by the linearization of $\underset{\approx}{\mathbf{K}}_g(\lambda)$ and the neglection of the difference between ${}^{m}\underset{\approx}{\mathbf{K}}_e$ and ${}^{*}\underset{\approx}{\mathbf{K}}_e$ vanishes and, hence, the estimate (eqn. (2.60)) becomes accurate when the lowest eigenvalue ${}^{m}\overline{\eta}_1$ approaches 1, i.e. the current configuration m approaches the critical one (${}^{m}\lambda \to {}^{*}\lambda$). This leads to the strategy of "accompanying eigenvalue analyses" in which the eigenvalue problem eqns. (2.59,2.60) is formulated and solved at several load levels during the incremental prebuckling analysis. ${}^{*}\overline{\lambda}(u_{ref})$, with u_{ref} being a properly chosen component of the displacement vector ${}^{m}\underset{\sim}{u}$, represents an "estimate curve" crossing the ${}^{m}\lambda(u_{ref})$ curve (the "load displacement path"), whenever the critical load situation is reached.

Very often a stability analysis due to thermal (or mechanical) loading only is needed with the mechanical loads (or the thermal load) being held constant. A procedure which accounts for this is implemented in the LFC-element:

The $\underset{\approx}{\mathbf{K}}_g^{(e)}$-matrix (eqn. (2.36)) can be broken down into a part which depends on mechanical loads and a part related to thermal loading. Considering the definition of the "stress-integrals", eqn. (2.39), the separation of $\underset{\approx}{\mathbf{K}}_g^{(e)}$ follows as:

$${}^{m}\underset{\approx}{\mathbf{K}}_g^{(e)} = {}^{m}\underset{\approx}{\mathbf{K}}_g^{(e)}{}_{mech} + {}^{m}\underset{\approx}{\mathbf{K}}_g^{(e)}{}_{th} \tag{2.62},$$

where ${}^{m}\underset{\approx}{\mathbf{K}}_g^{(e)}{}_{mech}$ is derived from eqn.(2.36) by evaluating $\underset{\approx}{\mathbf{S}}i$ (eqn. (2.39)) for $\overline{\vartheta} = \widehat{\vartheta} = 0$, and ${}^{m}\underset{\approx}{\mathbf{K}}_g^{(e)}{}_{th}$ is also obtained from eqn. (2.36), but using only the ϑ-dependent terms of $\underset{\approx}{\mathbf{S}}i$ (i.e. setting $\overline{\underset{\sim}{\varepsilon}} = \widehat{\underset{\sim}{\varepsilon}} = \widetilde{\underset{\sim}{\varepsilon}} = 0$).

After assembling the global stiffness matrices, the procedure for detection of buckling loads, described previously, can be applied either for constant thermal loading or constant mechanical loads by rewriting eqn. (2.59):

- for buckling due to mechanical loads only:

$$({}^{m}\underset{\approx}{\mathbf{K}}_{e} + {}^{m}\underset{\approx}{\mathbf{K}}_{g\ th} + {}^{m}\overline{\eta}\,{}^{m}\underset{\approx}{\mathbf{K}}_{g\ mech})\delta\underset{\sim}{\overline{u}} = 0 \qquad (2.63),$$

with a resulting estimate for the critical load (${}^{m}\vartheta$ being held constant)

$${}^{*}\underset{\sim}{r} \approx {}^{m}\overline{\eta}\,{}^{m}\underset{\sim}{r} \qquad (2.64);$$

- for buckling due to thermal loading only:

$$({}^{m}\underset{\approx}{\mathbf{K}}_{e} + {}^{m}\underset{\approx}{\mathbf{K}}_{g\ mech} + {}^{m}\overline{\eta}\,{}^{m}\underset{\approx}{\mathbf{K}}_{g\ th})\delta\underset{\sim}{\overline{u}} = 0 \qquad (2.65),$$

with a resulting estimate for the critical temperature load (${}^{m}\underset{\sim}{r}$ being held constant)

$$\begin{aligned} {}^{*}\overline{\vartheta} &\approx {}^{m}\overline{\eta}\,{}^{m}\overline{\vartheta} \\ {}^{*}\widehat{\vartheta} &\approx {}^{m}\overline{\eta}\,{}^{m}\widehat{\vartheta} \end{aligned} \qquad (2.66).$$

Another approach which is even more efficient with some respects is presented in another contribution to this book, see [22].

2.6 Numerical Examples

The following numerical examples show the applicability of the algorithms described above and explain some features typical for layered composite shells.

a) Two Layer Tension Specimen

By considering a simple two-layer strip under pure tension loading the bending-stretching coupling typical for cross-ply laminates and the torsion-stretching coupling typical for angle-ply laminates is shown. The material parameters and the finite element model are given by Fig. 2.5.

Lay-up: [0/90] ...... two layer cross-ply, layer thickness = 1.6 mm

[+45/-45] ... two layer angle-ply, layer thickness = 1.6 mm

Material:

$E_l = 31100.\ N/mm^2$

$E_q = 7600.\ N/mm^2$

$\nu_{ql} = 0.303$

$G_{lq} = 2900.\ N/mm^2$

$G_{lt} = 2900.\ N/mm^2$

$G_{qt} = 2600.\ N/mm^2$

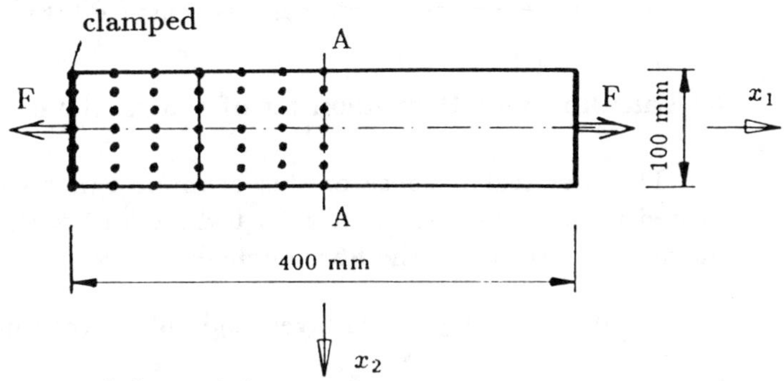

Fig. 2.5 Geometrical and material data of the tensile specimens

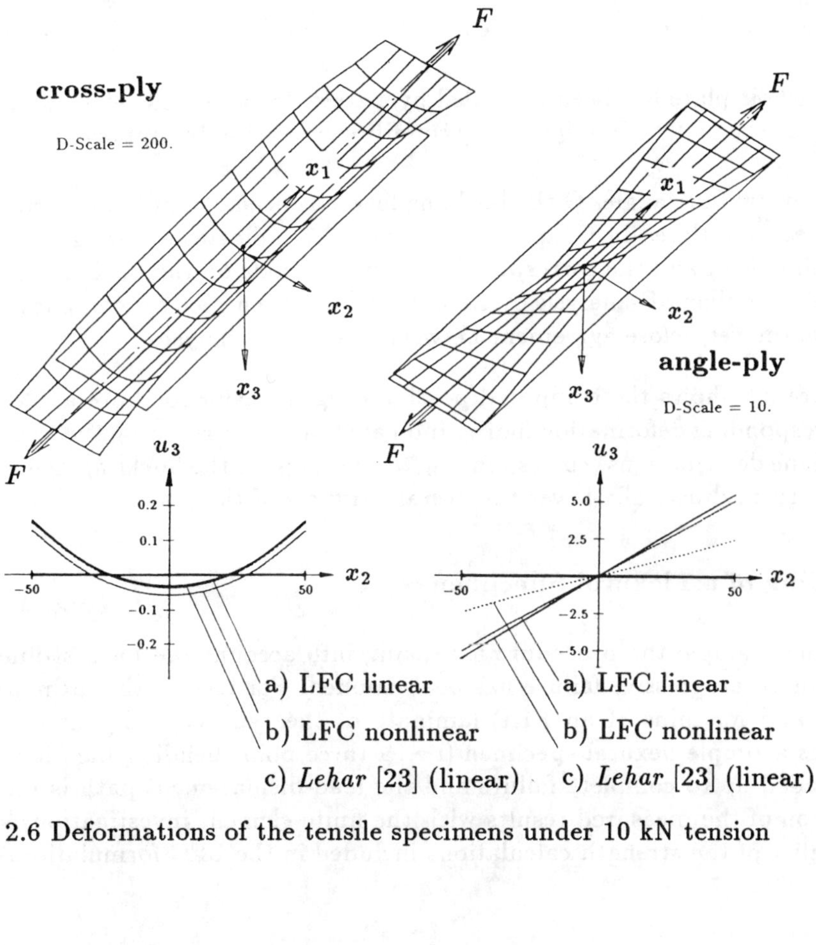

Fig. 2.6 Deformations of the tensile specimens under 10 kN tension

In Fig. 2.6 the deformations of the specimens at a tensile load of 10 kN are shown. Despite pure in-plane loading transversal displacements can be recognized due to the above mentioned coupling effects. In [23] the same problem is treated in a linear analysis. the results obtained by the LFC-model are in good agreement with that of [23] as long as linear behavior is concerned. However, Fig. 2.6 shows that even in this simple case nonlinearities must be taken into account.

b) Buckling and Postbuckling of a Multilayer Square Plate

The practical purpose of this example can be seen in optimizing an in-plane loaded plate (compression in $\Theta = 0$ direction) with regard to buckling. The optimization parameter is the fibre angle Θ.

Lay-up: $[(+\Theta/-\Theta)_{6s}]$... 24-layer angle-ply, layer thickness = 0.0529 mm

Material: Graphite/Epoxy

E_l = 127.5 kN/mm^2 E_q = 11.0 kN/mm^2 ν_{ql} = 0.35

G_{lq} = 5.5 kN/mm^2 G_{lt} = 5.5 kN/mm^2 G_{qt} = 4.6 kN/mm^2

The whole plate has been modelled by sixteen 16-node LFC-elements, the width of the square being 900 mm; boundary conditions: (a) simply supported, (b) clamped.

For several fibre angles Θ the buckling load has been computed by using buckling procedures as outlined in Chapter 2.5. Figure 2.7 indicates a strong dependence of the buckling load and the corresponding buckling mode on the fibre angle. To verify this result a similar FE-analysis by *Nemeth* [24] has been used for comparison. Figure 2.7 shows the very close agreement between these two analyses.

Figure 2.8 shows the computed postbuckling behavior for fibre angle $\Theta = 60°$. The corresponding deformation figures indicate that, after reaching the buckling load, out-of-plane deformations (corresponding to the shape of the buckling mode) develop, which, in turn, drastically lower the overall stiffness of the plate.

c) Bending of a Flexural Specimen

In this example the procedure for taking into account the local stiffness degradation due to progressive failure has been applied. For testing the ultimate flexural strength and modulus of an FRP laminate of the type typically used in aircraft structures a simple flexural specimen (i.e. a three point bending bar) is used. This bar is loaded up to complete failure and the load-displacement path is recorded. A comparison of the measured results with the finite element investigations shows the applicability of the strength calculations included in the LFC formulation.

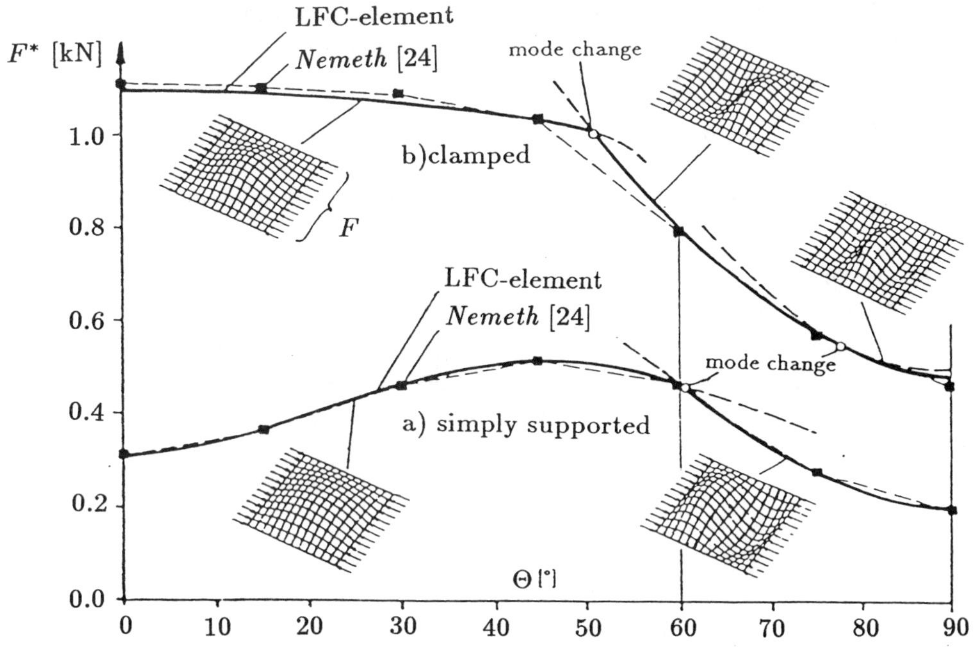

Fig. 2.7 Influence of fibre angle on buckling loads and buckling modes

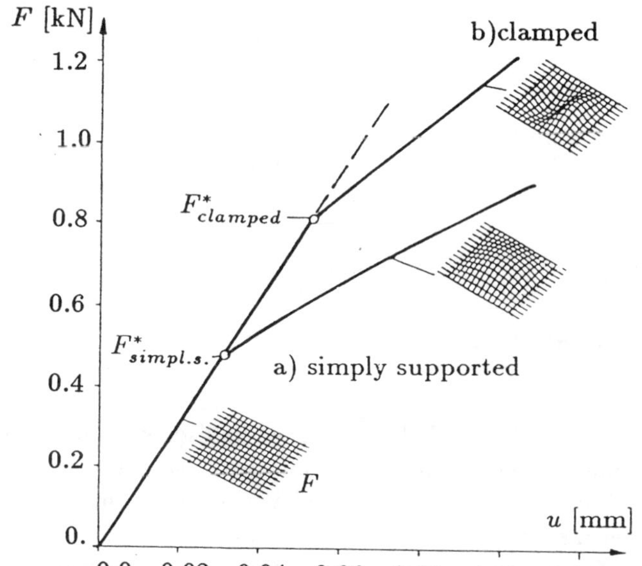

Fig. 2.8 Pre- and post-buckling behavior of the plate with $\Theta = 60°$

Lay-up: 30 layers of Kevlar29 fabric, effective layer thickness = 0.09 mm
For the FE model each fabric layer is subdivided into two sublayers in order to approximate the woven reinforcement by UD-layers.

→ assumed lay-up: $[(0/90)_{15s}]$... cross-ply, layer thickness = 0.045 mm

Model: one quarter of the bar has been modeled by eight 16-node elements

Material: Kevlar29 (UD)

$E_l = 57.2\ \mathrm{kN/mm^2}$	$E_q = 3.9\ \mathrm{kN/mm^2}$	$\nu_{ql} = 0.35$
$G_{lq} = 2.3\ \mathrm{kN/mm^2}$	$G_{lt} = 2.3\ \mathrm{kN/mm^2}$	$G_{qt} = 2.3\ \mathrm{kN/mm^2}$
$\sigma_{lTu} = 1300.\ \mathrm{N/mm^2}$	$\sigma_{lCu} = 227.\ \mathrm{N/mm^2}$	$\tau_{lqu} = 34.\ \mathrm{N/mm^2}$
$\sigma_{qTu} = 12.\ \mathrm{N/mm^2}$	$\sigma_{qCu} = 53.\ \mathrm{N/mm^2}$	$\beta_E = 0.2$

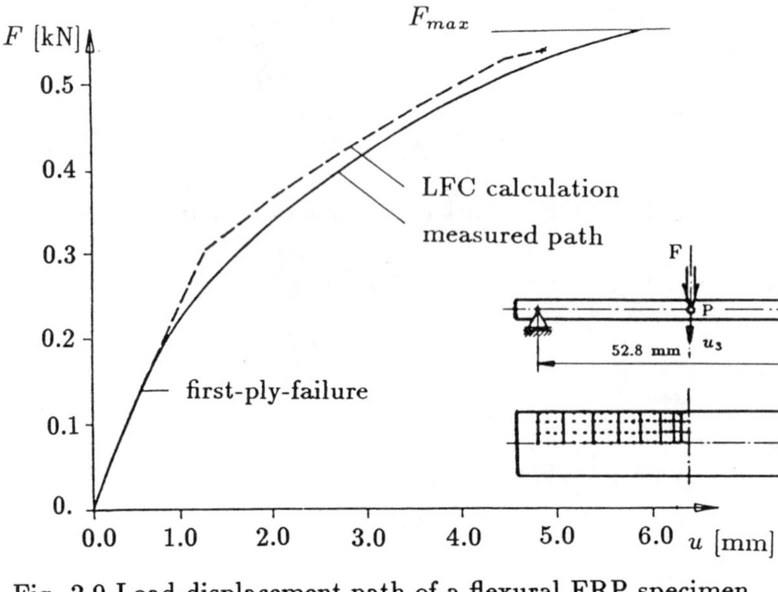

Fig. 2.9 Load-displacement path of a flexural FRP specimen

Figure 2.9 shows the calculated and the measured load-displacement path of the flexural specimen. Reasonably good agreement is obtained. First-ply-failure occurs at a relatively low load level and might be interpreted as the beginning of the nonlinear deformations of the experimentally obtained curve. However, in the FE calculations a noticable deviation from the linear path is observed only at a much higher load level. The calculated deformations and the calculated ultimate load are slightly lower than the measured values but for estimating the nonlinear behavior of the material the procedure appears to be satisfactory.

d) Thermally Loaded Cross-Ply Square Plate

In order to demonstrate the capabilities with respect to thermal loading let us now consider a simply supported composite square plate loaded by a uniform temperature rise. The following properties have been chosen:

Lay-up: [0/90]... two layer cross-ply, layer thickness = 0.2 mm

Material: Graphite/Epoxy

$E_l = 127.5\,\text{kN/mm}^2$	$E_q = 11.0\,\text{kN/mm}^2$	$\nu_{ql} = 0.35$
$G_{lq} = 5.5\,\text{kN/mm}^2$	$G_{lt} = 5.5\,\text{kN/mm}^2$	$G_{qt} = 4.6\,\text{kN/mm}^2$
$\alpha_l = -0.08\times10^{-5}\ {}^\circ\text{C}^{-1}$	$\alpha_q = 2.90\times10^{-5}\ {}^\circ\text{C}^{-1}$	

Model: The whole plate has been modelled by sixteen 16-node LFC-elements, the width of the square being 300 mm; all edges were restricted to remain straight.

Figure 2.10 shows the nonlinear load-displacement path as well as an estimate curve for the buckling load. The deformed shape of the plate is in good agreement with an analytically derived solution, see [25].

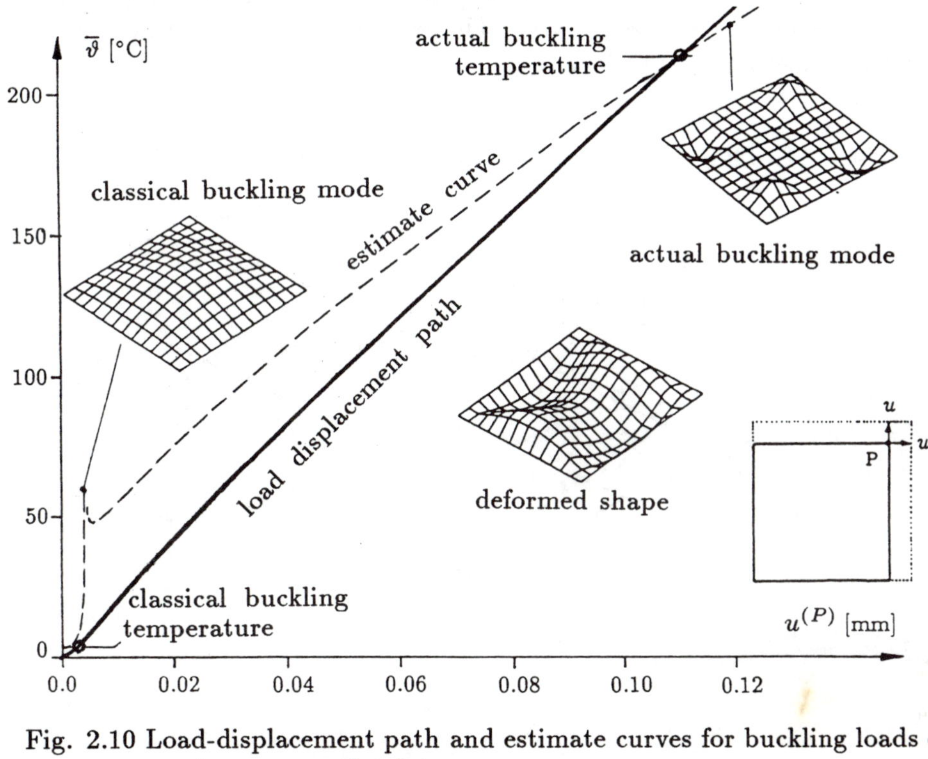

Fig. 2.10 Load-displacement path and estimate curves for buckling loads of a two-layer cross-ply plate

Because of the constant temperature rise all over the plate and the simply supported edges, no stability problem would be expected for a homogeneous isotropic plate. But, due to anisotropies occuring in the laminate, a linear buckling analysis (i.e. a buckling analysis after a very small load step, $\overline{\vartheta} = 1°\mathrm{C}$) yields a bifurcation point at $^*\overline{\vartheta} = 4°\mathrm{C}$. While in case of a symmetrical layer stacking classical buckling analysis could be applied in the example under consideration, however, a detailed analysis of the buckling behavior indicates that the nonlinear pre-buckling deformations make the results of the linear buckling analysis rather meaningless. As can be seen from the estimate curve in Fig. 2.10, the estimates of the relevant buckling mode change during the incremental increase of the temperature and the critical temperature is approximately 50 times higher than the corresponding value of the linear buckling analysis!

e) Shear loaded cylindrical shell panel

As an example of computation of the unstable post-buckling behavior let us consider a cylindrical shell panel under shear loads. The panel has the following properties given in dimension consistent values to allow a direct comparison with analytically derived results given in [26]. The geometrical and material parameters (in consistent dimensions) are given in Fig. 2.11.

Lay-up: [0/90]... two layer cross-ply, layer thickness = 0.79375

Material:

$E_l = 10.$

$E_q = 1.$

$\nu_{ql} = 0.22$

$G_{lq} = 0.333$

$G_{lt} = 0.333$

$G_{qt} = 0.3$

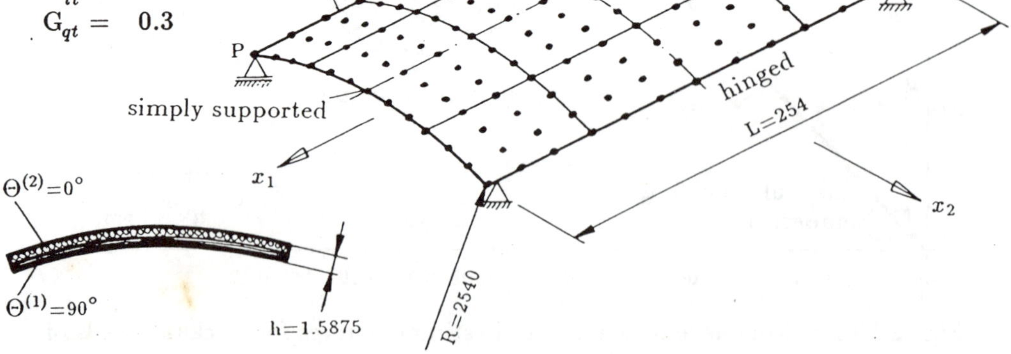

Fig. 2.11 Model of a cylindrical shell segment

Nondimensionalized geometry parameters and shear loads are used:

simplified flatness parameter [26]: $\theta_s = \dfrac{b}{\sqrt{R\,h}},$

aspect ratio: $L/B,$

nondimensional shear load [26]: $\overline{\tau} = \dfrac{F}{L}\dfrac{R}{E_q h^2}.$

First, critical shear loads $\overline{\tau}^*$ for several shell thicknesses h, cylinder radii R, and panel widths B have been calculated. By solving the eigenvalue problem eqn. (2.59) following a small linear load step, linear buckling loads have been obtained. The results are in very close agreement with the analytically derived solutions, see Fig. 2.12.

For comparing with the results for $N \to \infty$ a lay-up with 200 layers has been used. Due to the analytical thickness integration the number of layers did not affect the CPU-time significantly!

Next, the panel given in Fig. 2.11 with $\theta_s = 2$, L/B = 2, has been investigated. Figure 2.13 shows the deformations after a linear load step $\overline{\tau} = 1.0$ (D-scale = 200) and the buckling mode at the corresponding buckling load $\overline{\tau}^* = 2.89$.

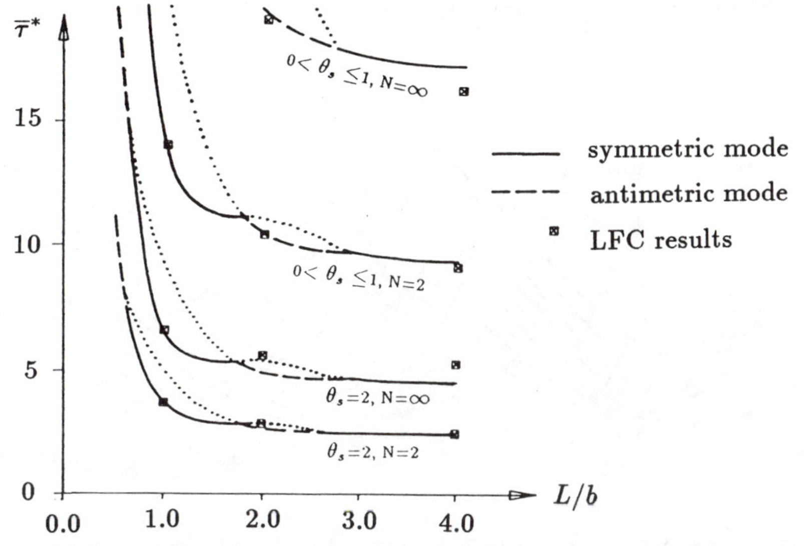

Fig. 2.12 Shear buckling load vs. aspect ratio for simply supported cross-ply cylindrical panels [26]

A slight variation of the boundary conditions or a change of the stacking sequence influence the critical shear load significantly:

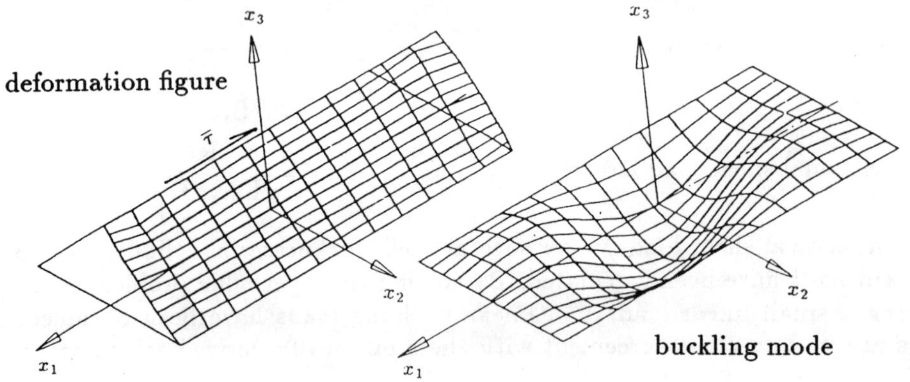

Fig. 2.13 Deformations and corresponding buckling mode for a cylindrical shell panel

By relaxing the boundary conditions on the loaded edge (see Fig. 2.11) in such a way that this edge is able to deform in the x_1–x_2 plane the buckling load is decreased by approx. 25% to $\bar{\tau}^* = 2.18$. In addition a qualitative change in the buckling mode becomes evident. Figure 2.14 shows the corresponding deformation figure and the buckling mode (compare with Fig. 2.13).

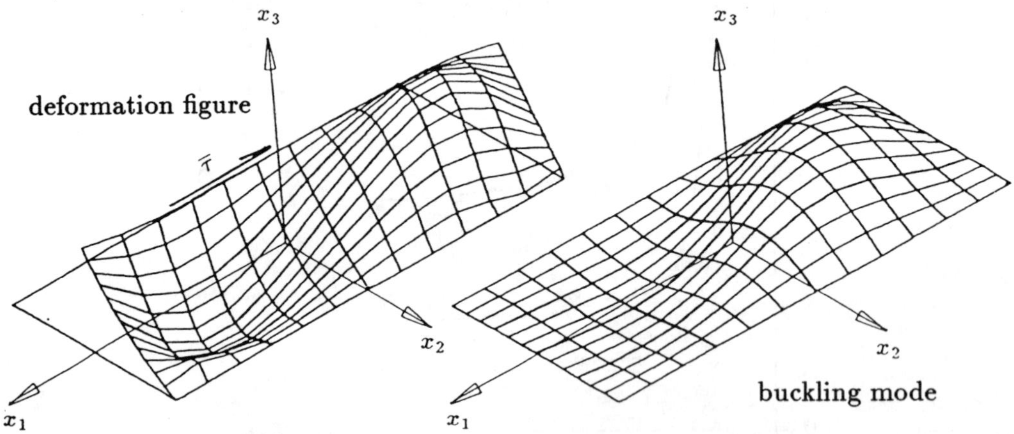

Fig. 2.14 Deformations and corresponding buckling mode for a cylindrical shell panel with relaxed boundary conditions

Reversing the stacking sequence, so that the outside layer is reinforced in circumferential direction and the inside layer in axial direction, renders an increase of the critical shear load of approx. 19% referred to the value $\overline{\tau}^*$ given in Fig. 2.13.

Relaxing the boundary conditions on the panel with the reversed stacking sequence leads to a small decrease of the buckling load, so that for these conditions the critical load is approx. 14% lower than $\overline{\tau}^*$ of the original panel.

Investigation of nonlinear prebuckling deformations in the buckling analysis of the panel shows interesting phenomena:

In Fig. 2.15 the load-displacement path and the "estimate curves" for the buckling loads are shown. On the abscissa of the diagram the displacements of the loaded edge in load direction are plotted.

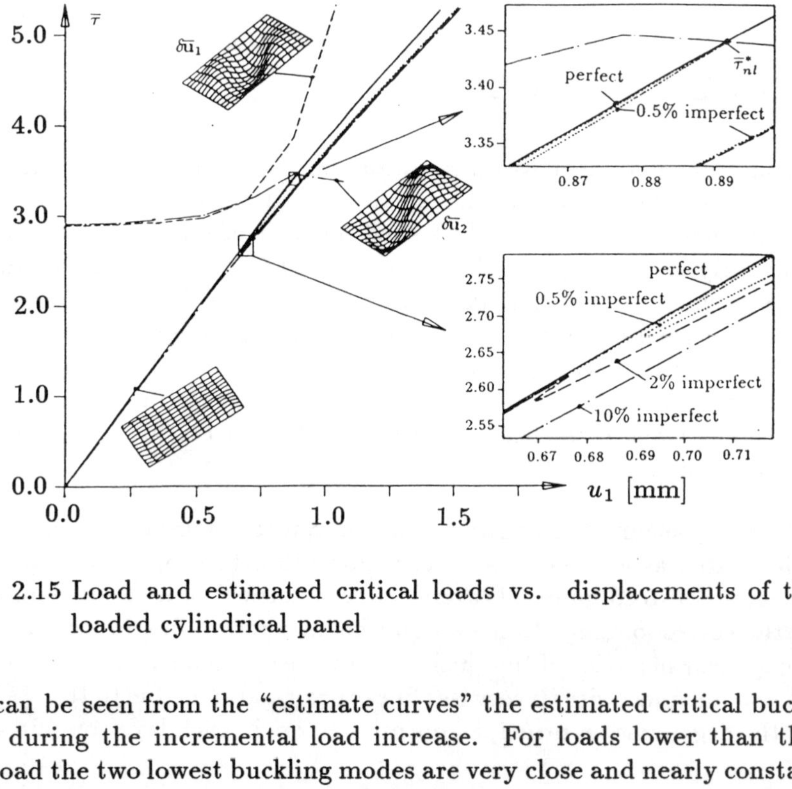

Fig. 2.15 Load and estimated critical loads vs. displacements of the shear loaded cylindrical panel

As can be seen from the "estimate curves" the estimated critical buckling mode changes during the incremental load increase. For loads lower than the effective critical load the two lowest buckling modes are very close and nearly constant. But on approaching the critical point the mode indicated by $\delta\underset{\sim}{\overline{u}_1}$ becomes uncritical, and the initially second mode finally corresponds to the relevant buckling figure. Compared with the linear analysis an increase of the critical shear load by approx. 20% due to the pre-buckling deformations can be observed.

In order to trace the post-buckling behavior the geometry of the panel has been slightly modified: the effective modeshape with an amplitude scaled to a certain fraction of the shell thickness (10%, 2%, 0.5%) has been superimposed on the original (perfect) geometry. Due to these imperfections the bifurcation point vanishes and the panel exhibits a snap-through behavior or no instability at all, depending on the amount of imperfection.

As can be seen from the close ups in Fig. 2.15, the 0.5% imperfect panel behaves like the perfect structure up to the bifurcation point but then the load-displacement path shows a snap through and after that a considerable deviation from the perfect panel. For the 2% imperfect panel the behavior is similar but the snap through is not as drastic. The load-displacement path for the 10% imperfect panel exhibits no stability limit but a continuously increasing deviation from the perfect structure can be observed.

3. THE LCSLFC-ELEMENT

The assumption of a linear strain variation over the shell's thickness in the LFC-element is justified as long as rather thin shells with a large number of different layers are considered. However, in case of thick shells or if the shell consists of a few layers only, with strongly different material behavior the LFC-element does not sufficiently represent the mechanical behavior. Especially its transverse shear behavior needs to be improved. Numerous different ways to overcome this problem have been developed [27,28,29]. One attempt that appears to be rather promising is to use first order shell elements for each layer and to combine them in a way that results in an element with layerwise constant transverse shear stresses. In [2] a successful application of this concept is presented.

In the LCSLFC-element (Layerwise Constant Shear Laminated Fiber Composite) each layer is treated as a seperate shell element with orthotropic material behavior. The layer degrees of freedom of all layers are transformed to a set of element DOFs along with the corresponding stiffnesses and nodal forces. This procedure improves the transverse shear behavior of the shell element since normal vectors of the shell are now able to undergo piecewise (layerwise) linear deformations (due to the assumptions included in the degeneration principle, normal vectors in each layer remain straight).

3.1 Stiffness Expressions for the Individual Layer

Each layer n is treated as a homogeneous shell with constant material properties over the layer thickness. Again the degeneration principle is employed to derive the shell element stiffness expressions from the 3/D continuum.

a) Description of Geometry

With the assumption that the normal vectors of the layer's reference surface remain straight and inextensible during deformation, the geometry of the layer can be described by one reference surface and its corresponding normal vectors. With appropriate shape functions the matrices $\underset{\approx}{\mathbf{B}}_l$ and $\underset{\approx}{\mathbf{B}}_{nl}$ (in eqns. (2.2–2.5)) can be constructed.

Due to the isoparametric element formulation these matrices are functions of natural coordinates r, s, t, see Fig. 3.1.

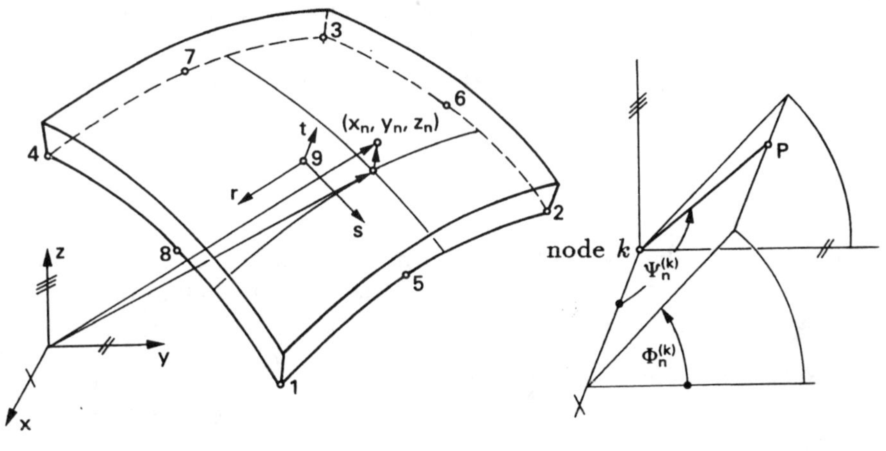

Fig. 3.1 Geometry of the degenerated shell (=layer) element

Interpolation of geometry (subscript n indicates the layer number):

$$
\begin{aligned}
x_n(r,s,t) &= \sum_{k=1}^{M} \phi^{(k)}(r,s)[x_n^{(k)} + t\,\mathrm{h}_n^{(k)} \cos\Psi_n^{(k)}] \\
y_n(r,s,t) &= \sum_{k=1}^{M} \phi^{(k)}(r,s)[y_n^{(k)} + t\,\mathrm{h}_n^{(k)} \sin\Psi_n^{(k)} \cos\Phi_n^{(k)}] \qquad (3.1), \\
z_n(r,s,t) &= \sum_{k=1}^{M} \phi^{(k)}(r,s)[z_n^{(k)} + t\,\mathrm{h}_n^{(k)} \sin\Psi_n^{(k)} \sin\Phi_n^{(k)}]
\end{aligned}
$$

with $\phi^{(k)}(r,s)$ being standard 2/D shape functions (i.e. Lagrangian polynomials); M is the number of nodes forming the element, $\mathrm{h}_n^{(k)}$ is the thickness of the layer at node

k, and $\Psi_n^{(k)}$ and $\Phi_n^{(k)}$ are used to determine the position of the layer's normal at node k, see Fig. 3.1.

Based on this kinematics and the further assumption that the contribution of the strain energy caused by stress components perpendicular to the reference surface is set to zero by using a modified material law, the stiffness contribution of the considered layer to the shell element stiffness matrix can be derived analogously to Chapter 2.

3.2 Assemblage of Element Stiffnesses

The procedure for establishing a connection between the individual layers is based on kinematic assumptions described below. Using this assumptions a relationship between conveniently chosen element DOFs and the layer DOFs (described in Chapter 3.1) can be derived.

a) Definition of Degrees of Freedom

In Figure 3.2 the definition of the element's geometry and of its DOFs is shown. To secure an easy and universal usage of the element, the definition of the element DOFs is based on the concept of "in-plane" and "out-of-plane" quantities:

- Displacements of the shell's midsurface (more precisely the shell's reference surface) represent the "in-plane" part (they are related to the membrane behavior).
- Rotations of shell's "normals" represent the "out-of-plane" part (they are related to the bending and transverse shearing behavior).

Basically, the same notation as used in the degeneration principle is employed here. The geometry of the entire shell is described in terms of:

- The coordinates of nodes (k) lying in the shell's midsurface, $x^{(k)}, y^{(k)}, z^{(k)}$,
- 2 angles at these nodes, $\Phi^{(k)}, \Psi^{(k)}$ which determine the position of the shell's "normal" (this vector is not necessarily exactly normal to the shell's midsurface),
- $2(N-1)$ additional angles at these nodes, $\varphi_n^{(k)}, \psi_n^{(k)}$ $(n = 1, \ldots, N-1;\ N$ $=$ number of layers), which determine the positions of the normal vectors of the layers, and
- the thickness of the shell at these nodes $h^{(k)}$, see Fig. 3.2.

Correspondingly, the following DOFs per node, expressed in increments, are defined:

- 3 displacement increments, $\Delta u^{(k)}, \Delta v^{(k)}, \Delta w^{(k)}$, and

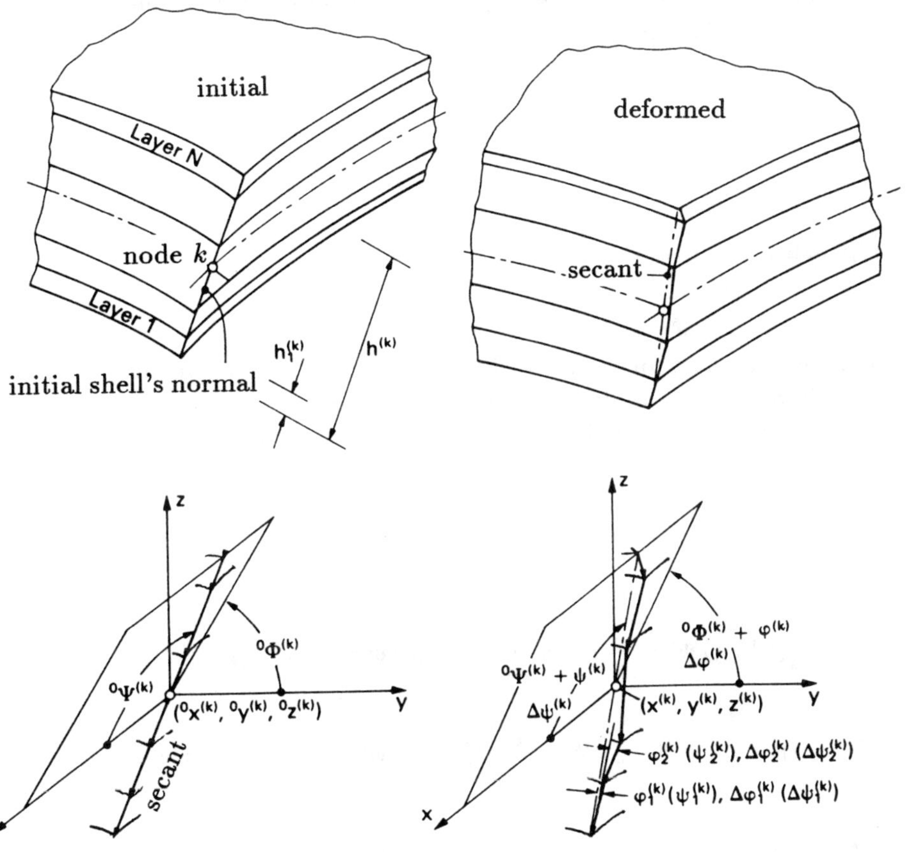

Fig. 3.2 Definition of element geometry and DOFs for the multilayer shell

– $2N$ angle increments, $\Delta\varphi^{(k)}, \Delta\psi^{(k)}, \Delta\varphi_n^{(k)}, \Delta\psi_n^{(k)}$ $(n = 1, \ldots, N-1)$.

Initially, the normal vectors of all layers lie on one straight line, namely the shell's normal, which is determined by the initial angles ${}^0\Phi^{(k)}, {}^0\Psi^{(k)}$, and can be computed from the nodal coordinates.

During loading, the normals of the layers deviate from the initial shell's normal. Due to the degeneration principle, these layer-normals remain straight, so that a piecewise linear deformation of the shell's normal occurs.

By defining a "secant" (which coincides with the shell's normal in the initial configuration) and angles $\varphi_n^{(k)}, \psi_n^{(k)}$ as corresponding deviations of the layer-normals from this secant (see Fig. 3.2) one can compute the coordinates $x_n^{(k)}$, $y_n^{(k)}$, $z_n^{(k)}$ and the total angles $\Phi_n^{(k)}$, $\Psi_n^{(k)}$ for the layers.

b) Description of Geometry

The nodal coordinates for layer n at beginning of the increment can be computed from the coordinates of the midsurface of the shell, the shell thickness, the angles of the "secant", and the angles of the layer-normals as follows:

$$
\begin{aligned}
x_n^{(k)} &= x^{(k)} - \tfrac{1}{2}\mathrm{h}^{(k)} \cos\Psi^{(k)} &&+ \sum_{j=1}^{n-1} \mathrm{h}_j^{(k)} \cos(\Psi^{(k)} + \psi_j^{(k)}) \\
y_n^{(k)} &= y^{(k)} - \tfrac{1}{2}\mathrm{h}^{(k)} \sin\Psi^{(k)} \cos\Phi^{(k)} &&+ \sum_{j=1}^{n-1} \mathrm{h}_j^{(k)} \sin(\Psi^{(k)} + \psi_j^{(k)}) \cos(\Phi^{(k)} + \varphi_j^{(k)}) \\
z_n^{(k)} &= z^{(k)} - \tfrac{1}{2}\mathrm{h}^{(k)} \sin\Psi^{(k)} \sin\Phi^{(k)} &&+ \sum_{j=1}^{n-1} \mathrm{h}_j^{(k)} \sin(\Psi^{(k)} + \psi_j^{(k)}) \sin(\Phi^{(k)} + \varphi_j^{(k)})
\end{aligned}
\qquad (3.2).
$$

$$
\Phi_n^{(k)} = \begin{cases} \Phi^{(k)} + \varphi_n^{(k)} & 1 \le n \le N-1 \\ \Phi^{(k)} - \arcsin\left(\sum\limits_{j=1}^{N-1} \sin\varphi_j^{(k)} \frac{\mathrm{h}_j^{(k)}}{\mathrm{h}_N^{(k)}}\right) & n = N \end{cases}
$$

$$
\Psi_n^{(k)} = \begin{cases} \Psi^{(k)} + \psi_n^{(k)} & 1 \le n \le N-1 \\ \Psi^{(k)} - \arcsin\left(\sum\limits_{j=1}^{N-1} \sin\psi_j^{(k)} \frac{\mathrm{h}_j^{(k)}}{\mathrm{h}_N^{(k)}}\right) & n = N \end{cases}
$$

The nodal displacements for layer n can be computed in terms of the shell's nodal displacements and nodal angles:

$$
\begin{aligned}
u_n^{(k)} &= u^{(k)} - \tfrac{1}{2}\mathrm{h}^{(k)}(\cos\Psi^{(k)} - \cos{}^0\Psi^{(k)}) \\
&\quad + \sum_{j=1}^{n-1} \mathrm{h}_j^{(k)}\left(\cos(\Psi^{(k)} + \psi_j^{(k)}) - \cos{}^0\Psi^{(k)}\right) \\
v_n^{(k)} &= v^{(k)} - \tfrac{1}{2}\mathrm{h}^{(k)}(\sin\Psi^{(k)}\cos\Phi^{(k)} - \sin{}^0\Psi^{(k)}\cos{}^0\Phi^{(k)}) \\
&\quad + \sum_{j=1}^{n-1} \mathrm{h}_j^{(k)}\left(\sin(\Psi^{(k)} + \psi_j^{(k)})\cos(\Phi^{(k)} + \varphi_j^{(k)}) - \sin{}^0\Psi^{(k)}\cos{}^0\Phi^{(k)}\right) \\
w_n^{(k)} &= w^{(k)} - \tfrac{1}{2}\mathrm{h}^{(k)}(\sin\Psi^{(k)}\sin\Phi^{(k)} - \sin{}^0\Psi^{(k)}\sin{}^0\Phi^{(k)}) \\
&\quad + \sum_{j=1}^{n-1} \mathrm{h}_j^{(k)}\left(\sin(\Psi^{(k)} + \psi_j^{(k)})\sin(\Phi^{(k)} + \varphi_j^{(k)}) - \sin{}^0\Psi^{(k)}\sin{}^0\Phi^{(k)}\right)
\end{aligned}
\qquad (3.3).
$$

The layer-coordinates from eqn. (3.2) and the layer-displacements from eqn. (3.3) are then used for computing the stiffness expressions of the layer (Chapter 3.1).

c) Transformation Layer → Shell

Equation (3.3) can be used for deriving the displacement increments. By assuming the angle increments $\Delta\varphi^{(k)}, \Delta\psi^{(k)}, \Delta\varphi_n^{(k)}, \Delta\psi_n^{(k)}$ to be small ($\ll 1$), the trigonometric functions can be linearized with respect to these angles, which results in a linear relationship between the layer DOFs ($\Delta u_n^{(k)}$, $\Delta v_n^{(k)}$, $\Delta w_n^{(k)}$, $\Delta\overline{\varphi}_n^{(k)}$, $\Delta\overline{\psi}_n^{(k)}$) and the shell DOFs ($\Delta u^{(k)}$, $\Delta v^{(k)}$, $\Delta w^{(k)}$, $\Delta\varphi^{(k)}$, $\Delta\psi^{(k)}$, $\Delta\varphi_n^{(k)}$, $\Delta\psi_n^{(k)}$, $n = 1, \ldots, N-1$):

$$
\Delta \underset{\sim}{u}_n^{(k)} = \underset{\approx}{\mathrm{G}}_n^{(k)}\, \Delta \underset{\sim}{u}^{(k)} \qquad (3.4).
$$

The detailed formulas for the transformation matrix $\underset{\approx}{\mathrm{G}}_n^{(k)}$ and the definitions of the vectors $\Delta\underset{\sim}{u}_n^{(k)}$ and $\Delta\underset{\sim}{u}^{(k)}$ can be found in [2].

The definition of the angles $\varphi_n^{(k)}$ and $\psi_n^{(k)}$ as deviations from the "secant" renders some distinct advantages:

- Enforcing a straight shell normal (e.g. as a boundary condition) can easily be done by setting the DOFs $\Delta\varphi_n^{(k)}$, $\Delta\psi_n^{(k)}$, $n = 1, \ldots, N-1$ to zero. This forces the total angles $\Phi_n^{(k)}$, $\Psi_n^{(k)}$ of all layers to be equal to the corresponding angles $\Phi^{(k)}$, $\Psi^{(k)}$ of the shell.

– The piecewise linear deformations of the shell's "normal" result directly from the analysis in terms of the deviation angles.

By applying the transformation from eqn. (3.1) to all nodes the following global transformation matrix can be derived:

$$\underset{\approx}{\mathbf{G}}_n = \begin{pmatrix} \underset{\approx}{\mathbf{G}}_n^{(1)} & \underset{\approx}{\mathbf{0}} & \cdots & \underset{\approx}{\mathbf{0}} & \cdots & \underset{\approx}{\mathbf{0}} \\ & \underset{\approx}{\mathbf{G}}_n^{(2)} & \cdots & \underset{\approx}{\mathbf{0}} & \cdots & \underset{\approx}{\mathbf{0}} \\ & & \ddots & \vdots & \vdots & \vdots \\ & sym. & & \underset{\approx}{\mathbf{G}}_n^{(k)} & \cdots & \underset{\approx}{\mathbf{0}} \\ & & & & \ddots & \vdots \\ & & & & & \underset{\approx}{\mathbf{G}}_n^{(M)} \end{pmatrix} \quad (3.5).$$

With this equation, the stiffness matrix and the vector of internal nodal forces can be transformed from the layer level to the element level:

$$\underset{\approx}{\mathbf{K}}^{(e)} = \sum_{n=1}^{N} \underset{\approx}{\mathbf{G}}_n^T (\underset{\approx}{\mathbf{K}}_e^{(n)} + \underset{\approx}{\mathbf{K}}_g^{(n)}) \underset{\approx}{\mathbf{G}}_n \quad (3.6),$$

$$\underset{\sim}{\mathbf{f}}_{th}^{(e)} = \sum_{n=1}^{N} \underset{\approx}{\mathbf{G}}_n^T (\underset{\sim}{\mathbf{f}}^{(n)} - \Delta \underset{\sim}{\mathbf{f}}_{th}^{(n)}) \quad (3.7).$$

Finally, after assembling the global tangent stiffness matrix and the global load vector, respectively, the incremental equilibrium equation, eqn. (2.1), can be formulated.

3.3 Verification Examples

The following examples show the effect of consideration of individual transverse shear deformations for each layer.

a) Pressure Loaded Thick Square Plate

A simply supported thick three-layer cross-ply square plate which is subject to a sinusoidal pressure load is used to demonstrate the accuracy of the LCSLFC-element. Both the exact solutions and FE results [27,28] have been used for comparison.

Lay-up: [0/90/0]...three layer cross-ply, shell thickness $h = 1$ in.

Material: as specified in [27]

E_l =25000 kpsi	E_q =1000 kpsi	ν_{ql} =0.25
G_{lq} = 500 kpsi	G_{lt} = 500 kpsi	G_{qt} = 200 kpsi

Model: A quarter of the plate has been modelled by four 16-node LCSLFC-elements. Width of the square $a = 4$ in. Symmetry conditions: along the $x-$symmetry line $\Delta v^{(k)}$, $\Delta\varphi^{(k)}$, $\Delta\varphi_n^{(k)}$ $(n = 1, \ldots, N-1)$ have been set to zero, and along the $y-$symmetry line $\Delta u^{(k)}$, $\Delta\psi^{(k)}$, $\Delta\psi_n^{(k)}$ $(n = 1, \ldots, N-1)$ have been set to zero. The 0° layers have been divided into two sub-layers to trace the variation of transverse shear within the layers.

Distributed load:

$$q(x, y) = q_0 \sin(\pi x/a) \sin(\pi y/a) \ ,$$
$$q_0 = (\pi/a)^2 \ .$$

For comparison purposes, the following normalized quantities have been used:

$$\overline{u} = \frac{100 E_q h^2}{q_0 a^3} u, \qquad \overline{w} = \frac{100 E_q h^3}{q_0 a^4} w, \qquad \overline{z} = z/h,$$
$$(\overline{\sigma}_x, \overline{\sigma}_y) = \frac{h^2}{q_0 a^2} (\sigma_x, \sigma_y), \qquad (\overline{\tau}_{xz}, \overline{\tau}_{yz}) = \frac{h}{q_0 a} (\tau_{xz}, \tau_{yz}).$$

Since nonlinear effects are not taken into account in [27], only a linear load step has been performed. Figures 3.3 to 3.5 show the through-the-thickness distribution of the normalized in-plane displacements $\overline{u}$ and the normalized stresses $\overline{\sigma}_x, \overline{\tau}_{xz}$, respectively, at specific locations. Analytical solutions and the results from classical laminate theory (CLT) are also given in these figures. Very good agreement can be observed between the LCSLFC-element and the analytical solution despite the fact, that this "plate" resembles a solid brick rather than a plate (the element length to thickness ratio is 1:1!).

The same computation has been performed with a different shell thickness ($h =$ 0.4) in order to investigate the importance of transverse shear. As one would expect, the differences between the classical lamination theory and more advanced methods (like the LCSLFC-element) become less pronounced as the thickness of the shell decreases, see Fig. 3.3(b). Table 3.1 lists some stresses and displacements for this case together with analytical results and other FE results.

These results clearly indicate that the LCSLFC-element is capable of handling thick laminated shell problems with high accuracy.

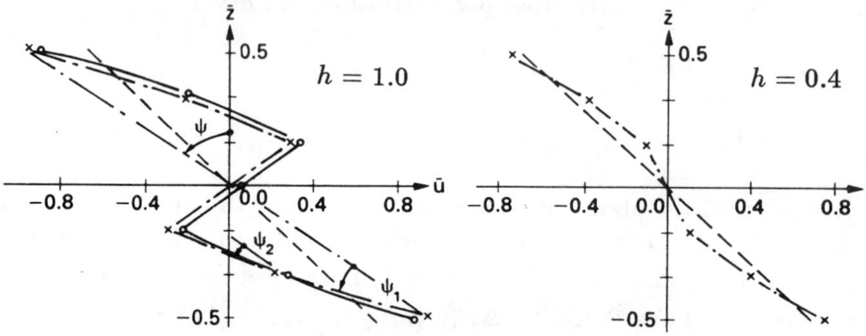

Fig. 3.3 Thickness distribution of normalized in-plane displacements $\overline{u}$ of a thick cross-ply square plate at $x = 0, y = \frac{a}{2}$ for case a) $h = 1.0$ and case b) $h = 0.4$

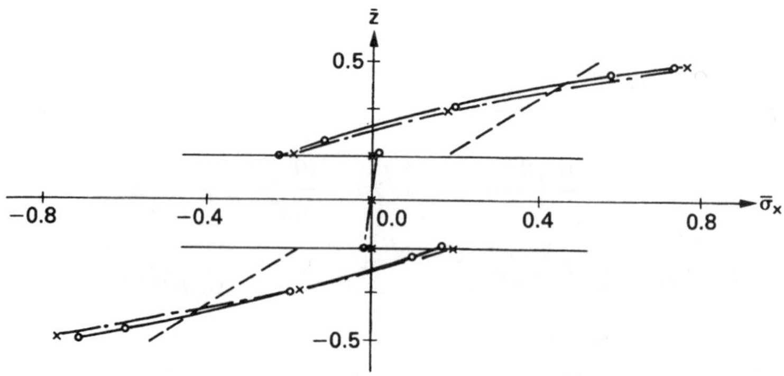

Fig. 3.4 Thickness distribution of normalized in-plane normal stresses $\overline{\sigma}_x$ at the center of a thick cross-ply square plate ($x = \frac{a}{2}, y = \frac{a}{2}$) for $h = 1.0$

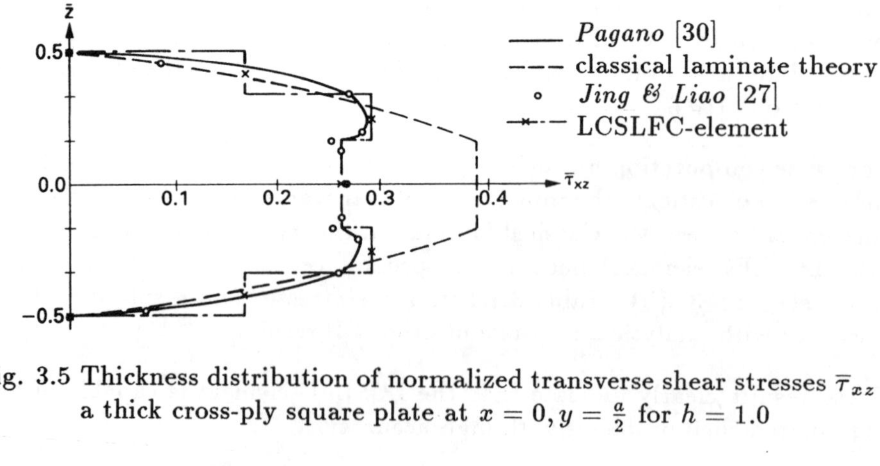

Fig. 3.5 Thickness distribution of normalized transverse shear stresses $\overline{\tau}_{xz}$ of a thick cross-ply square plate at $x = 0, y = \frac{a}{2}$ for $h = 1.0$

Table 3.1: Normalized stresses and displacements for $h = 0.4$

	$\overline{\sigma}_x(\frac{a}{2},\frac{a}{2},\pm\frac{h}{2})$	$\overline{\sigma}_y(\frac{a}{2},\frac{a}{2},\pm\frac{h}{6})$	$\overline{\tau}_{yz}(\frac{a}{2},0,0)$	$\overline{\tau}_{xz}(0,\frac{a}{2},0)$	$\overline{w}(\frac{a}{2},\frac{a}{2},0)$
Pagano [30]	±0.590	$^{+0.285}_{-0.288}$	+0.1228	+0.357	+0.7530
Jing & Liao [27]	$^{+0.5884}_{-0.5879}$	$^{+0.2834}_{-0.2873}$	+0.1284	+0.3627	+0.7531
LCSLFC-element	$^{+0.590}_{-0.586}$	$^{+0.285}_{-0.283}$	+0.1079	+0.364	+0.7559

b) Cylindrical Bending of a Plate Strip

A two-layer cross-ply plate strip subject to a distributed normal load [29] has been used to investigate the influence of stacking sequence, geometrical nonlinearities and boundary conditions on the mechanical behavior of laminated plates.

Lay-up: [0/90]... two layer cross-ply, layer thickness = 0.2 in.

Material: as specified in [29]

E_l =20000 kpsi	E_q =1400 kpsi	$\nu_{ql} = 0.3$
G_{lq} = 700 kpsi	G_{lt} = 700 kpsi	G_{qt} =700 kpsi

Model: The plate strip has been modelled by four 16-node LCSLFC-elements; length of the strip l = 18 in., width b = 3 in. Both ends of the strip have been pinned. Each layer has been divided into two sub-layers to ensure an accurate representation of the variation of transverse shear within the layers.

Different types of meshes have been tested to evaluate the robustness of the LCSLFC-element. Even with an aspect ratio (length to width) of 11:1 very accurate results have been obtained.

For comparison purposes the shell normals have been forced to remain straight. This was accomplished by setting the rotational DOFs $\Delta\varphi_1^{(k)}$ $\Delta\psi_1^{(k)}$ to zero. Excellent agreement with the investigations in [29] have been obtained, see Figure 3.6.

For simply supported ends, the magnitude of the displacements is independent of the sign of the applied load, because the plate strip is essentially in pure bending, see [29]. Due to the pinned ends of the strip in this study, in-plane stiffnesses are activated as soon as out of plane deflections occur (geometrically nonlinear effect) and, therefore, quite different results are obtained for positive and negative load directions (which can be translated into a reversed stacking sequence), respectively, see Fig. 3.6. For a detailed explanation of this phenomenon see [29].

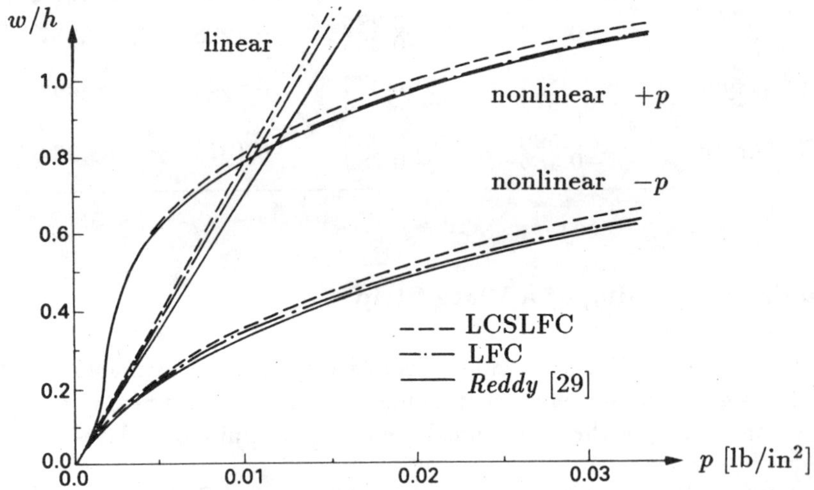

Fig. 3.6 Load-displacement path of a pinned two-layer cross-ply plate strip subject to a distributed normal load

Relaxing the boundary condition on the shell normals, which allows them to deform layerwise linearly (therefore having a more realistic model), results in only slightly larger deformations, as can be seen in Fig. 3.6. This indicates that transverse shear does not play a significant role in this rather thin shell application.

4. LM-ELEMENTS FOR SANDWICH SHELLS

Sandwich plates and shells becomes more and more important in lightweight constractions. Frequently, combinations of layered composite shells with sandwich shells are used to achieve optimized configurations.

4.1 Modifications to the LFC-Element

Under the assumption of *antiplane core* conditions a finite element formulation is derived for a layer model (LM-model) of the sandwich shell. Based on the LFC-element described in Chapter 2 the features typical for sandwich structures are introduced into the finite element formulation: The local material matrix $\underset{\approx}{C}$ is defined

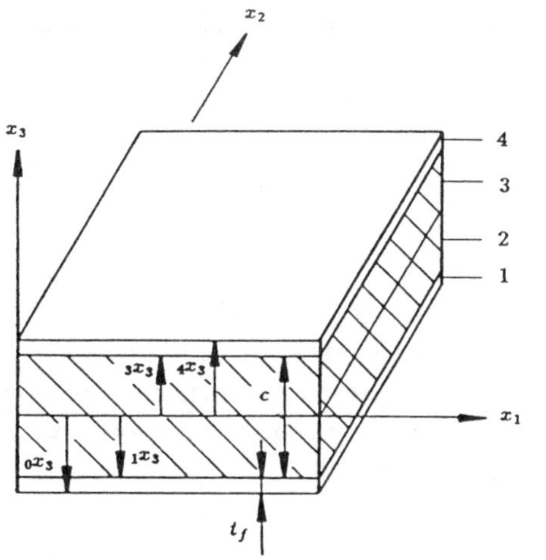

$$
\begin{aligned}
{}^0t &= \quad -1.0 \\
{}^1t &= -\frac{c}{2\ (c+2\ t_f)} \\
{}^2t &= \quad 0.0 \\
{}^3t &= \frac{c}{2\ (c+2\ t_f)} \\
{}^4t &= \quad 1.0
\end{aligned}
$$

Fig. 4.1 Definition of the sandwich layer model

independently for each layer.

Due to the assumptions of antiplane core conditions the material matrix $\underset{\approx}{\mathbf{C}}_L$ for the isotropic face layer material

$$
{}^1\underset{\approx}{\mathbf{C}}_L = {}^4\underset{\approx}{\mathbf{C}}_L = \underset{\approx}{\mathbf{C}}_f = \begin{pmatrix} \frac{E_f}{1-\nu_f^2} & \frac{\nu_f\ E_f}{1-\nu_f^2} & 0 & 0 & 0 & 0 \\ \frac{\nu_f\ E_f}{1-\nu_f^2} & \frac{E_f}{1-\nu_f^2} & 0 & 0 & 0 & 0 \\ 0 & 0 & 0 & 0 & 0 & 0 \\ 0 & 0 & 0 & G_{f,LW} & 0 & 0 \\ 0 & 0 & 0 & 0 & 0 & 0 \\ 0 & 0 & 0 & 0 & 0 & 0 \end{pmatrix} \tag{4.1},
$$

and the material matrix $\underset{\approx}{\mathbf{C}}_c$ for the orthotropic core material are defined with respect to the orthotropy axes:

$$
{}^2\underset{\approx}{\mathbf{C}}_L = {}^3\underset{\approx}{\mathbf{C}}_L = \underset{\approx}{\mathbf{C}}_c = \begin{pmatrix} 0 & 0 & 0 & 0 & 0 & 0 \\ 0 & 0 & 0 & 0 & 0 & 0 \\ 0 & 0 & 0 & 0 & 0 & 0 \\ 0 & 0 & 0 & 0 & 0 & 0 \\ 0 & 0 & 0 & 0 & G_{c,LT} & 0 \\ 0 & 0 & 0 & 0 & 0 & G_{c,WT} \end{pmatrix} \tag{4.2}.
$$

Since the out-of-plane shear stresses can be assumed to remain constant over the core thickness, no correction factors are required for the shear rigidities. If the local coordinate system defined for each layer does not coincide with the x_1',x_2',x_3'-coordinate system a transformation of the material matrices is needed leading to $\underset{\approx}{C}'$ matrices, see Chapter 2.

4.2 Accounting for Local Instability Phenomena

As the stiffness of the sandwich structure may considerably be changed by local buckling phenomena, the influence of regions with buckled face layers on the global deformation and stability behavior has to be considered for further postcritical loading. The following steps describe the procedure in accounting for local instability phenomena; see also [4,6,38]:

- Calculation of membrane forces in the face layers,
- Determination of critical loads inducing local buckling, i.e. buckling of the face layers at short wavelcngths including shear buckling and intracell buckling in the case of honeycomb cores,
- Check for the occurence of local instability: The smaller of the two critical loads is taken for checking the safety against local buckling:

$$P_{crit}^{buckl} = Min\left(P_{crit}^{wrinkl}, P_{crit}^{icb}\right) \qquad (4.3),$$

- Update of the vector of out-of-balance loads and of the tangential stiffness matrices in case of appearance of local buckling.

Since the determination of local stability limits is influenced by the actual loading state (biaxial loading, bending moments, ...), the critical loads have to be calculated for each state m.

a) Wrinkling of the Face Layers

Under the assumption of linear elastic isotropic face layers and linear elastic orthotropic core material the critical face layer force per unit width P_{crit}^{buckl} inducing short wavelength buckling of the face layers is derived in [4,31].

In the derivations of the wrinkling load the follwing conventions have been used: The $\tilde{x}$-direction of the local coordinate system ($\tilde{x}$, $\tilde{y}$, $\tilde{z}$) follows the direction of the larger of the two compressive principal force components, $\tilde{z} = 0$ denotes the face layer in which the maximum compressive principal force component of both face layers is found. As a consequence of the small wavelengths of the wrinkles as compared to the dimensions of a typical sandwich structure the analysis of critical wrinkling forces can be reduced to the analysis of the infinite plate. In the case of shells the influence of curvature on the local instability limit can be neglected, if the radii of shell curvature are essentially larger than the wavelengths of the local buckles. With regard to *St. Venant's* principle the influence of global boundary conditions of the structure can be neglected on analyzing wrinkling. For the same reasons axial forces in the face layers can be treated as constant in the local region of interest.

If sinusoidal wrinkling deformations of the face layers are assumed the nontrivial core deformations may be described by:

$$u(\tilde{x},\tilde{y},\tilde{z}) = \tilde{u}(\tilde{z})\cos\frac{\pi\tilde{x}}{a_{\tilde{x}}}\sin\frac{\pi\tilde{y}}{a_{\tilde{y}}} \qquad (4.4),$$

$$v(\tilde{x},\tilde{y},\tilde{z}) = \tilde{v}(\tilde{z})\sin\frac{\pi\tilde{x}}{a_{\tilde{x}}}\cos\frac{\pi\tilde{y}}{a_{\tilde{y}}} \qquad (4.5),$$

$$w(\tilde{x},\tilde{y},\tilde{z}) = \tilde{w}(\tilde{z})\sin\frac{\pi\tilde{x}}{a_{\tilde{x}}}\sin\frac{\pi\tilde{y}}{a_{\tilde{y}}} \qquad (4.6).$$

In order to characterize the wrinkling deformation pattern the following parameters are introduced:

$$\alpha = \frac{\pi}{a_{\tilde{x}}}, \qquad \beta = \frac{\pi}{a_{\tilde{y}}} \qquad (4.7).$$

The elasticity matrix $\underset{\approx}{C}_{orth}$ for orthotropic core materials (axes of orthotropy L, W, T) is generally described by:

$$\underset{\approx}{C}_{orth} = \begin{pmatrix} C_{LL} & C_{LW} & C_{LT} & 0 & 0 & 0 \\ C_{LW} & C_{WW} & C_{WT} & 0 & 0 & 0 \\ C_{LT} & C_{WT} & C_{TT} & 0 & 0 & 0 \\ 0 & 0 & 0 & 2G_{WT} & 0 & 0 \\ 0 & 0 & 0 & 0 & 2G_{LT} & 0 \\ 0 & 0 & 0 & 0 & 0 & 2G_{LW} \end{pmatrix} \qquad (4.8),$$

$C_{LL}, C_{LW}, \ldots, C_{TT}$ being defined by the direction dependent moduli of elasticity and the direction dependent Poisson ratios; see [4].

In the present analysis the local coordinate system ($\tilde{x}$,$\tilde{y}$,$\tilde{z}$) is assumed to coincide with the axes of orthotropy L, W, T. By using *Hooke's Law* the assumed wrinkling pattern in combination with the local equilibrium condition leads to the following set of coupled linear differential equations dependent on $\tilde{z}$ ($'$ defining derivatives with respect to $\tilde{z}$):

$$\underset{\approx}{\mathbf{U}}_1 \underset{\sim}{\mathrm{y}}'' - \underset{\approx}{\mathbf{U}}_2 \, \underset{\sim}{\mathrm{y}}' - \underset{\approx}{\mathbf{U}}_3 \, \underset{\sim}{\mathrm{y}} = \underset{\sim}{0} \qquad (4.9),$$

where $\underset{\sim}{\mathrm{y}}$ represents the amplitudes $\underset{\sim}{\mathrm{y}}^T = (\tilde{u}, \tilde{v}, \tilde{w})$ depending on $\tilde{z}$. The coefficient matrices $\underset{\approx}{\mathbf{U}}_i$ are:

$$\underset{\approx}{\mathbf{U}}_1 = \begin{pmatrix} G_{\tilde{x}\tilde{z}} & 0 & 0 \\ 0 & G_{\tilde{y}\tilde{z}} & 0 \\ 0 & 0 & C_{\tilde{z}\tilde{z}} \end{pmatrix} \qquad (4.10),$$

$$\underset{\approx}{\mathbf{U}}_2 = \begin{pmatrix} 0 & 0 & -\alpha(C_{\tilde{x}\tilde{z}} + G_{\tilde{x}\tilde{z}}) \\ 0 & 0 & -\beta(C_{\tilde{y}\tilde{z}} + G_{\tilde{y}\tilde{z}}) \\ \alpha(C_{\tilde{x}\tilde{z}} + G_{\tilde{x}\tilde{z}}) & \beta(C_{\tilde{y}\tilde{z}} + G_{\tilde{y}\tilde{z}}) & 0 \end{pmatrix} \qquad (4.11),$$

$$\underset{\approx}{\mathbf{U}}_3 = \begin{pmatrix} \alpha^2 C_{\tilde{x}\tilde{x}} + \beta^2 G_{\tilde{x}\tilde{y}} & \alpha\beta(C_{\tilde{x}\tilde{y}} + G_{\tilde{x}\tilde{y}}) & 0 \\ \alpha\beta(C_{\tilde{x}\tilde{y}} + G_{\tilde{x}\tilde{y}}) & \alpha^2 G_{\tilde{x}\tilde{y}} + \beta^2 C_{\tilde{y}\tilde{y}} & 0 \\ 0 & 0 & \alpha^2 G_{\tilde{x}\tilde{z}} + \beta^2 G_{\tilde{y}\tilde{z}} \end{pmatrix} \qquad (4.12).$$

The following substitution can be used for solving this set of differential equations:

$$\underset{\sim}{\mathrm{y}} = \underset{\sim}{\mu} \, e^{\varrho \tilde{z}} \qquad (4.13).$$

After inserting eqn. (4.13) into eqns. (4.10–4.12) one obtains

$$\underset{\approx}{\mathbf{U}} \, \underset{\sim}{\mu} \, e^{\varrho \tilde{z}} = \underset{\sim}{0} \qquad (4.14),$$

where $\underset{\approx}{\mathbf{U}}$ is defined by the eigenvalue problem:

$$\underset{\approx}{\mathbf{U}}\ \underset{\sim}{\mu} = \left(\underset{\approx}{\mathbf{U}}_1\ \varrho^2 - \underset{\approx}{\mathbf{U}}_2\ \varrho - \underset{\approx}{\mathbf{U}}_3 \right)\ \underset{\sim}{\mu} = \underset{\sim}{0} \qquad (4.15),$$

and $\underset{\sim}{\mu}$, ϱ are an eigenvector and the corresponding eigenvalue, respectively.

With

$$\Lambda = \varrho^2 \qquad (4.16)$$

the characteristic equation of the eigenvalue problem can be found to be

$$det\ \underset{\approx}{\mathbf{U}} = \Theta_6 \Lambda^3 + \Theta_4 \Lambda^2 + \Theta_2 \Lambda + \Theta_0 = 0 \qquad (4.17),$$

with parameters Θ_i depending on the material properties and the buckling wavelength parameters α and β. The roots of this cubic polynomial can be found analytically.

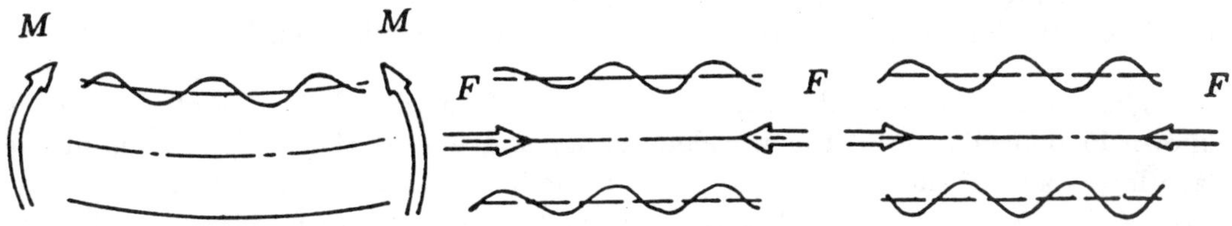

a) b) c)

Fig. 4.2 a) One-face layer wrinkling, b) antisymmetrical, c) symmetrical wrinkling

The six free coefficients in the individual solutions for $\underset{\sim}{\mathrm{y}}$ can be determined by the boundary conditions which depend on the local loading situation, i.e. bending or pure compression of the shell, and on the wrinkling mode, i.e. symmetrical or antisymmetrical wrinkling; see Fig. 4.2.

In any case fife of the six boundary conditions are:

$$u(\tilde{x},\tilde{y},\tilde{z}=0) = v(\tilde{x},\tilde{y},\tilde{z}=0) = u(\tilde{x},\tilde{y},\tilde{z}=c) = v(\tilde{x},\tilde{y},\tilde{z}=c) = 0 \qquad (4.18),$$

$$\sigma_z(\tilde{x},\tilde{y},\tilde{z}=0) = -\sigma_0 \sin\alpha\tilde{x} \sin\beta\tilde{y} \qquad (4.19).$$

The sixth boundary condition depends on the wrinkling mode; see Fig. 4.2.

In the case of pure bending (see Fig. 4.2(a)) the $\tilde{z}$-displacements are supposed to vanish completely at the opposite side of the core, because the resulting tensile forces do not induce buckling at the opposite face layer:

$$w(\tilde{x}, \tilde{y}, \tilde{z} = c) = 0 \qquad (4.20).$$

In the case of pure compressive forces the sixth boundary condition for antisymmetrical wrinkling is

$$\sigma_z(\tilde{x}, \tilde{y}, \tilde{z} = c) = \sigma_0 \sin \alpha\tilde{x} \sin \beta\tilde{y} \qquad (4.21).$$

If a symmetrical wrinkling pattern is assumed, the remaining boundary condition is determined by the condition of symmetry:

$$w(\tilde{x}, \tilde{y}, \tilde{z} = c/2) = 0 \qquad (4.22).$$

The verification of eqns. (4.18–4.22) yields to a system of 6 linear equations for the determination of the six free coefficients in the nontrivial solution $\underset{\sim}{y}$.

The vector $\underset{\sim}{y}$ is now determined and the deformation behavior of the core is known in dependence on the wavelengths $a_{\tilde{x}}$ and $a_{\tilde{y}}$. If the deformations of the face layers and of the core are supposed to be identical at the interface $\tilde{z} = 0$, i.e. $w_f(\tilde{x}, \tilde{y}, \tilde{z} = 0) = w(\tilde{x}, \tilde{y}, \tilde{z} = 0)$, corresponding to a perfect interface between the face layers and the core, the equivalent stiffness of the core C_{core} can be calculated as:

$$C_{core} = \frac{\sigma_0 \sin \dfrac{\pi\tilde{x}}{a_{\tilde{x}}} \sin \dfrac{\pi\tilde{y}}{a_{\tilde{y}}}}{\tilde{w}(\tilde{z} = 0) \sin \dfrac{\pi\tilde{x}}{a_{\tilde{x}}} \sin \dfrac{\pi\tilde{y}}{a_{\tilde{y}}}} = \frac{\sigma_0}{\tilde{w}(\tilde{z} = 0)} \qquad (4.23).$$

If the loading conditions differ from pure bending and pure compression the following approximation for C_{core} is proposed in analogy to [4.2]:

$$C_{core} = \frac{(1+\eta)\, C_{core}^{compr} + (1-\eta)\, C_{core}^{bending}}{2} \qquad (4.24),$$

where

$$\eta = \frac{P_{\tilde{x}}^{lf}}{P_{\tilde{x}}^{uf}} \qquad (4.25),$$

with $P_{\tilde{x}}^{uf}$ being the membrane force in the "upper" (i.e. $\tilde{z} = 0$) face layer relevant for wrinkling, and $P_{\tilde{x}}^{lf}$ is the membrane force in the "lower" (i.e. $\tilde{z} = c$) face layer.

Under general loading conditions the equivalent stiffness of the core depends on the parameter η and on the wavelengths $a_{\tilde{x}}$ and $a_{\tilde{y}}$.

Now the equation for the considered face layer modelled as a plate on an elastic foundation can be formulated:

$$K_f^{local} \nabla\nabla w_f + P_{\tilde{x}} \frac{\partial^2 w_f}{\partial \tilde{x}^2} + P_{\tilde{y}} \frac{\partial^2 w_f}{\partial \tilde{y}^2} + C_{core}(a_{\tilde{x}}, a_{\tilde{y}})\, w_f = 0 \qquad (4.26).$$

Since the axes $\tilde{x}$ and $\tilde{y}$ are assumed to coincide with the principal stress axes at the considered point on the face layer, the shear force term is not present in eqn. (4.26). Introduction of the definition

$$\xi = \frac{P_{\tilde{y}}}{P_{\tilde{x}}} \qquad (4.27)$$

allows the reduction of the description of biaxial loading by one parameter, and one obtains:

$$P_{\tilde{x}}^c = \frac{\left(\alpha^2 + \beta^2\right)^2 K_f^{local} + C_{core}(\alpha, \beta, \eta)}{\alpha^2 + \xi \beta^2} \qquad (4.28).$$

This equation for $P_{\tilde{x}}^c$ still contains the wavelength parameters α and β. The smallest value of $P_{\tilde{x}}^c$ defined as P_{crit}^{wrinkl} has to be determined by minimization in the parameter space $(\alpha \times \beta)$. Since the equivalent stiffness of the core C_{core} is not an explicit function of these parameters, the extremum cannot be determined by a closed formula but by a numerical procedure, see [31].

As a typical example Fig. 4.3 shows the results of the wrinkling analysis for a uniaxially loaded sandwich with $E_f = 70000 N/mm^2, E_c = 10 N/mm^2, \nu_f = \nu_c = 0.3, t_f = 0.1 mm$, the core thickness c is varied. For comparison reasons in this figure also the results obtained by approximations according to [33] are shown.

b) Intracell Buckling of the Face Layers

In honeycomb cores a further local instability phenomenon can occur: intracell buckling of the face layers. Supported by the cell walls the face layers can show a pattern of buckles (dimples) in the size of the honeycomb cells. Since the out-of-plane

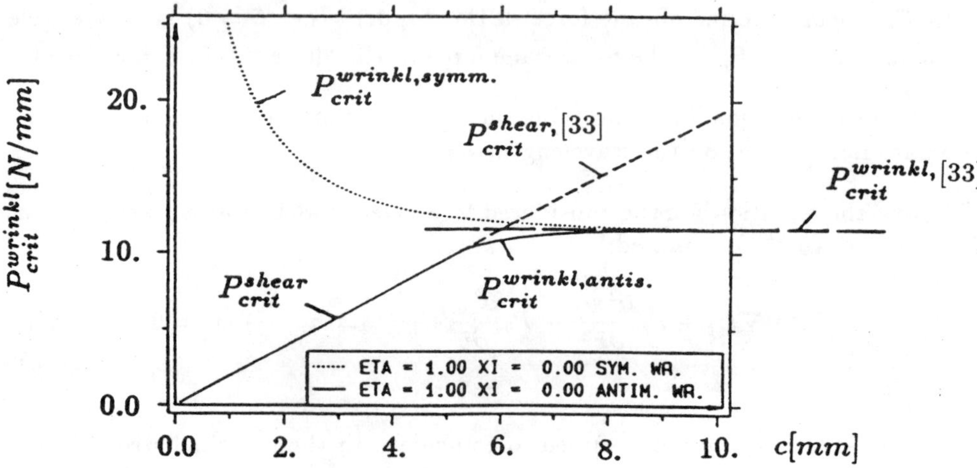

Fig. 4.3 P^{wrinkl}_{crit} over core height c

buckling deformations have to vanish at the cell walls, the analysis can be confined to the hexagonal area described by the supporting cell walls.

Stamm and Witte [32] investigated the phenomenon of intracell buckling for honeycomb cores with a nonregular hexagonal shape of the cells under biaxial loading conditions: In order to include the various forms of hexagons used in practical design, the following cell size parameters are introduced (see Fig. 4.4) and characterized by γ_{icb}:

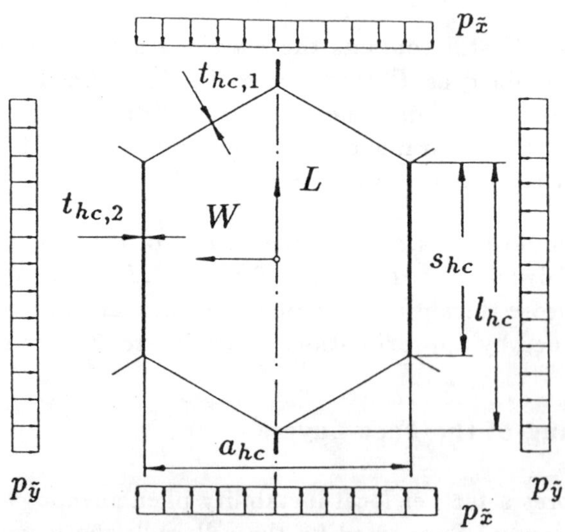

Fig. 4.4 Cell size parameters for nonregular hexagonal honeycomb cells

$$\gamma_{icb} = \frac{a_{hc}}{l_{hc}} \tag{4.29}$$

Assuming a local buckling pattern

$$w_f = \sin \frac{\pi \, x \, m}{l_{hc}} \, \sin \frac{\pi \, y \, n}{a_{hc}} \tag{4.30}$$

the following equation can be obtained for the critical force:

$$P_{crit}^{icb} = \Gamma_{icb} \, \frac{K_f^{local}}{a_{hc}^2} \tag{4.31}$$

where

$$\Gamma_{icb} = \pi^2 \, \frac{\left(\gamma_{icb}^2 \, m^2 + n^2\right)^2}{\gamma_{icb}^2 \, m^2 + \xi^p \, n^2} \qquad m, n = 1, 2, \ldots$$

$$\xi^p = \frac{p_{\bar{y}}}{p_{\bar{x}}} \tag{4.32}$$

Since eqn. (4.32) still contains the buckling wavenumbers (m, n), the critical value has to be determined by minimization with respect to these parameters.

c) Postcritical Analysis

The values of the principal forces and their directions can be calculated for each face layer. After the evaluation of the critical loads, P_{crit}^{buckl}, which are dependent on the deformation state m, the safety against local buckling is checked. If local buckling occurs, the stiffness matrix of the locally buckled shell element has to be updated. The postbuckling behavior of the wrinkled face layer is approximated by the assumption of a constant post-critical membrane compression force perpendicular to the wrinkles.

As in the case of local buckling the tangential membrane rigidity of the face layer is reduced, the elasticity parameters have to be adapted. Thus, in effect, a nonlinear material law is introduced for the shell element. By setting - as an approximation based on the stable post-buckling behavior of plates - the tangential Young's modulus of elasticity of the facing material E_f^ℓ to zero in the direction ℓ, where local buckling occurred, the material matrix can be updated. The shear terms in the material matrix are assumed not to be influenced by local buckling.

In the case of local buckling in direction ℓ the updated tangential material matrix ${}^m\breve{\underset{\approx}{C}}_f$ for the buckled face layer takes the form

$$\breve{\underset{\approx}{C}}{}^{\ell}_f = \begin{pmatrix} 0 & 0 & 0 & 0 & 0 & 0 \\ 0 & \frac{E_f}{1-\nu_f^2} & 0 & 0 & 0 & 0 \\ 0 & 0 & 0 & 0 & 0 & 0 \\ 0 & 0 & 0 & G_{f,LW} & 0 & 0 \\ 0 & 0 & 0 & 0 & 0 & 0 \\ 0 & 0 & 0 & 0 & 0 & 0 \end{pmatrix} \qquad (4.33).$$

Since this matrix is defined in a local coordinate system of the face layer with the main axis ℓ pointing into the direction of the local buckling load (i.e. the extension of wrinkles is perpendicular to this direction ℓ), a transformation into the shell's local coordinate system x'_1, x'_2, x'_3 is necessary.

In the case of biaxial local buckling in one face layer (i.e. buckling in both principal directions, which is only possible for pure biaxial compression: $\xi = 1.0$) all tangential stiffness material parameters are set to zero, except the shear modulus $G_{f,LW}$.

The updated tangential material matrix $\underset{\approx}{C}'$ for an integration point, at which local buckling occurred, then takes the form

$$\breve{\underset{\approx}{C}}'(r,s,t) = \begin{cases} {}^1\breve{\underset{\approx}{C}}'(r,s) & {}^0t \le t \le {}^1t \\ {}^2\underset{\approx}{C}'(r,s) & {}^1t < t \le {}^2t \\ {}^3\underset{\approx}{C}'(r,s) & {}^2t < t \le {}^3t \\ {}^4\breve{\underset{\approx}{C}}'(r,s) & {}^3t < t \le 1 \end{cases} \qquad (4.34).$$

By this approach stress redistribution during progressive wrinkling can be approximated. A more detailled consideration of this issue can be found in [34], where the stiffness loss due to core thickness reduction caused by wrinkling is taken into account, too.

4.3 Accounting for Thickness Variation Induced by Loading

If the shell is curved (by its original geometry or due to loading) bending moments lead - as a higher order effect - to normal stresses acting in the core perpendicularly to the shell's midsurface, see Fig. 4.5, [4,38].

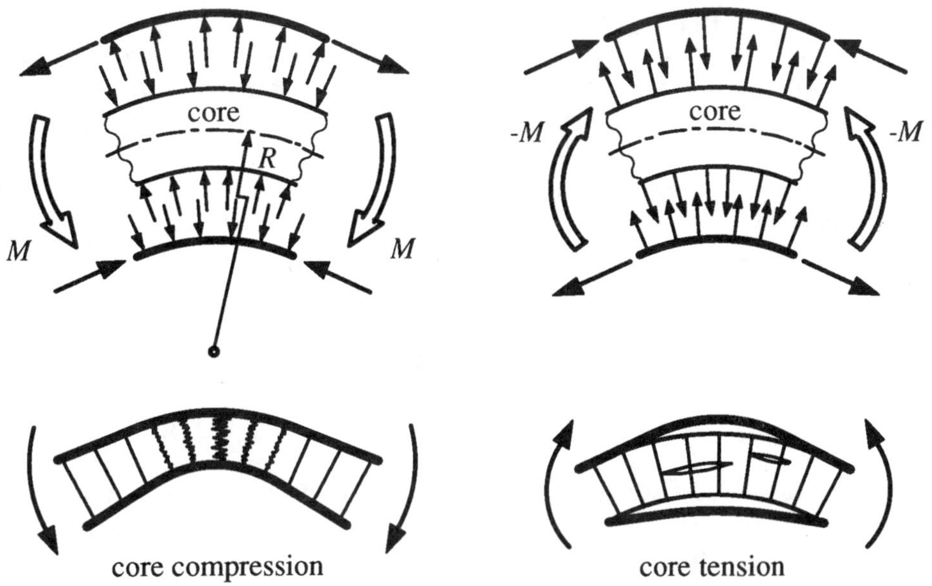

Fig. 4.5 Development of transverse normal stresses

Due to the limited resistance of the sandwich core to deformations perpendicular to the face layers transverse tensile stresses may lead to a thickness increase, to tensile cracking of the core or to debonding between face layers and core, and transverse compressive stresses may cause non-negligible local thickness reduction resulting in a loss in stiffness or buckling of core cell walls. This local phenomenon is similar to the *Brazier effect*. In order to investigate the influence of this effect in sandwich shells, the strains in the direction of the shell's normal have to be evaluated:

With M being the global bending moment of the shell and R the radius of shell curvature in the deformed state one obtaines the normal stress in the core in the direction of the current shell's normal

$$\sigma_n = -\frac{M}{R\ (c + t_f)} \qquad (4.35).$$

Since there exist two principal curvatures in each shell point, the resulting stress in the shell's normal direction can be approximated by their sum

$$\sigma_n^{bend} = \sigma_{n,1} + \sigma_{n,2} \qquad (4.36).$$

Thus, these normal stresses in the core can be evaluated at any point in the shell. As the determination of the stresses σ_n^{bend} requires the knowledge of the current bending moments acting on the shell, the thickness of the core is altered by an *a-posteriori*-procedure after each iteration. The evaluation of the radii of curvature has to take into account the effect of initial curvature of a structure as well as the influence of deformations induced by loading.

In the case of a sandwich structure loaded by external pressure additional normal stresses appear in the core. Due to the constant shear stresses in the core these normal stresses decrease linearly from the loaded face layer ($\hat{z} = 0$) to the unloaded one ($\hat{z} = c$):

$$^{m}\sigma_{z}^{pr} = {}^{m}p\left(\frac{\hat{z}}{c} - 1\right) \qquad (4.37).$$

The compressive core stress in the direction of the shell's normal is determined by superposition of the contributions due to bending and pressure loading. The updated core thickness results from

$$^{m+1}c = (1 + {}^{m}\epsilon_n)\ {}^{0}c \qquad (4.38),$$

with

$$^{m}\epsilon_n = \frac{^{m}\sigma_n^{bend} + {}^{m}\sigma_z^{pr,ave}}{C_{TT}} \qquad (4.39),$$

where $^{m}\sigma_z^{pr,ave} = -{}^{m}p/2$.

The thickness variation induced by loading has then to be taken into account in the next state $m+1$ to be considered. Thus, the update of the core thickness c lags behind the other kinematical quantities.

4.4 Numerical Examples

The following examples show the applicability of the above discussed approaches, see also [4,6,38].

a) Stability of a Clamped Beam

The analysis of the stability behavior of a clamped sandwich beam under compression shows the applicability of the algorithms derived in the present work to the case of global as well as local instabilities. Since there exist formulas for the analytical approximative determination of the global buckling load accounting for the shear effects, the results of the numerical analysis can be checked, provided the sandwich parameters fulfill the conditions assumed for the derivation of the analytical method.

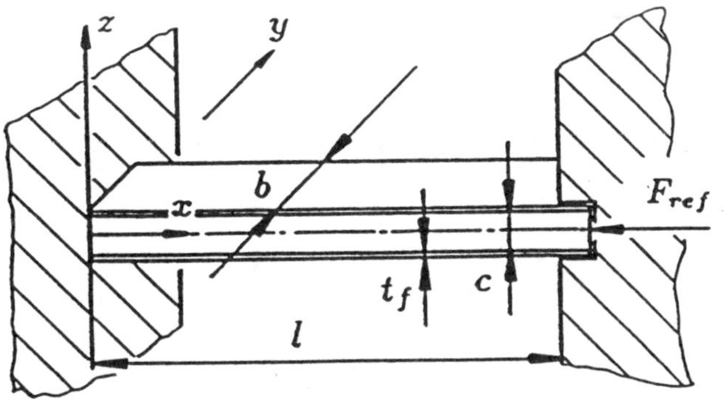

$l = 500\ mm$

$b = 60\ mm$

$c = 30\ mm$

$t_f = 0.5\ mm$

$E_f = 70kN/mm^2$

E_c: to be varied

$\nu_f = 0.3$

$F_{ref} = 120\ N$

$G_c = 0.385E_c$

Fig. 4.6 Clamped sandwich beam

The sandwich beam modelled for the FE analysis is shown schematically in Fig. 4.6.

In order to analyze the influence of an isotropic low-density core on the stability behavior, the Young's modulus of the core is varied in a series of computations. In Fig. 4.7 the critical loads corresponding to the individual kinds of stability loss are plottet in dependence on the ratio of the Young's moduli E_c/E_f.

Figure 4.7 shows how increased core stiffness, i.e. increased E_c, leads to transitions from shear buckling of the core (point E to D) to antisymmetrical wrinkling of the face layers (D to C), global buckling influenced by the small shear stiffness of the core (C to B) and, finally, to pure Euler buckling (B to A). The good correspondence with some approximative analytical formulas taken from [33] becomes obvious.

b) Wrinkling in a Square Sandwich Plate with a Central Circular Hole

In this example the influence of local instabilities on the deformation behavior of a square sandwich plate with a central circular hole is investigated computationally and the results are compared to laboratory experiments [35]. The plate is uniaxially loaded by compressive in-plane loads.

Due to the central cylindrical hole a stress concentration is induced near the hole, where the local instability will occur first. The plate configuration is shown schematically in Fig. 4.8.

The material for the test specimens and the material data have been provided by AIREX AG., Sims, Switzerland. The specimens for the particular example were

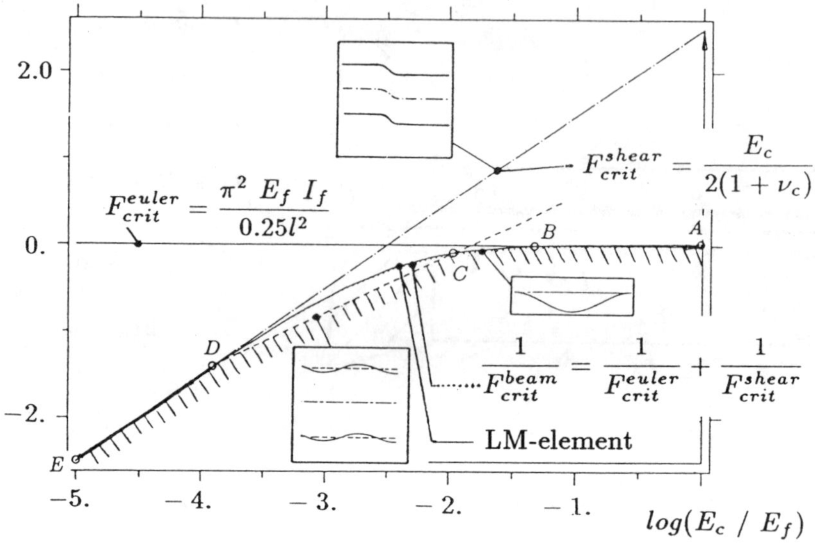

Fig. 4.7 Critical load ratio $F_{crit}/F^{euler}_{crit}$ in dependence on E_c/E_f

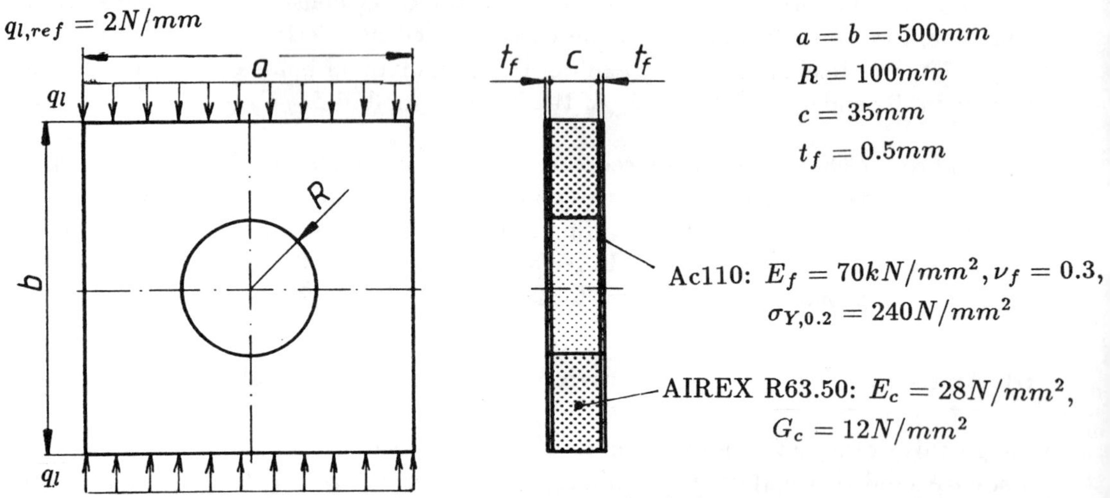

Fig. 4.8 Sandwich plate with a central circular hole

produced from sandwich plates with material properties as shown in Fig. 4.8[1].

---

[1] A series of tests performed at the Institute of Lightweight Structures and Aerospace Engineering at the Vienna Technical University leads to the presumption that the real G_c-value is approximately 80% of that one which is presented here and on which the computations are based, see [35].

Due to the double symmetry conditions only one quarter of the sandwich shell is modelled; 36 16-noded LM sandwich shell elements are used.The computed load-displacement developments for the edge at which the compressive loads are applied are plotted in Fig. 4.9. Local wrinkling is computed to appear at first at a load level $^*\lambda = 35$, corresponding to a critical distributed load $^*q_l = 70N/mm$. Of course, wrinkling leads to a considerable stiffness loss as can be seen from the behavior of the load displacement path. The collapse load multiplier is computed as $\lambda^{coll} = 67$.

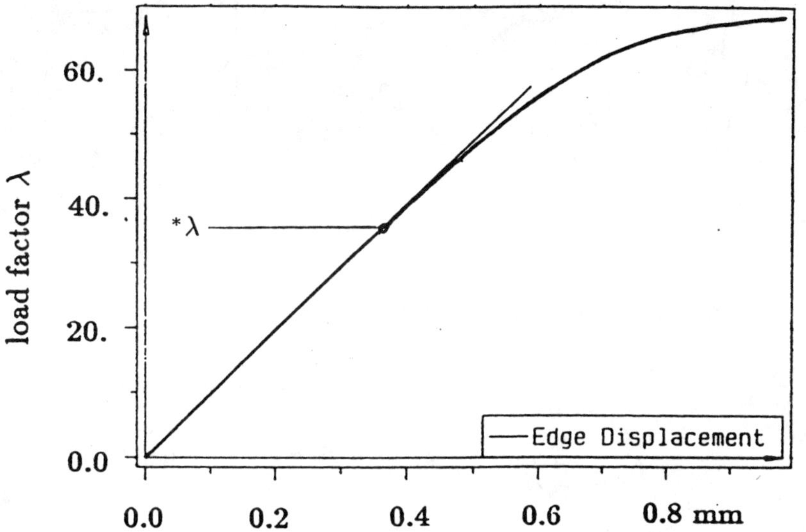

Fig. 4.9 Load-displacement development for a square sandwich plate with a central cylindrical hole

In Fig. 4.10 the development and growth of the wrinkled zones during increase of the compressive load is shown.

Wrinkling starts at the point of maximum face layer compressive stress, i.e. at the apex of the hole (Fig. 4.10a), and this wrinkled zone grows towards the outer edge (Fig. 4.10b). Immediately before the collapse load is reached wrinkling appears at the corners, too. (Fig. 4.10c).

For the above material data Fig. 4.11 shows the dependence of the compressive face layer membrane force being critical with respect to local buckling and the corresponding buckling mechanisms as functions of the core thickness. From this figure one can conclude that for the core thickness of the considered specimen (c = 35 mm) a distinction between symmetrical and antisymmetrical buckling of the face layers hardly can be made.

All the above described features were varified experimentally. Figure 4.12 shows

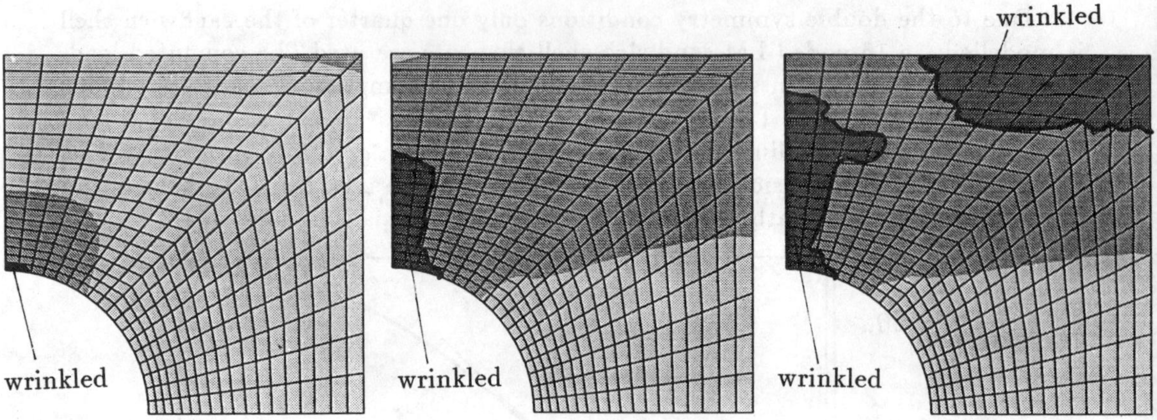

a) $\lambda = 35$ b) $\lambda = 60$ c) $\lambda = 67$

Fig. 4.10 Development of wrinkled zones

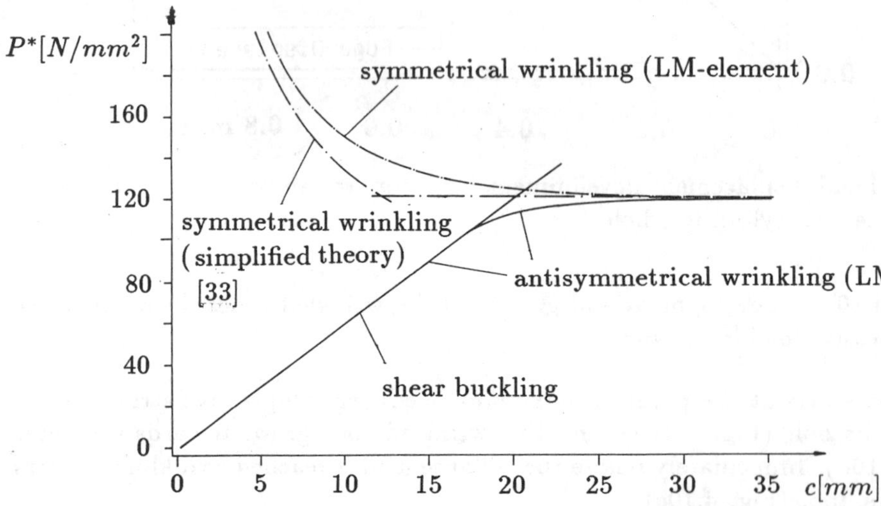

Fig. 4.11 Buckling mode and buckling load as functions of core thickness

the experimental set up including the specimen after collapse. Collapse is reached at a load of 51 kN which corresponds to $\lambda_{exp}^{coll} \approx 51$ being lower than the computed value. As mentioned above (see footnote 2) the foam stiffness must be expected to

Fig. 4.12 The experimental set up

be approximately 80% of the given value. Under this assumption the correspondence between precalculated and experimentally determined collapse load is reasonably good. Of course, the wide spread wrinkled area, compare Fig. 4.10, cannot be expected to be visible. The reason for this is the fact that after the central local buckle has approached the outer edge its depth increased rapidly forming a single plastic fold while the other elastic wrinkles were unloaded and disappeared.

Figure 4.13 shows this post-collapse fold at the apex of the hole. As computed and mentioned above, no distinction can be given whether symmetrical or antisymmetrical wrinkling appeared. A thinner plate with $c_f = 15$ mm and the same material data showed typical antisymmetrical buckling, see Fig. 4.14, as also theoretically expected (compare Fig. 4.11).

Figure 4.15 shows the edge of the 35 mm thick specimen with the central hole after collapse. In correspondence with Fig. 4.10c also the wrinkle at the corner could be verified experimentally.

c) Stability Analysis of a Cylindrical Sandwich Shell

In this example the stability behavior of a cylindrical sandwich shell under external pressure is analyzed. As the core consists of honeycomb material, local instability

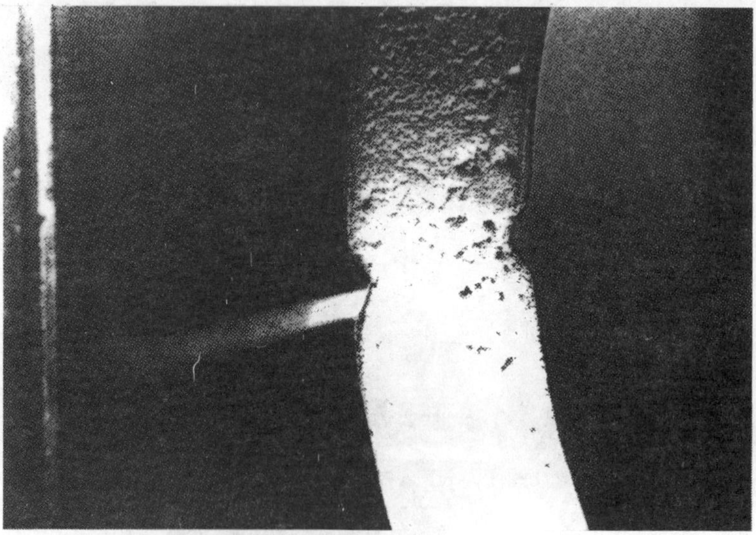

Fig. 4.13 Plastic post-collapse deformation

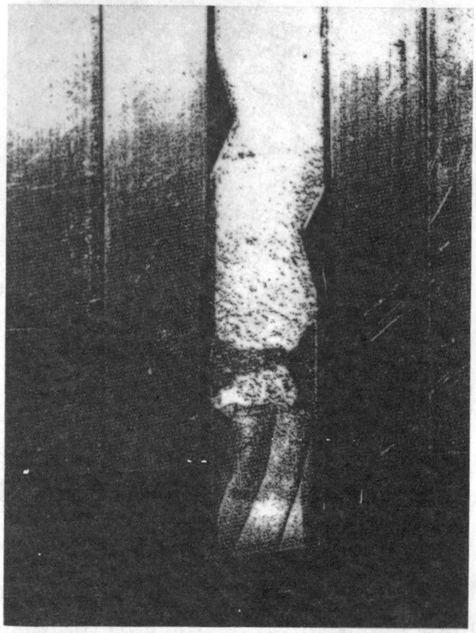

Fig. 4.14 Shear buckling of a specimen with $c_f = 15$ mm

phenomena including intracell buckling are taken into account.

The geometrical and the material parameters of the shell are shown in Fig. 4.16. The major orthotropy axis L of the honeycomb material is assumed to be parallel to

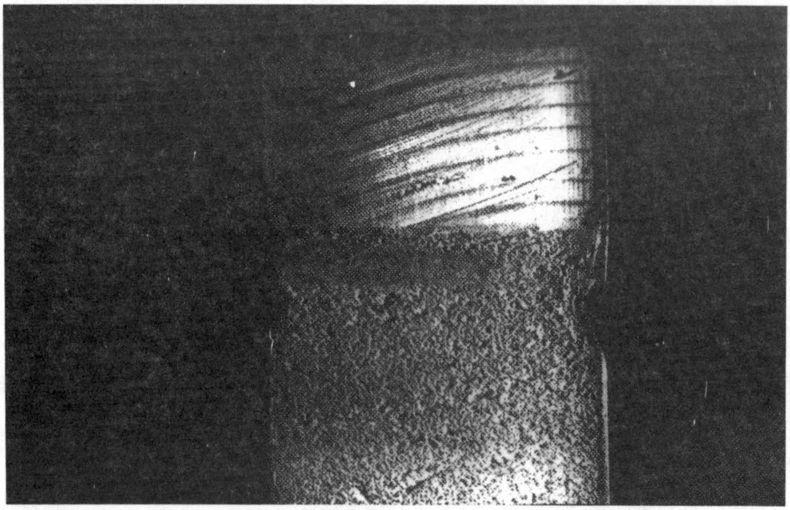

Fig. 4.15 Plastic post-collapse fold at the edge of the plate

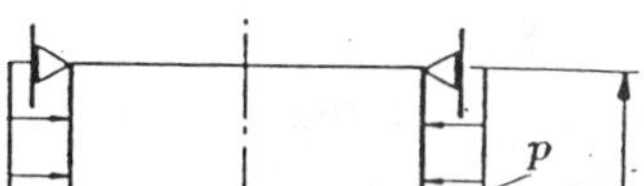

$p_{ref} = 0.002\ N/mm^2$

$R = 2000\ mm \quad t_f = 0.2\ mm$

$L = 4000\ mm \quad t_{hc,min} = 0.2\ mm$

$c = 35.6\ mm \quad a_{hc} = s_{hc} = 9.525\ mm$

$l_{hc} = 14.29\ mm$

$E_f = 70000\ N/mm^2$

$\nu_f = 0.3$

$E_{c,L} = E_{c,W} = 1\ N/mm^2$

$E_{c,T} = 69\ N/mm^2$

$\nu_{c,LW} = \nu_{c,LT} = \nu_{c,WT} = 0.3$

$G_{c,LT} = 82.7\ N/mm^2$

$G_{c,WT} = 69.0\ N/mm^2$

$G_{c,LW} = 1.0\ N/mm^2$

$E_c^{orig} = 70000\ N/mm^2$

$\nu_c^{orig} = 0.3$

Fig. 4.16 Cylindrical Sandwich Shell

the axis of revolution.

In the nonlinear finite element analyses only one quarter of the cylindrical sandwich shell is modelled; 30 LM-elements are used. First, no local phenomena are taken into consideration in the nonlinear analyses, while the pressure is continuously increased.

According to the method of accompanying eigenvalue analyses the critical buckling factors and modes are determined for several load steps. The corresponding crit-

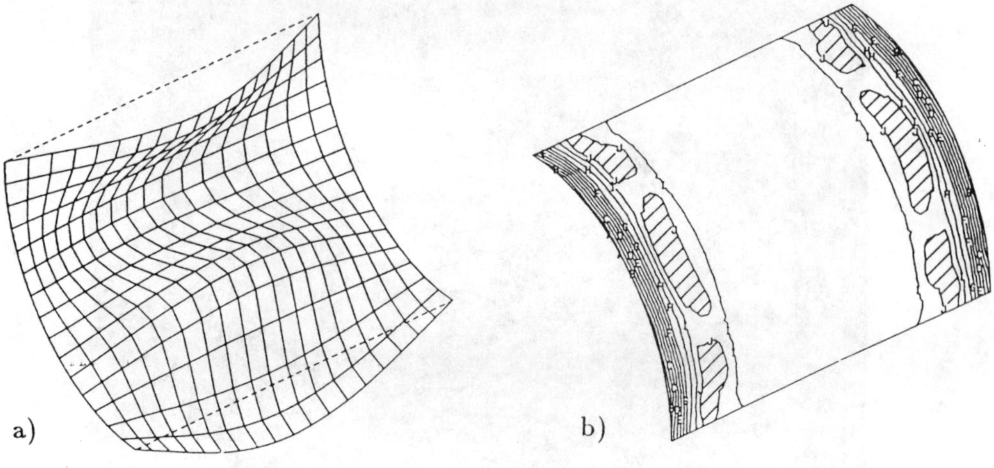

Fig. 4.17 a) Global Buckling Mode, b) Intracell Buckled Areas

ical buckling pressures, however, remain constant and are: $p_{crit}^{FE} = 0.05556 N/mm^2$. Figure 4.17(a) shows the buckling mode of the quarter cylinder.

If local buckling phenomena are taken into account, intracell buckling occurs before the buckling load of the previous analysis is reached. In Fig. 4.17(b) the locally buckled regions for a local postbuckling state are marked by hatching. The local intracell buckling is initiated in the region of the maximum bending moments. The load inducing local buckling cannot be surpassed, since the stiffness reduction is performed all over the circumference of the model.

5. LAYERED METAL MATRIX COMPOSITE SHELLS

In the above Chapters the nonlinearities in the composite material behavior were either due to progressive damage (FRP shells) or due to local instabilities and thickness reductions in the case of sandwich shells. If the shells are made of layers of fibre reinforced metals (typical MMCs) then plastification of the metallic matrix must be taken into account. In [3] the corresponding modifications and extensions to the LFC-element (see Chapter 2) are described in detail.

Based on micromechanical considerations a macromechanical description of the anisotropic elastic-plastic material behavior for each layer can be formulated. For example, the vanishing fibre diameter (VFD) approach developed by *Dvorak and*

Bahei-El-Din [36] uses stress or strain concentration factor matrices which are functions of the current stress and strain states to calculate the local stresses or strains in the metallic matrix of the composite. On this basis the tangential material matrix at each integration point can be adopted during the incremental-iterative analysis. A finite element implementation of the VFD-model can be found in [37]. A specialization of this approach based on assumptions as described in Chapter 2 leads to the above mentioned D3MMC-shell element [3].

Acknowledgements

The developments and results described here are partially derived in the course of research projects financially supported by the Federal Republic of Austria (Ministry of Science and Research), the Fonds zur Förderung der wissenschaftlichen Forschung (by Schrödinger Stipendium) and the Christian Doppler Gesellschaft of the Austrian Industries.

References

1. Dorninger, K.: Entwicklung von nichtlinearen FE-Algorithmen zur Berechnung von Schalenkonstruktionen aus Faserverbundstoffen, Doctoral Thesis, Vienna Technical University, Austria; Fortschritt-Berichte, VDI Reihe 18, Nr. 65, VDI Verlag, Düsseldorf, FRG, 1989.

2. Dorninger, K.: A Nonlinear Layered Shell Finite Element With Improved Transverse Shear Behavior, Composites Engineering **1**(1991), 211-224.

3. Svobodnik, A.J.: Numerical Treatment of Elastic-Platic Macromechanical Behavior of Longfiber-Reinforced Metal Matrix Composites, Doctoral Thesis, Vienna Technical University, Austria; Fortschritt-Berichte, VDI Reihe 18, Nr. 90, VDI Verlag, Düsseldorf, FRG, 1990.

4. Starlinger, A.: Development of Efficient Finite Shell Elements for the Analysis of Sandwich Structures Under Large Deformations and Global as Well as Local Instabilities, Doctoral Thesis, Vienna Technical University, Austria, 1990; Fortschritt-Berichte, VDI Reihe 18, Nr. 93, VDI Verlag, Düsseldorf, FRG, 1991.

5. CARINA - Computer **A**ided **R**esearch **In** **N**onlinear **A**nalysis; a Finite Element Program of the Institute of Lightweight Structures and Aerospace Engineering, Vienna Technical University, Vienna, Austria.

6. Rammerstorfer, F.G., K.Dorninger and A.Starlinger: Nonlinear Finite Element Analysis of Composite and Sandwich Shell Structures, in: Nonlinear Computational Mechanics - State of the Art (Ed. P. Wriggers and W. Wagner), Springer-Verlag, Berlin, Heidelberg, 1991.

7. Ramm, E.: A Plate/Shell Element for Large Deflections and Rotations, in: Formulations and Computational Algorithms in Finite Element Analysis (Ed. K.J. Bathe, J.T. Owen, W. Wunderlich), Proc. U.S.-German Symp., MIT, Cambridge, 1977.

8. Andelfinger U. and E.Ramm: An Assessment of Hybrid-Mixed Four-Node Shell Elements, in this book.

9. Dvorkin, E.N.: On Nonlinear Analysis of Shells Using Finite Elements Based on Mixed Interpolation of Tensorial Components, in this book.

10. Dorninger, K. and F.G. Rammerstorfer: A Layered Composite Shell Element for Elastic and Thermoelastic Stress and Stability Analysis at Large Deformations, Int. J. Num. Meths. Eng. **30**(1990), 833–858.

11. Anderson, R.S.: The Mechanical Properties of Fibre Reinforced Composite Plates, Doctoral Thesis, University of Aston, UK, 1977.

12. Niederstadt G., J.Block, B.Geier, K.Rohwer and R.Weiß: Leichtbau mit kohlenstoffaserverstärkten Kunststoffen, Expert Verlag, Sindelfingen, 1985.

13. Huang, C.-M. and D.C. Lagoudas: A Fortran Program for Effective Properties of Composite Materials Based on the Mori-Tanaka Scheme, Department of Mechanical Engineering and Mechanics and Department of Civil Engineering, Rensselaer Polytechnic Institute, Troy, NY, 1990.

14. Noor, A.K. and J.M. Peters: A Posteriori Estimates for Shear Correction Factors in Multilayered Composite Cylinders, J. Eng. Mechs. **115**(1989), 1225–1244.

15. Stanley, G.M.: Continuum-Based Shell Elements, Doctoral Thesis, Stanford University, Stanford, CA, 1985.

16. Ramm, E. and A. Matzenmiller: Large Deformation Shell Analysis Based on the Degeneration Concept, in: State-of-the-Art Texts on FEM for Plate and Shell Structures (Eds. T.J.R. Hughes and E. Hinton), Pineridge Press, Swansea, UK, 1986.

17. Laschet, G., J.P. Jeusette and P. Beckers: Homogenization and Pre-Integration Techniques for Multilayer Composites and Sandwich Finite Element Models, Int. J. Num. Meths. Eng. **27**(1989), 257–269.

18. Chawla, K.K.: Composite Materials, Springer Verlag, New York, NY, 1987.

19. Rammerstorfer, F.G.: Jump Phenomena Associated with the Stability of Geometrically Nonlinear Structures, in: Recent Advances in Non-Linear Computational Mechanics (Eds. E. Hinton, D.R.J. Owen and C. Taylor), Pineridge Press, Swansea, UK, 1982.

20. Ziegler, F. and F.G. Rammerstorfer: Thermoelastic Stability, in: Thermal Stresses III (Ed. R.B. Hetnarski), North-Holland, Amsterdam, 1989.

21. Ramm, E.: Geometrisch nichtlineare Elastostatik und finite Elemente, Habilitationsschrift, University of Stuttgart, FRG, 1975.

22. Wagner W.: Nonlinear Stability Analysis of Shells with the Finite Element Method, in this book.

23. Lehar, H.: Beitrag zur numerischen Behandlung ebener, anisotroper Schichtverbunde mittels Finite Elemente Methoden, Doctoral Thesis, University of Innsbruck, Austria, 1984.

24. Nemeth, M.P.: Importance of Anisotropy on Buckling of Compression-Loaded Symmetric Composite Plates, AIAA Journal **24**(1986), 1831–1835.

25. Tauchert, T.R.: Thermal Stresses in Plates – Statical Problems, in: Thermal Stresses I (Ed. R.B. Hetnarski), North-Holland, Amsterdam, 1986.

26. Hui, D.: Effects of Shear Loads on Vibration and Buckling of Antisymmetric Cross-Ply Cylindrical Panels, Int. J. Non-Linear Mechs. **23**(1988), 177-187.

27. Jing, H.-S. and M.-L. Liao: Partial Hybrid Stress Element for the Analysis of Thick Laminated Composite Plates, Int. J. Num. Meths. Eng. **28**(1989), 2813–2827.

28. Li, Z.H. and D.R.J. Owen: Elastic-Plastic Analysis of Laminated Anisotropic Shells by a Refined Finite Element Laminated Model, Computers & Structures **32**(1989), 1005–1024.

29. Reddy, J.N.: On Refined Computational Models of Composite Laminates, Int. J. Num. Meths. Eng. **27**(1989), 361–382.

30. Pagano, N.J.: Exact Solutions for the Rectangular Bidirection Composites and Sandwich Plates, J. Appl. Mech. **51**(1970), 20–34.

31. Starlinger, A. and F.G. Rammerstorfer: Berücksichtigung von lokalem kurzwelligem Deckschichtbeulen in speziellen finiten Sandwich-Schlenelementen, ZAMM

71(1991),T619-623.

32. Stamm K. and H. Witte: Sandwichkonstruktionen - Berechnung, Fertigung, Ausführung, Springer-Verlag, Wien, 1974.

33. Wiedemann J.: Leichtbau: Band 1 - Elemente, Springer-Verlag, Berlin, 1986.

34. Kühhorn, A.: Geometrisch nichtlineare Theorie für Sandwichschalen - Knitterphänomen, VDI-Verlag, Düsseldorf, FRG, 1991.

35. Pehn,W.: Entwicklung eines Versuchsaufbaus und experimentelle Untersuchungen zum lokalen Beulverhalten von Sandwichplatten, Heft ILFB-3/91 der Berichte aus dem Institut für Leichtbau und Flugzeugbau der TU Wien, Vienna Technical University, Vienna, 1991.

36. Dvorak, G.J. and Y.A. Bahei-El-Din: Plasticity Analysis of Fibrous Composites, J.Appl.Mech. **49**(1982), 327–335.

37. Svobodnik, A.J., H.J.Böhm and F.G.Rammerstorfer: A 3/D Finite Element Approach for Metal Matrix Composites Based on Micromechanical Models, Int. J. Plasticity **7**(1991), 781–802.

38. Starlinger A. and F.G.Rammerstorfer: A Finite Element Formulation for Sandwich Shells Accounting for Local Failure Phenomena, in: Proc. Second Int. Conf. on Sandwich Construction, March 9-12, 1992, Univ. of Florida, EMAS, Warley, UK, Vol.1.

NON LINEAR FINITE ELEMENT ANALYSIS OF CONCRETE SHELLS

E. Oñate
Polytechnic University of Cataluña, Barcelona, Spain

ABSTRACT

This chapter deals with the non linear analysis of concrete shells using the finite element method. The finite element formulation is based on small displacements Reissner-Mindlin facet shell theory. Shear locking in dealt with by using an assumed shear strain approach. A layered model is used to take into account material non linearities in the plain concrete and the reinforcing steel over the shell thickness. A constitutive model for concrete based on plastic damage theory including stiffness degradation effects is presented. Details of the general non linear finite element solution are also given. A full section is devoted to the treatment of beam stiffeners using simple two noded layered Timoshenko beam elements and an example of this formulation to the analysis of a slab-beam bridge is also presented. The last section includes a number of examples of application of the finite element method to the non linear analysis of different plain and reinforced concrete shell-type structures like a deep beam, a slab-beam bridge, a cylindrical shell and a cryogenic concrete tank for storage of liquid gas.

1. INTRODUCTION

Concrete shell structures are very common in civil engineering practice. Typical examples are reinforced and pre–stressed slabs, shell roofs, water tanks, nuclear reactors cylindrical walls and buildings, bridges, etc. Some examples of these structures are shown in Figure 1. The importance of all these structures requires an adequate design based on an accurate evaluation of the structural response both at service and ultimate loading levels.

The analysis and design of reinforced and prestressed concrete structures has been based on simple equilibrium conditions and empirical rules for almost a century. The traditional methods generally result in safe designs, but they frecuently contain inherent inconsistencies and often do not reflect a clear understanding of the actual composite action of the material. Present–day design codes continue, in many respects, to be based on empirical approaches and rely heavily on the results of a considerable amount of experimental data. This situation is largely attributable to the complex behaviour of reinforced concrete components and structures. Concrete cracking, tension stiffening, nonlinear multiaxial material properties and complex interface behaviours were previously ignored or treated in a very approximate manner. Numerical methods, and particularly the finite element technique, now permits a more rational analysis of these complexities.

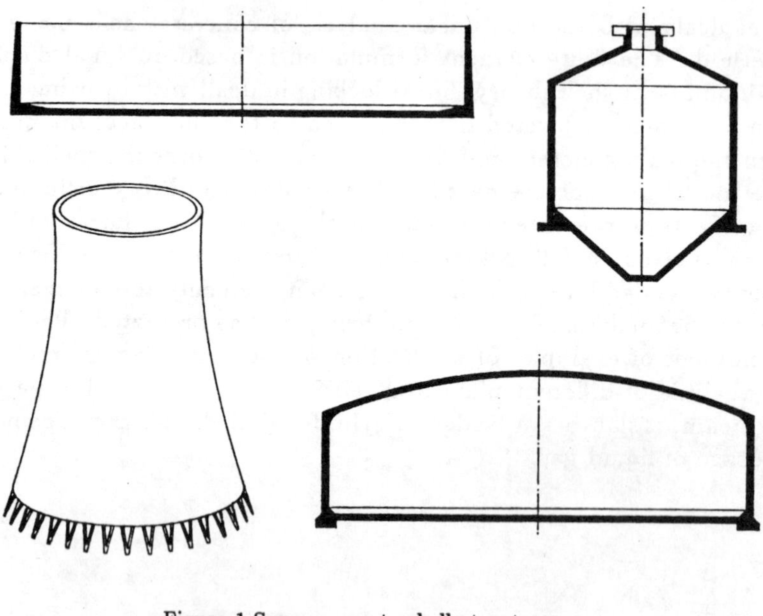

Figure 1 Some concrete shell structures.

The last two decades have witnessed rapid advances in the use of finite element methods for the analysis of reinforced concrete structures as reported in several comprehensive review articles [1–5]. Some numerical approaches have been developed to mainly study local behaviours, such as bond effects craking, interface shear, and dowel action [6–9], while other numerical studies have been directed at the analysis and design of components and structures [10–13]. The primary objectives in the latter case are the accurate prediction of the overall deformation characteristics and limit loads. A layered approach is generally employed to simulate steel reinforcement with the crack effects being assumed to be distributed (smeared) within each concrete layer. Full bond is assumed at the steel–concrete interfaces.

In this chapter we present the basis of the finite element formulation for the analysis of reinforced concrete shells using layered shell theory and facet shell elements. Only small displacements will be considered.

The layout of the chapter is the following. In next section a description of the finite element formulation using layered facet shell elements is presented. Next, the constitutive model for concrete based on plastic damage theory, is described, together with details of the general non linear solutions process. The treatment of beam stiffners is presented next. Finally, some examples of applications of non linear analaysis of several concrete structures is presented.

2. FORMULATION OF FACET SHELL ELEMENTS FOR ANALYSIS OF REINFORCED CONCRETE SHELLS

2.1 Basic theory

The simplest approach for deriving finite elements for shell analysis is to assume that the mid–surface of each shell element is flat. It is well known that in this case membrane and bending kinematics are uncoupled at element level. We will assume the standard hypothesis of Reissner-Mindlin thick plate theory [22], [23], i.e. the normals to the element mid–surface before deformation remain straight but not necessarly normal to the mid–surface after deformation. Therefore the displacement field can be written as (Figure 2)

$$\begin{aligned} u'(x',y',z') &= u'_o(x',y') - z'\theta_{x'}(x',y') \\ v'(x',y',z') &= v'_o(x',y') - z'\theta_{y'}(x',y') \\ w'(x',y',z') &= w'_o(x',y') \end{aligned} \tag{1}$$

The local displacement vector is then defined as

$$\mathbf{u}' = \left[u'_o, v'_o, w'_o, \theta_{x'}, \theta_{y'}\right]^T \tag{2}$$

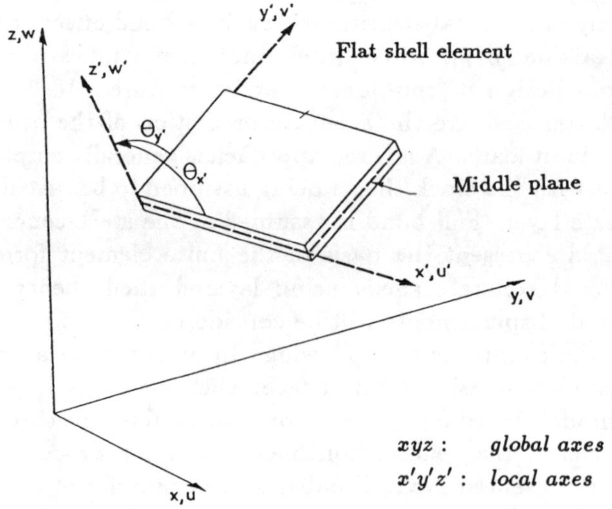

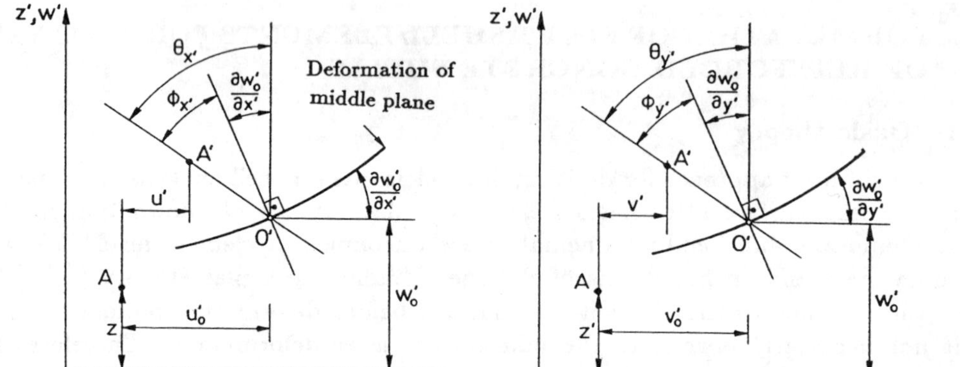

Figure 2 Definition of local displacements and rotations in flat shell theory.

where u'_o, v'_o are the in–plane (membrane) displacements and $w'_o, \theta_{x'}$ and $\theta_{y'}$ are the transverse displacements and local rotations of the normal (flexural displacements)

Assuming now the standard plane stress hypothesis ($\sigma_{z'} = 0$) allows to elliminate the thickness strain in local axes. The local strain vector is thus obtained using (1) as

$$\boldsymbol{\varepsilon}' = \left\{ \begin{array}{c} \varepsilon_{x'} \\ \varepsilon_{y'} \\ \gamma_{x'y'} \\ \cdots\cdots \\ \gamma_{x'z'} \\ \gamma_{y'z'} \end{array} \right\} = \left\{ \begin{array}{c} \frac{\partial u'}{\partial x'} \\ \frac{\partial v'}{\partial y'} \\ \frac{\partial u'}{\partial y'} + \frac{\partial v'}{\partial x'} \\ \cdots\cdots \\ \frac{\partial u'}{\partial z'} + \frac{\partial w'}{\partial x'} \\ \frac{\partial v'}{\partial z'} + \frac{\partial w'}{\partial y'} \end{array} \right\} = \left\{ \begin{array}{c} \frac{\partial u'_o}{\partial x'} \\ \frac{\partial v'_o}{\partial y'} \\ \frac{\partial u'_o}{\partial y'} + \frac{\partial v'_o}{\partial x'} \\ \cdots\cdots\cdots \\ 0 \\ 0 \end{array} \right\} + \left\{ \begin{array}{c} -z' \frac{\partial \theta_{x'}}{\partial x'} \\ -z' \frac{\partial \theta_{y'}}{\partial y'} \\ -z' (\frac{\partial \theta_{x'}}{\partial y'} + \frac{\partial \theta_{y'}}{\partial x'}) \\ \cdots\cdots\cdots \\ \frac{\partial w'_o}{\partial x'} - \theta_{x'} \\ \frac{\partial w'_o}{\partial y'} - \theta_{y'} \end{array} \right\} \quad (3)$$

Eq. (3) can be rewritten as

$$\boldsymbol{\varepsilon}' = \left\{ \begin{array}{c} \hat{\boldsymbol{\varepsilon}}'_m \\ \cdots\cdots \\ 0 \end{array} \right\} + \left\{ \begin{array}{c} z' \hat{\boldsymbol{\varepsilon}}'_b \\ \cdots\cdots \\ \hat{\boldsymbol{\varepsilon}}'_s \end{array} \right\} \quad (4)$$

where

$$\hat{\boldsymbol{\varepsilon}}'_m = \left[\frac{\partial u'_o}{\partial x'}, \frac{\partial v'_o}{\partial y'}, (\frac{\partial u'_o}{\partial y'} + \frac{\partial v'_o}{\partial x'}) \right]^T \quad (5a)$$

$$\hat{\boldsymbol{\varepsilon}}'_b = \left[-\frac{\partial \theta_{x'}}{\partial x'}, -\frac{\partial \theta_{y'}}{\partial y'}, -(\frac{\partial \theta_{x'}}{\partial y'} + \frac{\partial \theta_{y'}}{\partial x'}) \right]^T \quad (5b)$$

$$\hat{\boldsymbol{\varepsilon}}'_s = \left[(\frac{\partial w'_o}{\partial x'} - \theta_{x'}), (\frac{\partial w'_o}{\partial y'} - \theta_{y'}) \right]^T \quad (5c)$$

are respectively the generalized ***membrane***, ***bending*** and ***shear*** local strain vectors.

For linear elastic analysis the local stresses are related to the local strains in the standard manner by

$$\boldsymbol{\sigma}' = \left\{ \begin{array}{c} \sigma_{x'} \\ \sigma_{y'} \\ \tau_{x'y'} \\ \cdots\cdots \\ \tau_{x'z'} \\ \tau_{y'z'} \end{array} \right\} = \left\{ \begin{array}{c} \boldsymbol{\sigma}'_f \\ \cdots \\ \boldsymbol{\sigma}'_s \end{array} \right\} = \left[\begin{array}{ccc} \mathbf{D}'_f & \vdots & \mathbf{0} \\ \cdots & \cdots & \cdots \\ \mathbf{0} & \vdots & \mathbf{D}'_s \end{array} \right] \left\{ \begin{array}{c} \varepsilon_{x'} \\ \varepsilon_{y'} \\ \gamma_{x'y'} \\ \cdots\cdots \\ \gamma_{x'z'} \\ \gamma_{y'z'} \end{array} \right\} = \mathbf{D}' \boldsymbol{\varepsilon}' \quad (6)$$

where for orthotropic material

$$\mathbf{D}'_f = \frac{1}{1-\nu_{x'y'}\nu_{y'x'}}\begin{bmatrix} E_{x'} & \nu_{x'y'}E'_x & 0 \\ \nu_{y'x'}E'_x & E'_y & 0 \\ 0 & 0 & (1-\nu_{x'y'}\nu_{y'x'})G_{x'y'} \end{bmatrix}$$

$$\mathbf{D}'_s = \begin{bmatrix} \alpha G_{x'z'} & 0 \\ 0 & \alpha G_{y'z'} \end{bmatrix} \tag{7}$$

and for isotropic material

$$E_{x'} = E_{y'} = E \;;\; \nu_{x'y'} = \nu_{y'x'} = \nu \;;\; G_{x'y'} = G_{x'z'} = G_{y'z'} = \frac{E}{2(1+\nu)} \tag{8}$$

In (6) and (8) $(\cdot)_f$ and $(\cdot)_s$ stand for *flexural* and transverse *shear* terms, respectively.
From (4) and (6) it can be obtained

$$\begin{aligned} \boldsymbol{\sigma}'_f &= \mathbf{D}'_f(\hat{\boldsymbol{\varepsilon}}'_m + z'\hat{\boldsymbol{\varepsilon}}'_b) \\ \boldsymbol{\sigma}'_s &= \mathbf{D}'_s\hat{\boldsymbol{\varepsilon}}'_s \end{aligned} \tag{9}$$

The *resultant stress* vector is now defined as

$$\hat{\boldsymbol{\sigma}}' = \begin{Bmatrix} \hat{\boldsymbol{\sigma}}'_m \\ \cdots \\ \hat{\boldsymbol{\sigma}}'_b \\ \cdots \\ \hat{\boldsymbol{\sigma}}'_s \end{Bmatrix} = \begin{Bmatrix} N_{x'} \\ N_{y'} \\ N_{x'y'} \\ \cdots \\ M_{x'} \\ M_{y'} \\ M_{x'y'} \\ \cdots \\ Q_{x'} \\ Q_{y'} \end{Bmatrix} = \int_{-\frac{t}{2}}^{+\frac{t}{2}} \begin{Bmatrix} \sigma_{x'} \\ \sigma_{y'} \\ \tau_{x'y'} \\ \cdots \\ z'\sigma_{x'} \\ z'\sigma_{y'} \\ z'\tau_{x'y'} \\ \cdots \\ \tau_{x'z'} \\ \tau_{y'z'} \end{Bmatrix} dz' = \int_{-\frac{t}{2}}^{+\frac{t}{2}} \begin{Bmatrix} \boldsymbol{\sigma}'_m \\ \cdots \\ z'\boldsymbol{\sigma}'_b \\ \cdots \\ \boldsymbol{\sigma}'_s \end{Bmatrix} dz' \tag{10}$$

where $\hat{\boldsymbol{\sigma}}'_m$, $\hat{\boldsymbol{\sigma}}'_b$, $\hat{\boldsymbol{\sigma}}'_s$ are respectively the local membrane, bending and shear resultant stress vectors. For sign convention see Figure 3.

The relationship between resultant stresses and generalized strains in local axes can be obtained by combining (10) and (9) as

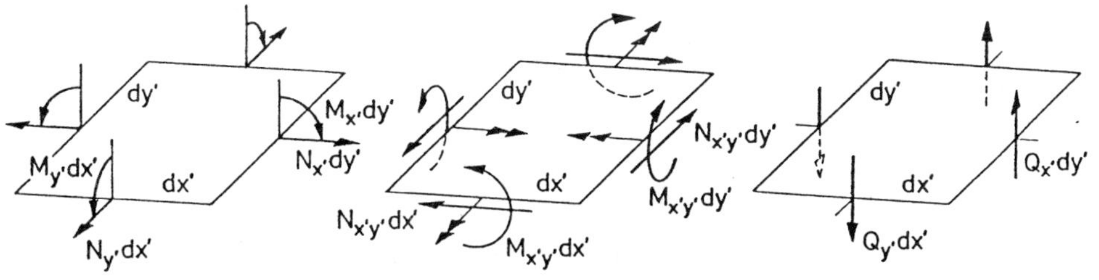

Figure 3 Sign convention for local resultant stresses in flat shell elements.

$$
\begin{aligned}
\hat{\boldsymbol{\sigma}}' &= \begin{Bmatrix} \hat{\boldsymbol{\sigma}}'_m \\ \cdots \\ \hat{\boldsymbol{\sigma}}'_b \\ \cdots \\ \hat{\boldsymbol{\sigma}}'_s \end{Bmatrix} = \int_{-\frac{t}{2}}^{+\frac{t}{2}} \begin{Bmatrix} \boldsymbol{\sigma}'_m \\ \cdots \\ z'\boldsymbol{\sigma}'_b \\ \cdots \\ \boldsymbol{\sigma}'_s \end{Bmatrix} dz' = \\
&= \int_{-\frac{t}{2}}^{+\frac{t}{2}} \begin{Bmatrix} \mathbf{D}'_b(\hat{\boldsymbol{\varepsilon}}'_m + z'\hat{\boldsymbol{\varepsilon}}'_b) \\ \cdots\cdots\cdots\cdots \\ z'\mathbf{D}'_f(\hat{\boldsymbol{\varepsilon}}'_m + z'\hat{\boldsymbol{\varepsilon}}'_b) \\ \cdots\cdots\cdots\cdots \\ \mathbf{D}'_s\hat{\boldsymbol{\varepsilon}}'_s \end{Bmatrix} dz' = \hat{\mathbf{D}}' \begin{Bmatrix} \hat{\boldsymbol{\varepsilon}}'_m \\ \cdots \\ \hat{\boldsymbol{\varepsilon}}'_b \\ \cdots \\ \hat{\boldsymbol{\varepsilon}}'_s \end{Bmatrix} = \hat{\mathbf{D}}'\hat{\boldsymbol{\varepsilon}}' \qquad (11)
\end{aligned}
$$

where $\hat{\mathbf{D}}'$ is given by

$$
\hat{\mathbf{D}}' = \int_{-\frac{t}{2}}^{+\frac{t}{2}} \begin{bmatrix} \mathbf{D}'_f & z'\mathbf{D}'_f & \mathbf{0} \\ z'\mathbf{D}'_f & z'^2\mathbf{D}'_f & \mathbf{0} \\ \mathbf{0} & \mathbf{0} & \mathbf{D}'_s \end{bmatrix} dz' = \begin{bmatrix} \hat{\mathbf{D}}'_m & \hat{\mathbf{D}}'_{mb} & \mathbf{0} \\ \hat{\mathbf{D}}'_{mb} & \hat{\mathbf{D}}'_b & \mathbf{0} \\ \mathbf{0} & \mathbf{0} & \hat{\mathbf{D}}'_s \end{bmatrix} \qquad (12a)
$$

with
$$
\begin{aligned}
\hat{\mathbf{D}}'_m &= \int_{-\frac{t}{2}}^{+\frac{t}{2}} \mathbf{D}'_f dz' \quad ; \quad \hat{\mathbf{D}}'_{mb} = \int_{-\frac{t}{2}}^{+\frac{t}{2}} z'\mathbf{D}'_f dz' \\
\hat{\mathbf{D}}'_b &= \int_{-\frac{t}{2}}^{+\frac{t}{2}} z'^2\mathbf{D}'_f dz' \quad ; \quad \hat{\mathbf{D}}'_s = \int_{-\frac{t}{2}}^{+\frac{t}{2}} \mathbf{D}'_s dz' \qquad (12b)
\end{aligned}
$$

where $\hat{\mathbf{D}}'_m$, $\hat{\mathbf{D}}'_b$ and $\hat{\mathbf{D}}'_s$ are respectively the generalized membrane, bending and shear constitutive matrices, and $\hat{\mathbf{D}}'_{mb}$ is the coupled bending–membrane constitutive matrix. Note that $\hat{\mathbf{D}}'_{mb} = \mathbf{0}$ in the case of homogeneous material, or if the material properties are symmetricaly distributed with respect to the element mid–surface. However, in the analysis of general reinforced concrete shells material heterogenity prevails and $\hat{\mathbf{D}}'_{mb}$ must be appropiately computed.

If an elasto–plastic material model is considered eq.(6) is defined in an incremental form as

$$d\boldsymbol{\sigma}' = \mathbf{D}'_{ep} d\boldsymbol{\varepsilon}' \tag{13a}$$

where $\mathbf{D}'_{ep}$ is the tangent elasto–plastic constitutive matrix which will be defined in a latter section. Integration of $\mathbf{D}'_{ep}$ across the element thickness allows to write eq.(11) also in an incremental form as

$$d\hat{\boldsymbol{\sigma}}' = \hat{\mathbf{D}}'_{ep} d\hat{\boldsymbol{\varepsilon}}' \tag{13b}$$

2.2 Layered model

In reinforced concrete shell problems a convenient representation of concrete and steel behaviour across the shell thickness is needed. This is of particular importance if the non linear behaviour of compressive concrete, concrete cracking and reinforcement response are to be appropiately modelled. The most popular computational approach is to use a layered model in which the shell thickness is divided into a series of plain (unreinforced) concrete layers and of reinforcing steel layers (Figure 4). Plain concrete layers can be either elastic, (singly or doubly) cracked, and yielded or crushed. Appropiate stress–strain relations must be used for each of these states of behaviour (see Section 3). On the other hand, the reinforcing steel is replaced by an equivalent smeared uniformly distributed steel layer with stiffness only in the direction of the reinforcement. The equivalent thickness of the steel layer is determined such that the corresponding area of reinforcement in the layer remains unchanged. Ordinarly, a concrete shell is reinforced by at least two sets of reinforcing bars. It is also usualy assumed that the reinforcing steel is arranged in layers forming grids intersecting each other at arbitrary angles. Any number of such layers can be accounted for and each layer is to be located exactly in space for the purpose of generating its stiffness properties. Perfect bond is assumed to exist between the reinforcing steel and the surrounding concrete. However, appropiate bond slip laws can also be incoporated into the analysis.

Layers are numbered sequentially, starting at the bottom surface of the shell element, and each layer contains stress points on its mid–surface. The stress components of the layer are computed at these stress points and are assumed to be constant over the thickness of each layer, so that the actual stress distribution over the shell thickness is modelled by a piecewise constant approximation [6] (Figure 4).

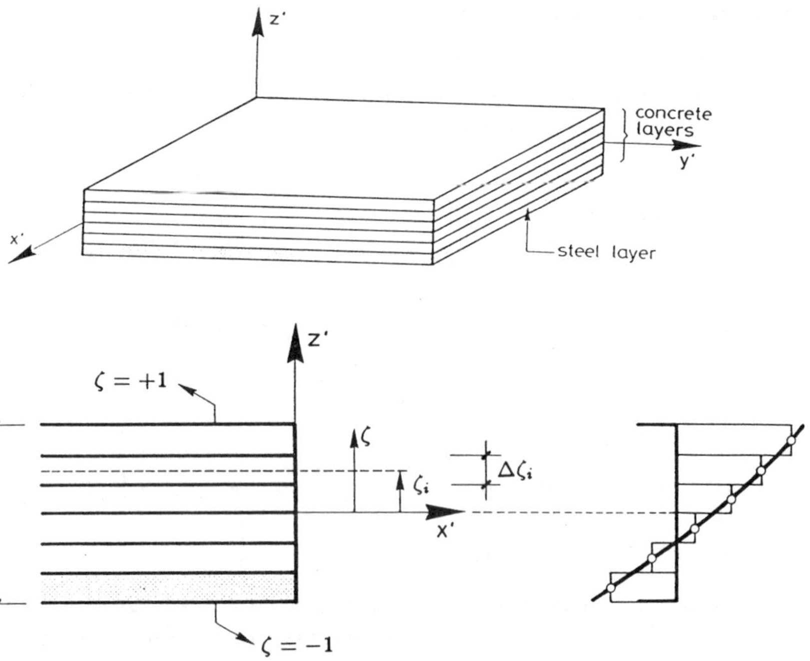

Figure 4 Layered discretization of reinforced concrete slab.

Layers of different thickness can be employed, as well as different number of layers per element. The specification of the layer thickness in terms of a normalized thickness coordinate $\zeta = \frac{2}{t}z'$, permits the variation of the layer thickness as the shell thickness varies [6] [7].

The stress resultants are obtained from eq.(11) by adequately integrating the constitutive matrices (12) across the element layers as

$$\begin{aligned}
\hat{\mathbf{D}}'_m &= \frac{t}{2}\sum_{i=1}^{l}\hat{\mathbf{D}}'_{f_i}\Delta\zeta_i \quad ; \quad \hat{\mathbf{D}}'_s = \frac{t}{2}\sum_{i=1}^{l}\hat{\mathbf{D}}'_{s_i}\Delta\zeta_i \\
\hat{\mathbf{D}}'_b &= \frac{t^3}{8}\sum_{i=1}^{l}\hat{\mathbf{D}}'_{f_i}\Delta\zeta_i^2\zeta_i \quad ; \quad \hat{\mathbf{D}}'_{m_b} = \frac{t^2}{4}\sum_{i=1}^{l}\hat{\mathbf{D}}'_{f_i}\zeta_i\Delta\zeta_i
\end{aligned} \tag{14}$$

where $\zeta_i = \frac{2z'_i}{t}$, l is the number of layers and $(\cdot)_i$ denotes values in the ith layer. For the non linear material case the elastic matrices in (14) will be substituted by the corresponding non linear operators via eq.(13b).

2.3 Finite element discretization

We will consider the discretization of the shell mid–surface in a mesh of isoparametric flat finite elements of n nodes (Figure 5). The local displacement field can be interpolated in the standard form

$$\mathbf{u}' = \sum_{i=1}^{n} \mathbf{N}_i \mathbf{a}_i'^{(e)} = [\mathbf{N}_1, \mathbf{N}_2, \cdots, \mathbf{N}_n] \begin{Bmatrix} \mathbf{a}_1'^{(e)} \\ \mathbf{a}_2'^{(e)} \\ \vdots \\ \mathbf{a}_n'^{(e)} \end{Bmatrix} = \mathbf{N}\mathbf{a}'^{(e)} \tag{15}$$

where

$$\mathbf{N}_i = \begin{bmatrix} N_i & 0 & 0 & 0 & 0 \\ 0 & N_i & 0 & 0 & 0 \\ 0 & 0 & N_i & 0 & 0 \\ 0 & 0 & 0 & N_i & 0 \\ 0 & 0 & 0 & 0 & N_i \end{bmatrix} \; ; \quad \mathbf{a}_i'^{(e)} = \left[u'_{o_i}, v'_{o_i}, w'_{o_i}, \theta_{x'_i}, \theta_{y'_i} \right]^T \tag{16}$$

are the shape function matrix and the local displacement vector of a node i.

From (5) and (15) the local generalized strains can be obtained as

$$\hat{\boldsymbol{\varepsilon}}' = \mathbf{B}'\mathbf{a}'^{(e)} \tag{17}$$

$$\text{with } \mathbf{B}' = [\mathbf{B}'_1, \mathbf{B}'_2, \cdots, \mathbf{B}'_m] \text{ and } \mathbf{B}'_i = \begin{Bmatrix} \mathbf{B}'_{m_i} \\ \mathbf{B}'_{b_i} \\ \mathbf{B}'_{s_i} \end{Bmatrix} \tag{18}$$

where $\mathbf{B}'_{mi}$, $\mathbf{B}'_{bi}$ and $\mathbf{B}'_{si}$ are respectively the local membrane, bending and shear generalized strain matrices of a node i, given by

$$\mathbf{B}'_{m_i} = \begin{bmatrix} \frac{\partial N_i}{\partial x'} & 0 & 0 & 0 & 0 \\ 0 & \frac{\partial N_i}{\partial y'} & 0 & 0 & 0 \\ \frac{\partial N_i}{\partial y'} & \frac{\partial N_i}{\partial x'} & 0 & 0 & 0 \end{bmatrix} \tag{19}$$

$$\mathbf{B}'_{b_i} = \begin{bmatrix} 0 & 0 & 0 & -\frac{\partial N_i}{\partial x'} & 0 \\ 0 & 0 & 0 & 0 & -\frac{\partial N_i}{\partial y'} \\ 0 & 0 & 0 & -\frac{\partial N_i}{\partial y'} & -\frac{\partial N_i}{\partial x'} \end{bmatrix} \tag{20}$$

$$\mathbf{B}'_{s_i} = \begin{bmatrix} 0 & 0 & \frac{\partial N_i}{\partial x'} & -N_i & 0 \\ 0 & 0 & \frac{\partial N_i}{\partial y'} & 0 & -N_i \end{bmatrix} \tag{21}$$

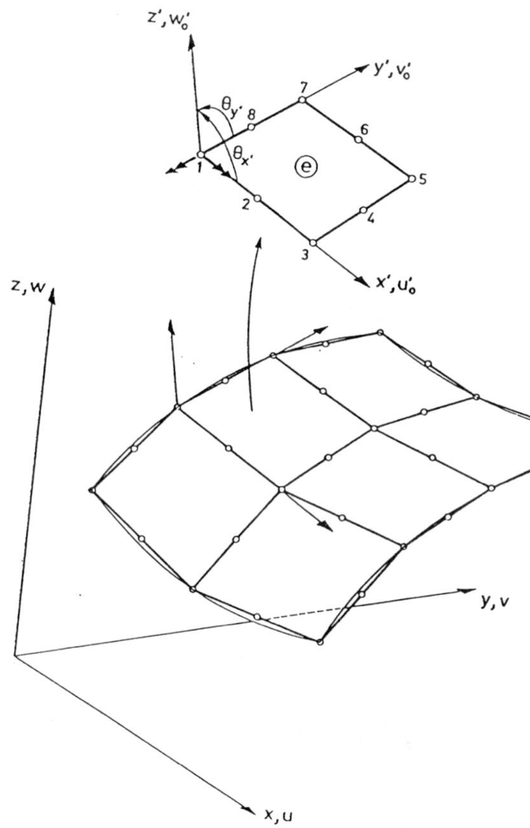

Figure 5 Discretization of a shell into flat shell finite elements.

The virtual work principle for a single element can be written as

$$\int\int_{A^{(e)}} \delta\hat{\boldsymbol{\varepsilon}}'^T \hat{\boldsymbol{\sigma}}' dA = \int\int_{A^{(e)}} \delta\mathbf{u}'^T \mathbf{t}' dA + \left[\delta\mathbf{a}'^{(e)}\right]^T \mathbf{q}'^{(e)} \tag{22}$$

where $\mathbf{t}'$ and $\mathbf{q}'^{(e)}$ are the distributed local vector and nodal point load vector, respectively.

Substituting eqs.(11), (15) and (17) in (22) the standard stiffness equilibrium equations for a single element can be obtained as

$$\mathbf{q}'^{(e)} = \mathbf{K}'^{(e)}\mathbf{a}'^{(e)} - \mathbf{f}'^{(e)} \tag{23}$$

where

$$\mathbf{K}'^{(e)}_{ij} = \int\int_{A^{(e)}} \mathbf{B}'^T_i \hat{\mathbf{D}}' \mathbf{B}'_j dx' dy' \tag{24}$$

$$\mathbf{f}'^{(e)}_i = \int\int_{A^{(e)}} \mathbf{N}^T_i \mathbf{t}' dx' dy' \tag{25}$$

are the element stiffness matrix and equivalent nodal load vector, respectively.

Matrix $\mathbf{K}'^{(e)}_{ij}$ can be rewritten using (12) and (18) as

$$\mathbf{K}'^{(e)}_{ij} = \mathbf{K}'^{(e)}_{m_{ij}} + \mathbf{K}'^{(e)}_{b_{ij}} + \mathbf{K}'^{(e)}_{s_{ij}} + \mathbf{K}'^{(e)}_{mb_{ij}} + \mathbf{K}'^{(e)}_{bm_{ij}} \tag{26}$$

where

$$\begin{aligned}
\mathbf{K}'^{(e)}_{m_{ij}} &= \int\int_{A^{(e)}} \mathbf{B}'^T_{m_i} \hat{\mathbf{D}}'_m \mathbf{B}'_{m_j} dx' dy' \\
\mathbf{K}'^{(e)}_{b_{ij}} &= \int\int_{A^{(e)}} \mathbf{B}'^T_{b_i} \hat{\mathbf{D}}'_b \mathbf{B}'_{b_j} dx' dy' \\
\mathbf{K}'^{(e)}_{s_{ij}} &= \int\int_{A^{(e)}} \mathbf{B}'^T_{s_i} \hat{\mathbf{D}}'_s \mathbf{B}'_{s_j} dx' dy' \\
\mathbf{K}'^{(e)}_{mb_{ij}} &= \int\int_{A^{(e)}} \mathbf{B}'^T_{m_i} \hat{\mathbf{D}}'_{mb} \mathbf{B}'_{b_j} dx' dy' = \left[\mathbf{K}'^{(e)}_{bm_{ij}}\right]^T
\end{aligned} \tag{27}$$

are respectively the membrane, bending, shear and membrane-bending coupling local stiffness matrices. Note that if $\hat{\mathbf{D}}'_{mb}$ is zero (which is the case for homogeneous material or when there is material symmetry with respect to the mid-plane) $\mathbf{K}'^{(e)}_{mb}$ and $\mathbf{K}'^{(e)}_{bm}$ are also zero and the *local* stiffness matrix can be directly obtained by simple addition of the membrane, bending and shear uncoupled contributions.

Note that in (23)–(27) we have assumed elastic material behaviour. The non linear case will be treated in a later section.

The global stiffness matrix and the global equivalent nodal load vector for the whole mesh are obtained by assembly of the individual element contributions in the standard manner [14]. This involves first a transformation of local degrees of freedom and forces to a common global cartesian coordinate system as

$$\mathbf{a}'^{(e)} = \mathbf{T}\mathbf{a}^{(e)} \quad \text{and} \quad \mathbf{f}'^{(e)} = \mathbf{T}\mathbf{f}^{(e)} \tag{28}$$

where $\mathbf{T}$ is the transformation matrix relating local and global nodal degrees of freedom and forces at element level [14], [15]. The global element stiffness matrix is then computed by the well known transformation

$$\mathbf{K}^{(e)} = \mathbf{T}^T \mathbf{K}'^{(e)} \mathbf{T} \tag{29}$$

If the shell has folds or kinks the transformations (28)–(29) involve and additional global rotation θ_z which plays the role of a sixth degree of freedom at each non-coplanar node, whereas the standard five degrees of freedom (three global displacements and two local rotations) can be kept at the coplanar nodes (Figure 6). Details of the treatment of co-planar and non coplanar nodes can be found in [14], [15].

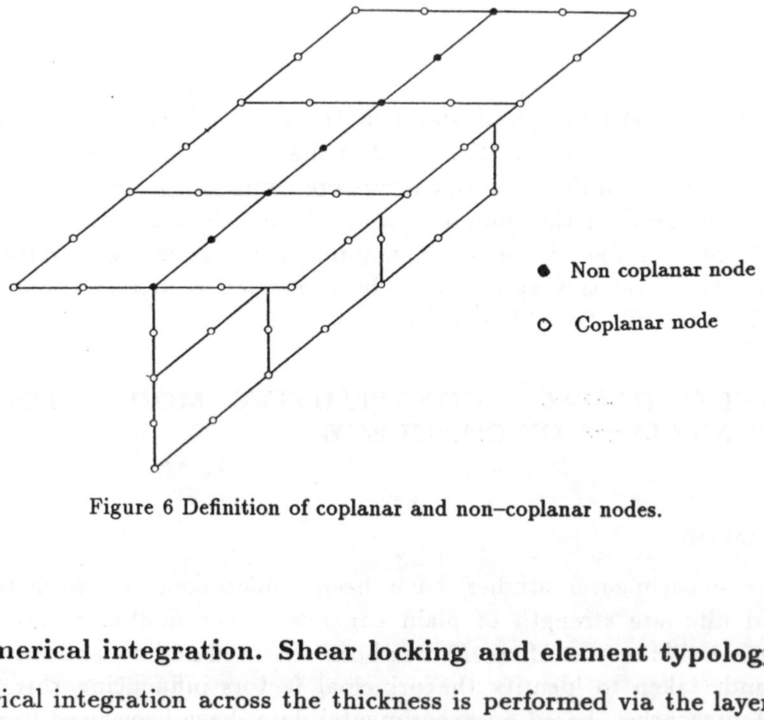

Figure 6 Definition of coplanar and non–coplanar nodes.

2.4 Numerical integration. Shear locking and element typology

Numerical integration across the thickness is performed via the layered model as described previously. In the shell plane the normal (full) integration rule consists of $m \times m$ Gauss points where m is the number of nodes along each element side. Nevertheless when flat shell elements are fully integrated they exhibit shear locking and over–stiff solutions are obtained in the majority of applications [14], [15].

The simplest procedure to overcome shear locking behaviour is to use a reduced integration quadrature for the shear stiffness [14], [15] whereas the rest of the stiffness

An alternative approach for derivation of *robust* shell elements is based on the use of an assumed shear strain field. In this method a shear strain field is "a priori" assumed over the element in the natural coordinate system.

$$\gamma_\xi = \mathbf{N}_\gamma \bar{\gamma}_\xi \tag{30}$$

where $\bar{\gamma}_\xi$ contains the values of the shear strains at some prescribed points within the element and $\mathbf{N}_\gamma$ are appropiate shear interpolating functions. The displacement and rotations are interpolated in the standard manner. However, a different interpolation for each displacement field must sometimes be used to satisfy the the requirements for the existence of the solution [14], [18].

By relating $\bar{\gamma}_\xi$ with the cartesian shear strains and these with the element nodal displacements $\mathbf{a}^{(e)}$ through eq.(17) a final relationship beteween $\boldsymbol{\gamma}'_s$ and $\mathbf{a}^{(e)}$ can be found in the form

$$\boldsymbol{\gamma}'_s = \hat{\mathbf{B}}'_s \mathbf{a}^{(e)} \tag{31}$$

where $\hat{\mathbf{B}}'_s$ is termed *substitute shear strain* matrix (or shear **B**-bar matrix). Matrix $\mathbf{K}'_s$ is computed by eq.(27) using $\hat{\mathbf{B}}'_s$ instead of the original shear strain matrix $\mathbf{B}'_s$, whereas the rest of the stiffness matrix terms are computed as shown in (27). Full integration is now used for the computation of *all* the element matrices.

Figure 7 shows some of the most popular rectangular and triangular flat shell elements based on this approach. The interested reader can found further information in [14], [19], [20], [21] and [24].

3. A PLASTIC DAMAGE CONSTITUTIVE MODEL FOR NON LINEAR ANALYSIS OF CONCRETE

3.1 Introduction

Extensive experimental studies have been undertaken to characterise the response and ultimate strength of plain concrete under multiaxial stress states [30,31]. Considerable scatter of results has been observed and collaborative studies have been undertaken to identify the principal factors influencing this variation [32]. Several approaches, based on experimental data, have been used to represent the constitutive relationship of concrete under multiaxial stress states and these can be categorised into the four following groups: (a) Linear and nonlinear elasticity theories, [33-36] (b) perfect and workhardening plasticity theories [37-39], (c) endochronic theory of plasticity [40] and (d) plastic fracturing theory [41].

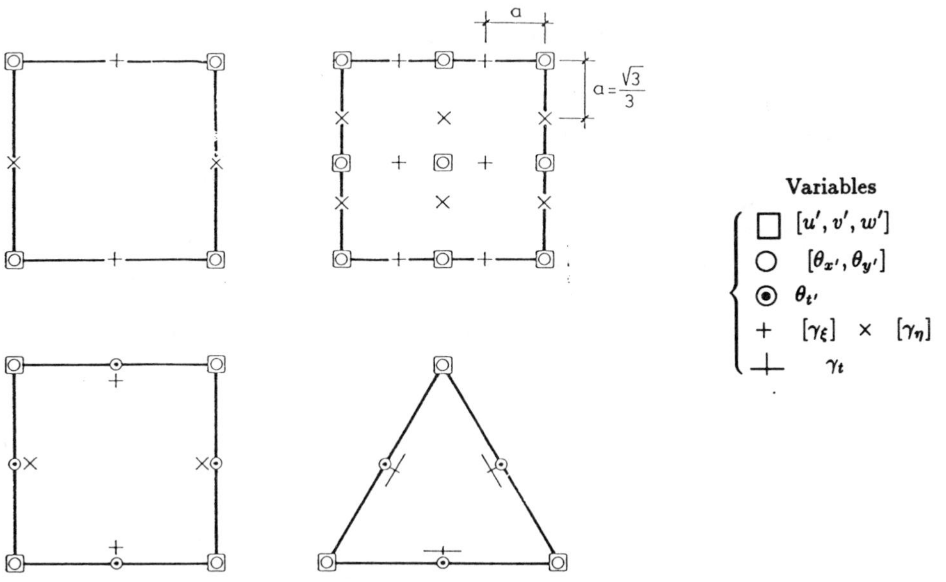

Figure 7 Some quadrilateral and triangular assumed shear strain shell elements [19–21], [24].

Experimental evidendence indicates that the nonlinear deformation in concrete is basically inelastic and therefore the stress-strain behaviour may be separated into recoverable and irrecoverable components.

A simple and popular model non linear finite element analysis of concrete assumes elasto–plastic (or viscoplastic) constitutive equations for compression behaviour, whereas a conceptually more simple elasto–brittle model is used for defining onset and progression of cracks at points in tension. Different versions of this model have been successfully used by different authors for non–linear analysis of plain and reinforced concrete structures [31-60].

The elasto–plastic–brittle model, in spite of its popularity, presents various controversial features such as the need for defining uncoupled behaviour along each principal stress (or strain) direction: the use of a shear retention factor to ensure some shear resistence along the crack; the lack of equilibrium at the cracking point when more than one crack is formed [62]; the difficulties in defining stress paths following the opening and closing of cracks under cycling loading conditions and the difficulty for dealing with the combined effect of cracking an plasticity at the damaged point.

It is well known that microcracking in concrete takes place at low load levels due to physical debonding between aggregate and mortar particles, or to simple microcracking in the mortar area. Cracking progresses following a non–homogeneous path which combines the two mentioned mechanisms with growth and linking between microckacks along different directions. Experiments carried out on mortar specimens shown that the distribution of microcracking is fairly discontinuos with arbitrary orientations [61]. This fact is supported by many experiments which shown that *cracking can be considered, at microscopic level, as a non–directional phenomenon* and that the propagation of microcraks at aggregate level follows an erratic path which depends on the size of the aggregate particles. Thus, the *dominant cracking directions can be interpreted at macroscopic level as the locus of trajectories of the damage points* (Figure 8).

The above concepts support *the idea that the nonlinear behavior of concrete can be modelled using concepts of classical plasticity theory* only provided an adequate yied functions is defined for taking into account the different response of concrete under tension and compression states. Cracking can, therefore, be interpreted as a *local damage effect*, defined by the evolution of known material parameters and by a single yield function which controls the onset and evolution of damage.

One of the advantages of such a model is the independence of the analysis with respect to crack directions which can be simply identified *a posteriori* from the converged values of the nonlineal solution. This allows to overcome the problems associated to most elastic–plastic–brittle smeared cracking models. In this section an elastoplastic model developed by the author's group for nonlinear analysis of concrete based on the concepts of *plastic damage* mentioned above is presented [63-67]. The model takes into account all the important aspects which should be considered in the nonlinear analysis of concrete, such as the different response under tension and compression, the effect of stiffness degradation and the problem of objectivity of the results with respect to the finite element mesh.

3.2 Basic concepts of the plastic damage model

The plastic damage model proposed can be considered as a general form of classical plasticity in which the standard hardening variable is replaced by a normalized plastic damage variable κ^p, such that $0 \leq \kappa^p \leq 1$. This variable is similar to the former in the sense that it never decreases and it only increases if plastic deformation takes place which is associated to the existence of microcraking. The limit value of $\kappa^p = 1$ denotes total damage at a point with complete loss of cohesion. This can be interpreted as the formation of a macroscopic crack.

If stiffness degradation effects are negleted (and this will be separately treated in a latter section) the basic equations of the model are:

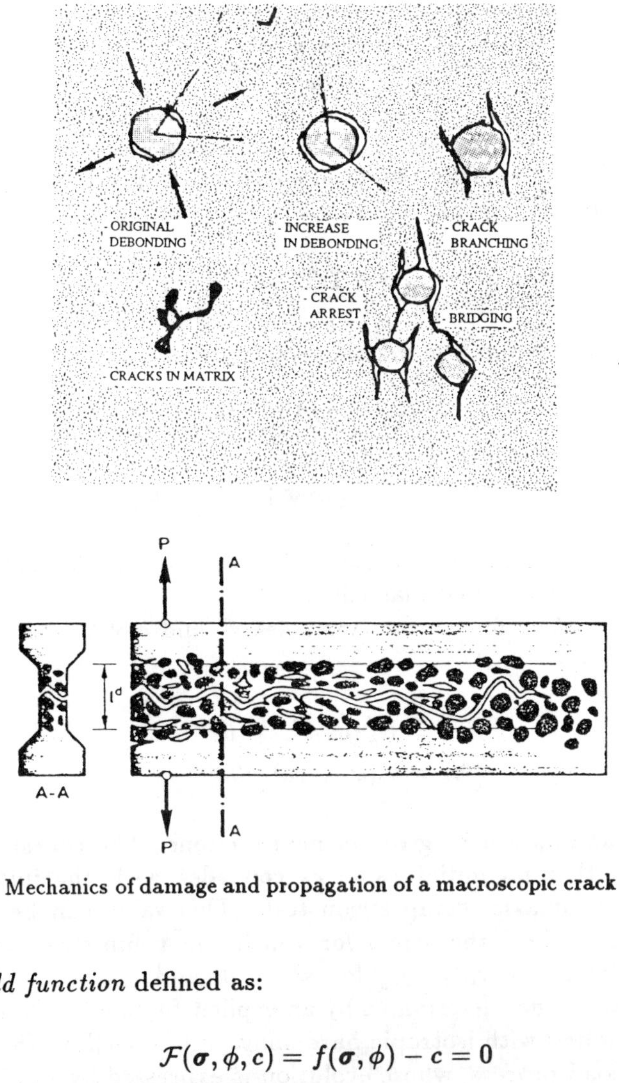

Figure 8 Mechanics of damage and propagation of a macroscopic crack in concrete.

(a) *The yield function* defined as:

$$\mathcal{F}(\boldsymbol{\sigma}, \phi, c) = f(\boldsymbol{\sigma}, \phi) - c = 0 \tag{32}$$

where c is a cohesion or some constant multiple thereof, and ϕ is an internal friction angle, $f(\boldsymbol{\sigma}, \phi)$ is a function of the stress components that is first degree homogeneous in the stresses $\boldsymbol{\sigma}$, given a physical meaning of scaled stress to the cohesion. The particular form of $\mathcal{F}$ used in this work are presented in next section.

(b) *The elasto–plastic strain descomposition* as:

$$d\boldsymbol{\varepsilon} = \mathbf{D}^{-1} d\boldsymbol{\sigma} + \boldsymbol{\varepsilon}^p = d\boldsymbol{\varepsilon}^e + d\boldsymbol{\varepsilon}^p \tag{33}$$

where $\mathbf{D}$ is the elastic constitutive matrix.

(c) *The flow rule* is defined for the general case of non–associated plasticity as:

$$d\boldsymbol{\varepsilon}^p = \lambda \frac{\partial \mathcal{G}(\boldsymbol{\sigma}, \psi, c)}{\partial \boldsymbol{\sigma}} = \lambda \mathbf{g} \tag{34}$$

where λ is the plastic loading factor, ψ is a dilatancy angle and $\mathbf{g}$ is a plastic flow vector, normal to the plastic potential surface $\mathcal{G}(\boldsymbol{\sigma}, \psi, c)$. From eqs.(32-34) the standard elastoplastic incremental constitutive equation can be obtained as:

$$d\boldsymbol{\sigma} = \mathbf{D}^{ep} \cdot d\boldsymbol{\varepsilon} \tag{35}$$

with the elastoplastic constitutive matrix given by:

$$\mathbf{D}^{ep} = \mathbf{D} - \frac{\left[\mathbf{D} \cdot \left\{\frac{\partial \mathcal{G}}{\partial \boldsymbol{\sigma}}\right\}\right] \otimes \left[\mathbf{D} \cdot \left\{\frac{\partial \mathcal{F}}{\partial \boldsymbol{\sigma}}\right\}\right]}{A + \left[\left\{\frac{\partial \mathcal{F}}{\partial \boldsymbol{\sigma}}\right\} \cdot \mathbf{D} \cdot \left\{\frac{\partial \mathcal{G}}{\partial \boldsymbol{\sigma}}\right\}\right]} \tag{36}$$

where A is the hardenening parameter. Note from eq.(36) that $\mathbf{D}^{ep}$ is only symmetric for $\mathcal{G} = \mathcal{F}$ (associated plasticity).

(d) *The evolution laws* for internal variables κ^p and c of the form:

$$d\kappa = \mathbf{h}_K^T(\boldsymbol{\sigma}, \kappa^p, c) \cdot d\boldsymbol{\varepsilon}^p \tag{37}$$

$$dc = h_c(\boldsymbol{\sigma}, \kappa^p, c) \cdot d\kappa^p \tag{38}$$

Functions $\mathbf{h}_K$ and h_c will be given in next sections. The cohesion c is a scaled uniaxial stress, so that its initial value c_o coincides with the initial yield stres f_{c_o} obtained from a uniaxial compression test. This value can be interpreted as a discontinuity stress, i.e. the stress for which the volumetric strain reaches a minimum. Therefore, $c = c_o = f_{c_o}$ for $\kappa^p = 0$, and $c = c_o = 0$ for $\kappa^p = 1$. Note, however, that c is not determined by an explicit fuction of κ^p, as is the case in simple plasticity models with isotropic hardening, but is itself an *internal variable*, depending on the load process, whose evolution is expressed by eq.(38)

3.3 Definition of the yield surface

Recent work has shown that the behaviour of concrete under triaxial compression states can be adequately modelled by yield criteria of the type of eq.(32) with $\mathcal{F}$ being a function with straight meridians, that is first degree homogeneous in the stress components.

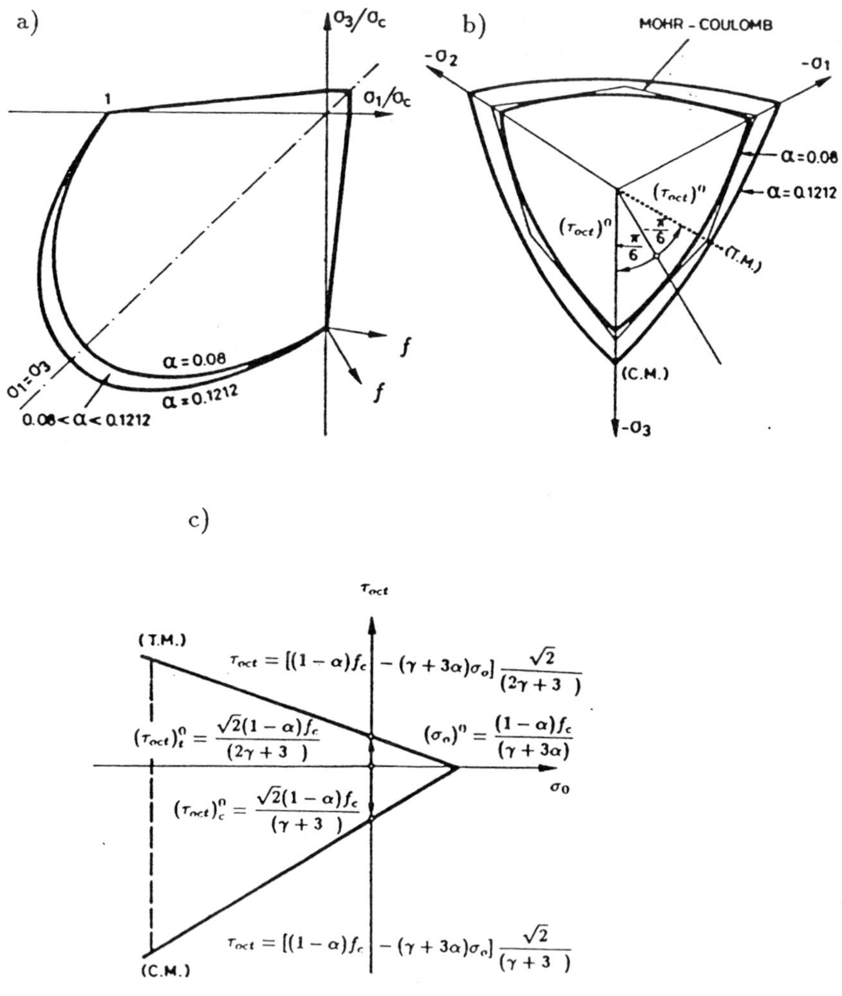

Figure 9 Proposed yield surface: (a) $(\sigma_{11} - \sigma_{33};\ \sigma_{22} = 0)$ plane; (b) octahedral plane; (c) meridian plane.

A more detailed study of the experimental work reported for biaxial and triaxial behaviour of concrete allowed the author's group to define a function of the form [63,64] (Figure 9)

$$\mathcal{F} = \mathcal{F}(\boldsymbol{\sigma}, \phi, c) = \frac{1}{(1-\alpha)}[\sqrt{3J_2} + \alpha I_1 + \beta < \sigma^{max} > -\gamma < -\sigma^{max} >] - c = 0 \quad (39)$$

where I_1 is the first invariant of stress, α, β and γ are dimensionless parameters that can be expressed as functions of the friction angle ϕ, σ^{max} is the maximum principal stress and $< \pm x > = \frac{1}{2}[x \pm |x|]$ is a ramp function. Note that when $\sigma^{max} = 0$ (biaxial compression) $\mathcal{F}$ is just the Drucker–Prager criterion, with the exception of parameter α. This can be obtained comparing the initial equibiaxial compression stress f_{b_o} with initial uniaxial compression stress f_{c_o}, yielding [63,64]:

$$\alpha = \frac{\left(\frac{f_{b_o}}{f_{c_o}-1}\right)}{\left(\frac{2f_{b_o}}{f_{c_o}-1}\right)} \tag{40}$$

Experimental values give $1.10 \leq f_{b_o}/f_{c_o} \leq 1.16$ which yields $0.08 \leq \alpha \leq 0.1212$.

Once α is known, β can be determined from the value of $R = f_{c_o}/f_{T_o}$, where f_{T_o} is the initial uniaxial tensile yield, as [63,64]:

$$\beta = (1-\alpha)R - (1+\alpha) \tag{41}$$

and for $R \simeq 10$ and $\alpha \simeq 0.10$ gives $\beta \simeq 7.50$

The parameter γ appears only in triaxial compression with $\sigma^{max} < 0$. Considering the equations of the straight tension (TM) and compression (CM) meridians of the yield surface it can be obtained [63,64]

$$\gamma = \frac{3(1 - r_{oct}^{max})}{2r_{oct}^{max} - 1} \tag{42}$$

where

$$r^{max}{}_{oct} = (\sqrt{J_2})_{TM}/(\sqrt{J_2})_{CM} \qquad \text{at a given} \quad I_1 \tag{43}$$

Experimental tests show thar r_{oct}^{max} has a constant mean values of $\simeq 0.65$ [63,64] which yields a value of $\lambda \simeq 3.5$

3.4 Definition of the plastic damage variable κ^p

Let us consider stress–plastic strain diagrams for uniaxial tension and compression tests (see Figure 10) for each test we define

$$\begin{aligned} \kappa^p &= \frac{1}{g_T^p}\int_{t=0}^{t} \sigma_T d\varepsilon_T^p, && \text{for uniaxial tension, and} \\ \kappa^p &= \frac{1}{g_C^p}\int_{t=0}^{t} \sigma_C d\varepsilon_C^p, && \text{for uniaxial compression} \end{aligned} \tag{44}$$

where g_T^p and g_C^p are the specific plastic works, defined by the areas under each of the curves $\sigma_T - \varepsilon_T^p$ and $\sigma_C - \varepsilon_C^p$ obtained from the tension and compression uniaxial tests,

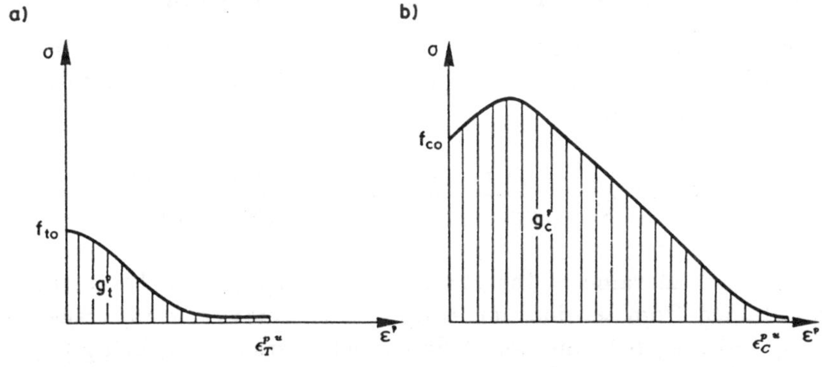

Figure 10 Uniaxial curves ($\sigma - \varepsilon^p$). (a) Tension; (b) compression.

respectively. The eqs.(44) allow the transformation of uniaxial diagrams: $\sigma = f(\varepsilon^p)$ in other: $\sigma = f(\kappa^p)$ such that (Figure 10)

$$\begin{array}{lll} \text{tension test:} & f_T(0) = f_{T_o} \quad \text{and} & f_T(1) = 0 \\ \text{compression test:} & f_C(0) = f_{C_o} \quad \text{and} & f_C(1) = 0 \end{array}$$

Starting from these concepts, the evolution law for κ^p can be generalized for a multiaxial stress state (written in terms of principal stress and plastic strain), as [63,64]

$$d\kappa^p = \mathbf{h}_K(\boldsymbol{\sigma}, \kappa^p, c)d\boldsymbol{\varepsilon}^p = \sum_{i=1}^{3}(h_{c_io}d\varepsilon_i^p) \tag{45}$$

with:

$$\begin{aligned} h_{\kappa_i} &= [(h_{\kappa_i})_T + (h_{\kappa_i})_C] = \frac{1}{g_T^{p*}} < \sigma_i > + \frac{1}{g_C^{p*}} < -\sigma_i > \\ g_T^{p*} &= g_T^p \frac{\sum_{i=1}^3 < \sigma_i >}{\sigma_T}; \quad g_C^{p*} = g_C^p \frac{\sum_{i=1}^3 < -\sigma_i >}{\sigma_C} \end{aligned} \tag{46}$$

where subscrits T and C denote values obtained from uniaxial tension and compression tests, respectively. In eq.(46) g_T^{p*} and g_C^{p*} are normalized values of the uniaxial specific plastic work for tension and compression processes, accordingly to the yield function chose and also to the uniaxial tension and compression stresses σ_T and σ_T. For further details see [63,64].

3.5 Evolution law for the internal variable of cohesion c

The evolution of the cohesion c must satisfy $c \to 0$ for $\kappa^p \to 1$. In this model the evolution law for the cohesion is given by eq.(38) with the evolution function $h_c(\boldsymbol{\sigma}, \kappa^p, c)$ defined by [63,64]:

$$h_c((\boldsymbol{\sigma}, \kappa^p, c) = \left[\frac{r(\boldsymbol{\sigma})}{c_T(\kappa^p)}\frac{dc_T(\kappa^p)}{d\kappa^p} + \frac{1-r(\boldsymbol{\sigma})}{c_C(\kappa^p)}\frac{dc_C(\kappa^p)}{d\kappa^p}\right] \tag{47}$$

where c is the actual value of the cohesion, $c_T(\kappa^p)$ and $c_C(\kappa^p)$ are the cohesion functions obtained from tension and compression uniaxial tests, respectively (see Figure 11a), and $r(\boldsymbol{\sigma})$ is a function defining the stress stated, being $0 \leq r(\boldsymbol{\sigma}) \leq 1$ with $r(\boldsymbol{\sigma}) = 1$ if $\sigma_i \geq 0$ over all $i = 1,2,3$ and $r(\boldsymbol{\sigma}) = 0$ if $\sigma_1 \leq 0$ over all $i = 1,2,3$.

We have taken

$$r(\boldsymbol{\sigma}) = \frac{\sum_{i=1}^{3} < \sigma_i >}{\sum_{i=1}^{3} |\sigma_i|} \tag{48}$$

For further details the reader is referred to [35,36,44,45,63,64].

3.6 Evolution law for the internal friction angle ϕ

It has been shown [35,44,45] that the loss of cohesion in concrete due to increase plastic damage affects the value of angle of internal friction, which ranges from $\phi \simeq 0$ for initial cohesion c_0 until $\phi = \phi^{max}$ for the ultimate value of cohesion $c = c_u = 0$. In this work the following evolution law for ϕ has been chosen

$$\sin\phi = \begin{cases} 2\frac{\sqrt{\kappa^p\kappa^L}}{\kappa^p+\kappa^L}\sin\phi^{max}; & \forall\kappa^p \leq \kappa^L \\ \sin\phi^{max}; & \forall\kappa^p > \kappa^L \end{cases} \tag{49}$$

where κ^L denotes the limit damage for which the value of ϕ remains constant (see Figure 11b).

3.7 Plastic potential function and dilatancy angle ψ

Granular materials like concrete exhibit dilatancy phenomenon. This can be modelled introducing an adequate plastic potential function G to mach the numerical values obtained for the inelastic volume change with experimental data. In this work we have chosen for G the modified Mohr-Coulomb yield function of Figure 9 with the angle of dilatancy ψ substituting the internal friction angle ϕ. The evolution law for ψ has been obtained via a simple modification of the general expression used by De Borst and Vermeer [68] as

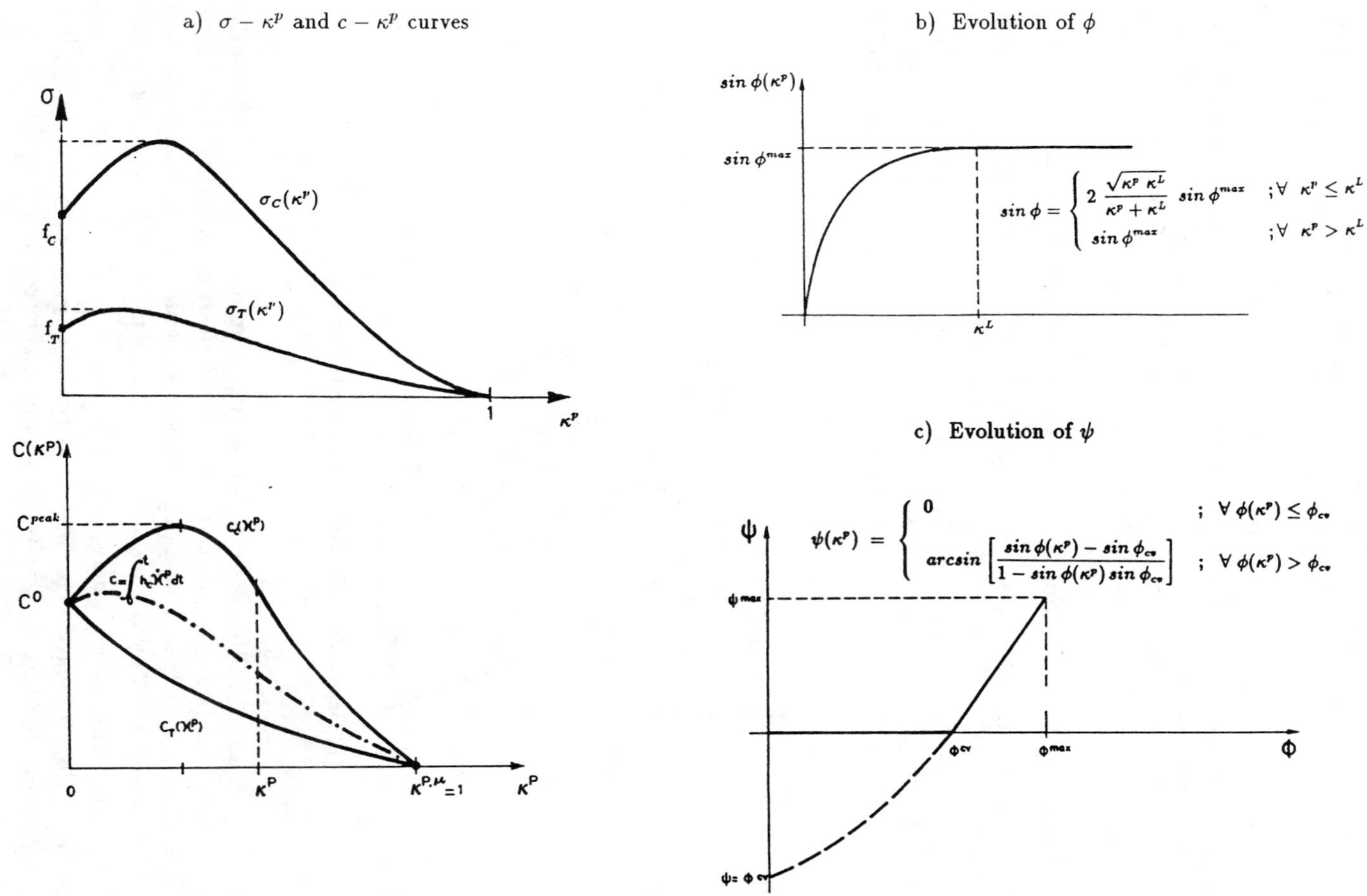

Figure 11 **a)Uniaxial curves $\sigma - \kappa^p$ and $c - \kappa^p$ for tension and compression tests. b) Evolution law for the internal friction angle ϕ c) Evolution law for the dilatance Ψ.**

$$\psi(\kappa^p) = \arcsin\left[\frac{\sin\phi(\kappa^p) - \sin\phi_{cv}}{1 - \sin\phi(\kappa^p)\sin\phi_{cv}}\right] \tag{50}$$

where ϕ_{cv} can be taken as a constant value. For concrete $\phi_{cv} \simeq 13^\circ$.

Eq.(50) gives for the initial stages of the process a negative dilatancy, which increase as plastic damage increases, takes a zero value for $\phi = \phi_{cv}$ and reaches a maximum for $\phi = \phi^{max}$. For concrete a negative value of ψ has not physical meaning and, therefore, it must be taken $\phi = 0$ for $\phi \leq \phi_{cv}$ (see Figure 11c).

3.8 Generalization of the model to include stiffness degradation

Experimental results show that near and beyond peak strength comented granular materials exhibit an increacing degradation of sttiffness due to microcracking (Figure 12). The consideration of the phenomenon of stiffness degradation makes it necessary some modification in the basic concepts of the theory of plasticity used in previous sections and, in particular, that of associated plasticity.

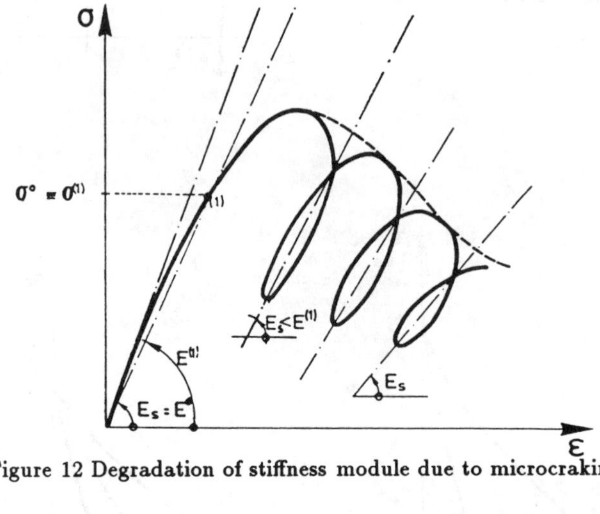

Figure 12 Degradation of stiffness module due to microcraking.

Taking into account the stiffness degradation effects implies modifying the elastic secant constitutive matrix $\mathbf{D}_s$ in terms of two sets of internal variables: the elastic degradation variables and the plastic degradation variables whose evolutions will be assumed to be governed by rate equations of the form: $\dot{d}_i^e = \phi_i < \mathbf{k}_i \cdot \dot{\boldsymbol{\varepsilon}} >$ and $\dot{d}_j^p = \mathbf{I}_j \cdot \dot{\boldsymbol{\varepsilon}}^p$ [63,64] repectively; where $\mathbf{k}_i$ and $\mathbf{I}_j$ are vectors in the stresses space denoting the directions of elastic and plastic degradation, respectively; and ϕ_i is a positive scalar factor (for further details the reader is referred to [63,64,67].

The simplest assumption for elastic degradation based on a simple isotropic degradation can be variable: d^e, such that the secant constitutive matrix is modified by:

$$\mathbf{D}_s(d^e) = (1 - d^e)\mathbf{D}^o \tag{51}$$

where $\mathbf{D}^o$ is the initial stiffness. Parameter d^e can be interpreted as the ratio between the area of degradated material and the total area, and it can be expressed [63,64] as

$$d^e = 1 - e^{-\phi w^e} \tag{52}$$

where $2w^e = \boldsymbol{\varepsilon}^e \cdot \mathbf{D}^o \cdot \boldsymbol{\varepsilon}^e$ is the square of the undamaged energy norm of the strain, $\boldsymbol{\varepsilon}^e$ is the elastic strain and ϕ is a constant given for this particular case by [63,64]:

$$\phi = \frac{2}{E^o(\varepsilon^1)^2} \ln \frac{E'}{E^o} \tag{53}$$

E^o is initial Young modulus; E' and ε^1 are the secant Young modulus and the elastic deformation at the limit stress point of elastic degradation, respectively. For further details the reader is referred to [63,64].

For the plastic degradation a simple one–parameter model can also be used [63,64,67]. This is based on the assumption that plastic degradation takes place only in the softening branch and that the stiffness is then is then proportional to the cohesion. The secant constitutive matrix is thus given by:

$$\mathbf{D}_s(d^p, \mathbf{d}^e) = (1 - d^p)\mathbf{D}_s(\mathbf{d}^e) \tag{54}$$

with the plastic degradation parameter d^p given by

$$d^p = 1 - \frac{c}{c^{peak}} \tag{55}$$

where c is the actual value of cohesion and c^{peak} is the maximum cohesion value reached [64,64].

3.9 Problem of objectivity response

It has been made abundantly clear over the past decade that the strain–softening branch of the stress–strain curves cannot represent a local physical property of the material. The argument have been advanced both on physical grounds and on the basis of the mesh–sensitivity of numerical solutions obtained by means of the finite–element method. The mesh–sensitivity can be largely eliminated if one defines

$g_T^p = G_T/l$ and $g_C^p = G_C/l$, where l is a characteristic length related to mesh size, and G_T and G_C are quantities with the dimensions of energy/area that are assumed to be material properties.

In problem involving tensile cracking, G_T may be identified with the specific fracture energy G_f, defined as the energy required for form a unit area of crack. It has generally been assumed that G_f is a true property, and methods have been developed for determining it. For the characteristic length l, various approaches have been proposed [69].

Not so much attention has been paid to the corresponding compressive problem. Compressive failure may occur through several mechanisms–crushing, shearing and transverse cracking–and consequently G_c, if indeed it is a material property, cannot be readily identified with any particular physical energy. Moreover, it must be kept in mind that it is only the descending portion of the stress-strain curve that is mesh–sensitive.

3.10 Determination of cracks by postprocessing the numerical results

The amount and directions of cracking at a point in the plastic–damage model is obtained *a posteriori*, once convergence of the non–linear solution has been reached, as follows:

(a) Cracking initiates at a point when the effective plastic strain, $\bar{\varepsilon}^p$, is greater than zero. The direction of cracking is assumed to be orthogonal to that of the maximum principal strain at the point (see Figure 13).

(b) The increment of plastic strains along the directions of the crack, $\Delta\boldsymbol{\varepsilon}^{cr}$, can be obtained as $\Delta\boldsymbol{\varepsilon}^{cr} = \mathbf{T}\cdot\Delta\boldsymbol{\varepsilon}^p$, where $\Delta\boldsymbol{\varepsilon}^p$ is the vector of plastic strain increment expressed in global Cartesian axes and $\mathbf{T}$ is a transformation matrix given by:

$$\mathbf{T} = \begin{bmatrix} \cos^2\theta & \sin^2\theta & \frac{\sin^2\theta}{2} \\ \sin^2\theta & \cos^2\theta & -\frac{\sin^2\theta}{2} \\ -\sin 2\theta & \sin 2\theta & \cos 2\theta \end{bmatrix} \tag{56}$$

where θ is the angle which the direction of the maximum principal strain forms with the global x axis (See Figure 13).

Vector $\Delta\mathbf{e}^{cr}$ is used to accumulate the plastic strain dissipated along the crack local axes.

(c) The energy dissipated in the structure due to cracking in a load increment is obtained as:

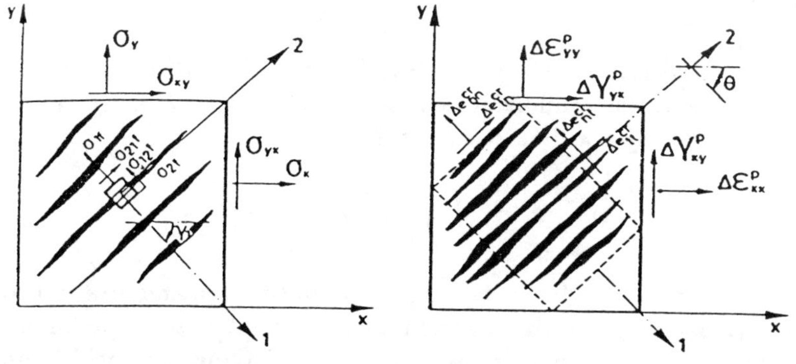

$$\Delta \boldsymbol{\varepsilon}^{cr} = [\Delta e_{nn}^{cr}, \Delta e_{tt}^{cr}, \Delta e_{nt}^{cr}]^T \quad ; \quad \Delta \boldsymbol{\varepsilon}^{p} = [\Delta \varepsilon_{xx}^{p}, \Delta \varepsilon_{yy}^{p}, \Delta \varepsilon_{xy}^{p}]^T$$

Figure 13 Direction of cracking at a damaged point.

$$\Delta W^p = \int_V \boldsymbol{\sigma}^T \cdot \Delta \boldsymbol{\varepsilon} \cdot dV \tag{57}$$

where V is the volume of the structure.

(d) The model also allows to obtain the shear retention factor a crack as $\beta = \tau / \tau^e$ where τ is the actual shear stress parallel to the direction of the crack and τ^e is the value of τ obtained from a linear elastic analysis.

Therefore, the elasto–plastic model proposed here allows the computations of all the necesary information for fully defining the state of cracking in the structure. However, the fact that all this information is obtained *a posteriori* can be considered a clear advantge with respect to other discrete or smeared cracking models, which involve detailed transformations during the non–linear numerical solution stage.

4 BEHAVIOUR OF REINFORCING STEEL IN TENSION AND COMPRESSION

The reinforcing bars are considered as steel layers of equivalent thickness in the present model. Each steel layer has an uniaxial behaviour resisting only the axial force in the bar direction. A bilinear or a trilinear idelization can be adopted in order to model the elasto–plastic stress-strain relationship. The basic relationships for uniaxial elasto -plastic behaviour and the corresponding numerical formulation can be found in reference [49].

5 GENERAL NUMERICAL PROCEDURE FOR NON LINEAR ANALYSIS

During the general stage of the incremental/iterative solution of a finite element elasto–plastic problem, the equilibrium equations will not be exactly satisfed and a system of residual forces $\boldsymbol{\Psi}$ will exist such that

$$\boldsymbol{\Psi}_i^n = \mathbf{p}_i^n - \mathbf{f}_i^n = \int\int_A \mathbf{B}'^T \hat{\boldsymbol{\sigma}}_i'^n dV - \mathbf{f}_i^n \neq 0 \tag{58}$$

in which $\mathbf{f}^n$ and $\mathbf{p}^n$ are respectively the external applied force and internal equivalent force vectores, $\mathbf{B}'$ is the generalized strain/displacement matrix (defined in (18)). $\hat{\boldsymbol{\sigma}}'^n$ is the current (local) stress field (see eq.(11)) satisfying the yield condition, A denotes the area of the shell, the superscript n denotes the load increment number, and subscript i the iteration cycle number within that increment.

An iteration sequence must be performed for each load increment in order to obtain a displacement fiel, $\mathbf{a}_i^n$, which provides a stress field $\boldsymbol{\sigma}_i'^n$ in (58) such that the residuals $\boldsymbol{\Psi}_i^n$ vanish. In particular, the displacements are updated at the end of each iteration according to

$$\mathbf{a}_i^n = \mathbf{a}_{i-1}^n + \Delta\mathbf{a}_i^n \tag{59}$$

where $\Delta\mathbf{a}_i^n$ denotes the displacement change occuring during the iteration. Several options exist for the choice of the displacement search directions. If the tangential stiffness approach is employed the iterative displacement change is evaluated according to [14,16]

$$\Delta\mathbf{a}_i^n = -\left[\mathbf{K}_{T_{i-1}}^n\right]^{-1} \boldsymbol{\Psi}_{i-1}^n \tag{60}$$

in which $\mathbf{K}_{T_{i-1}}^n$ is the tangential stiffness matrix of the structure evaluated at the beginning of the ith iteration. as

$$\mathbf{K}_T = \frac{d\mathbf{p}}{d\mathbf{a}} = \int_V \mathbf{B}'^T \frac{d\boldsymbol{\sigma}'}{d\boldsymbol{\varepsilon}'}\frac{d\boldsymbol{\varepsilon}'}{d\mathbf{a}'} dV = \int_V \mathbf{B}'^T \mathbf{D}^{ep} \mathbf{B}' dV \tag{61}$$

The updated displacements $\mathbf{a}_i^n$ obtained from (59) are used to evaluate the current stresses $\boldsymbol{\sigma}_i'^n$ and hence the resultant stress field and the residual forces by (58). The iteration process is repeated until these residual forces are deemed to be sufficiently close to zero. Detalis of the stress computation are not given and can be found elsewhere [14,16].

It should be noted that assembly and inversion of the full equation system is required for each iteration. A variant on the above algorithm is offered by the initial stiffness scheme in which the original structural stiffness matrix $\mathbf{K}_{T_o}^{o}$ is employed at each stage of the iteration process. This reduces the computational cost per iteration but unfortunately also reduces the rate of convergence of the process. In practice the optimum algorithm is generally provided by updating the stiffnesses at selected iterative intervals only. As an example we could consider two typical possibilities (a) the structural stiffness matrix is updated at the beginning of a load increment and maintained constant during iteration to equilibrium, so that $\mathbf{K}_{T_i}^{n}$ in (60) is replaced by $\mathbf{K}_{T_o}^{n}$. (b) the stiffnesses are updated after the first iteration of each load incement only (i.e. $\mathbf{K}_{T_1}^{n}$ is used in (60)).

6 STIFFENED SHELLS. MODELLING OF ECCENTRIC BEAM STIFFENERS

Beam stiffened shells are very common in practice. Typical examples of application are found in slab–beam bridges, edge beams in shell roofs and beam stiffened ship hulls (Figure 14).

The finite element analysis of these structures precises an adequate modelling of both the beam and the shell structures. For the shell any of the flat or degenerate shell elements formulations presented in previous section can be used. On the other hand the analysis of the beam can be based on standard straight two node beam elements, more elaborated curved beam elements or even more sophisticated beam elements developed from a degeneration of 3D solid elements. Also for each of these element types the hypothesis of orthogonality of the transverse normal sections after deformation can or can not be assumed yielding the so called Euler–Bernoulli or Timoshenko beam theories, respectively.

In this section we will concentrate on the study of beam elements based on Timoshenko beam theory. Only straight elements will be considered. The study is focused on the analysis of reinforced concrete beams for use as eccentric stiffners in concrete shells, and also in slab beam bridges. Therefore, the non linear material behaviour will be treated using a layered approach, similar to that studied for shells. Also non linear geometrical effects will be negleted in the analysis.

6.1 3D Timoshenko beam elements. Basic theory

Let's consider a beam defined in a global coordinate system xyz by the center line S and the different transverse sections of area $A(S)$ (Figure 15). A local coordinate system $\bar{x}\bar{y}\bar{z}$ is defined at the centroid of each section G so that $\bar{x}$ is normal to the transverse section and tangent to the center line in G and $\bar{y}$ and $\bar{z}$ coincide with the two principal directions of the transverse section. For simplicity we will assume that the shear center coincides with G.

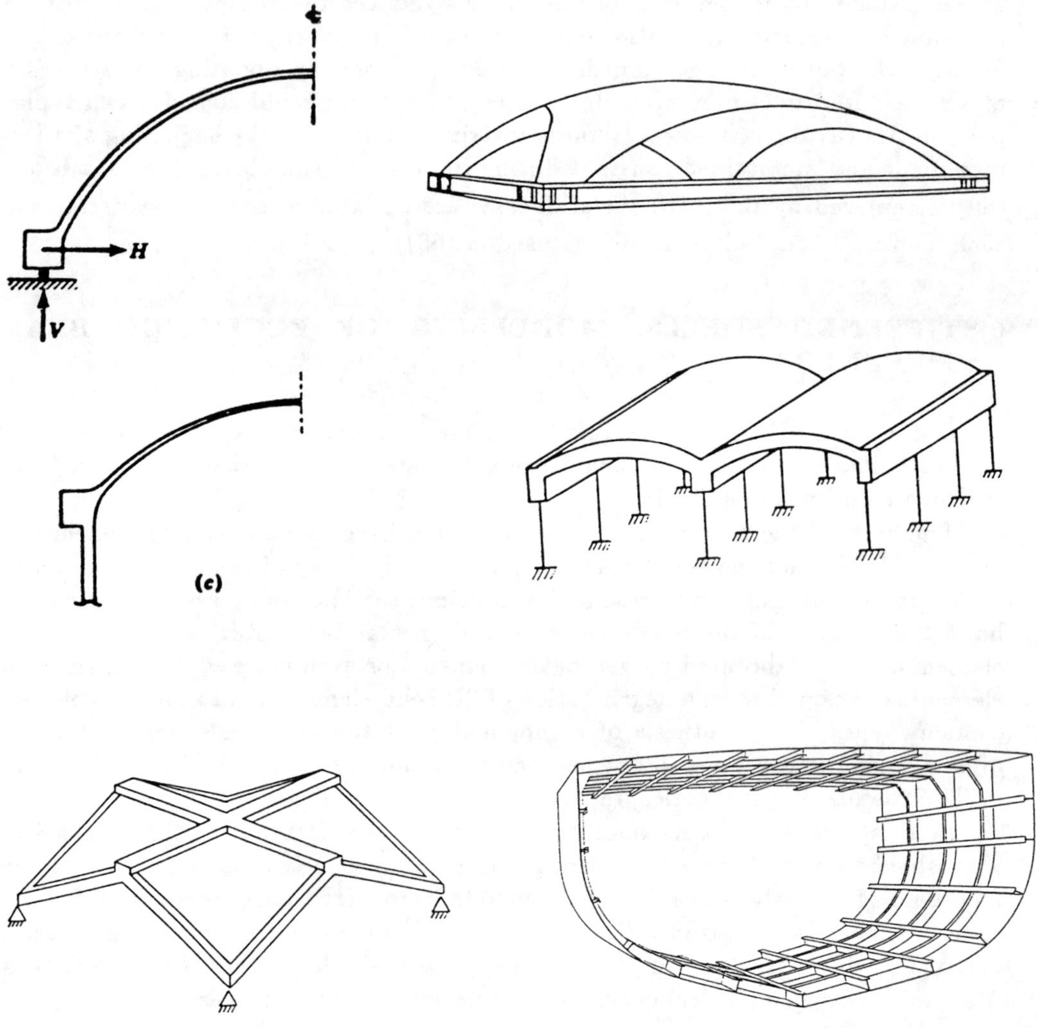

Figure 14 Some beam stiffened structures.

The kinematic description is based on Timoshenko beam theory, i.e. the transverse sections remain plane, but not necessarily normal to the center line after deformation. This assumption is analogous to that of Reissner-Mindlin for shells used in a previous sections.

With this assumption the displacement field can be written as

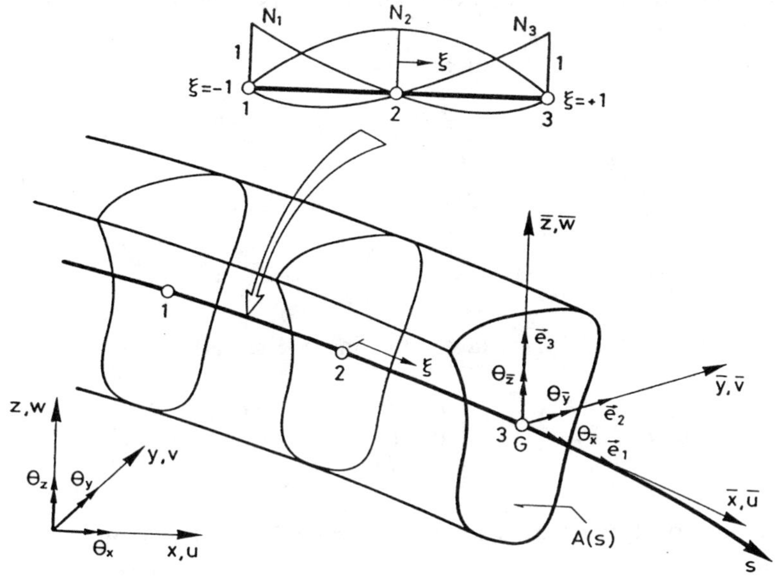

Figure 15 Geometric description of an straight beam element.

$$
\begin{aligned}
\bar{u} &= \bar{u}_G + \bar{z}\theta_{\bar{y}} - \bar{y}\theta_{\bar{z}} \\
\bar{v} &= \bar{v}_G - \bar{z}\theta_{\bar{x}} \\
\bar{w} &= \bar{w}_G + \bar{y}\theta_{\bar{y}}
\end{aligned}
\tag{62}
$$

where $(\cdot)_G$ denotes displacements of the centroid, $\theta_{\bar{x}}$ is the torsional rotation and $\theta_{\bar{y}}, \theta_{\bar{z}}$ are the two rotations about $\bar{y}$ and $\bar{z}$ axes, respectively (see Figure 15).

The local and displacement vectors of a point are related by

$$
\mathbf{u}' = \mathbf{L}\mathbf{u} \tag{63}
$$

with

$$
\begin{aligned}
\mathbf{u}' &= [\bar{u}_G, \bar{v}_G, \bar{w}_G, \bar{\theta}_{\bar{x}}, \bar{\theta}_{\bar{y}}, \bar{\theta}_{\bar{z}}]^T \\
\mathbf{u} &= [u_G, v_G, w_G, \theta_x, \theta_y, \theta_z]^T
\end{aligned}
\tag{64}
$$

$$
\mathbf{L} = \begin{bmatrix} \mathbf{T} & \mathbf{0} \\ \mathbf{0} & \mathbf{T} \end{bmatrix} \quad ; \quad \mathbf{T} = [\mathbf{e}_1, \mathbf{e}_2, \mathbf{e}_3]
$$

where $\mathbf{e}_1, \mathbf{e}_2$ and $\mathbf{e}_3$ are the cartesian components of the unit vectors in local directions $\bar{x}, \bar{y}$ and $\bar{z}$ respectively.

The three non–zero local strains are

$$
\begin{aligned}
\varepsilon_{\bar{x}} &= \frac{\partial \bar{u}}{\partial \bar{x}} = \frac{\partial \bar{u}_G}{\partial \bar{x}} + \bar{z}\frac{\partial \theta_{\bar{y}}}{\partial \bar{x}} - \bar{y}\frac{\partial \theta_{\bar{z}}}{\partial \bar{x}} \\
\gamma_{\bar{x}\bar{y}} &= \frac{\partial \bar{u}}{\partial \bar{y}} + \frac{\partial \bar{v}}{\partial \bar{x}} = \frac{\partial \bar{v}_G}{\partial \bar{x}} - \theta_{\bar{z}} - \bar{z}\frac{\partial \theta_{\bar{x}}}{\partial \bar{x}} \\
\gamma_{\bar{x}\bar{z}} &= \frac{\partial \bar{u}}{\partial \bar{z}} + \frac{\partial \bar{w}}{\partial \bar{x}} = \frac{\partial \bar{w}_G}{\partial \bar{x}} + \theta_{\bar{y}} + \bar{y}\frac{\partial \theta_{\bar{x}}}{\partial \bar{x}}
\end{aligned} \tag{65}
$$

The stress–strain constitutive law is for linear–elasticity

$$
\boldsymbol{\sigma}' = \begin{Bmatrix} \sigma_{\bar{x}} \\ \tau_{\bar{x}\bar{y}} \\ \tau_{\bar{x}\bar{z}} \end{Bmatrix} = \begin{bmatrix} E & 0 & 0 \\ 0 & G & 0 \\ 0 & 0 & G \end{bmatrix} \begin{Bmatrix} \varepsilon_{\bar{x}} \\ \gamma_{\bar{x}\bar{y}} \\ \gamma_{\bar{x}\bar{z}} \end{Bmatrix} = \mathbf{D}'\boldsymbol{\varepsilon}' \tag{66}
$$

If a layer approach is used as shown in Figure 16 (66) will be written for each individual layer. Also, if material non–linearities are considered, eq.(66) will be expressed in the standard incremental form as

$$
d\boldsymbol{\sigma}' = \mathbf{D}'_{ep} d\boldsymbol{\varepsilon}'
$$

where $\mathbf{D}'_{ep}$ is the elasto-plastic constitutive matrix obtained using non–linear material model for concrete explained in Section 3.

The virtual work principle is written as

$$
\int_l \delta\hat{\boldsymbol{\varepsilon}}'\hat{\boldsymbol{\sigma}} = \int_l \delta\mathbf{u}^T \mathbf{t}\, ds + \sum_i \delta\mathbf{u}_i^T \mathbf{p}_i \tag{67}
$$

where

$$
\begin{aligned}
\mathbf{t} &= [t_x, t_y, t_z, m_x, m_y, m_z]^T \\
\mathbf{p}_i &= [P_{x_i}, P_{y_i}, P_{z_i}, M_{x_i}, M_{y_i}, M_{z_i}]^T
\end{aligned} \tag{68}
$$

are the distributed and point load vectors in global axes, respectively.
where

$$
\hat{\boldsymbol{\varepsilon}}' = \left[\frac{\partial \bar{u}_G}{\partial s}, \frac{\partial \theta_{\bar{y}}}{\partial s}, -\frac{\partial \theta_{\bar{z}}}{\partial s}, \frac{\partial \theta_{\bar{x}}}{\partial s}, \left(\frac{\partial \bar{v}_G}{\partial s} - \theta_{\bar{z}}\right), \left(\frac{\partial \bar{w}_G}{\partial s} + \theta_{\bar{y}}\right)\right]^T \tag{69}
$$

is the local generalized strain vector in which $\frac{\partial \bar{w}_G}{\partial s}$ is the axial elongation, $\frac{\partial \theta_{\bar{y}}}{\partial s}$ and $\frac{\partial \theta_{\bar{z}}}{\partial s}$ are the beam curvatures, $\frac{\partial \theta_{\bar{x}}}{\partial s}$ the gradient of torsional rotation and $\frac{\partial \bar{v}_G}{\partial s} - \theta_{\bar{z}}$ and $\frac{\partial \bar{w}_G}{\partial s} + \theta_{\bar{y}}$ the shear deformations.

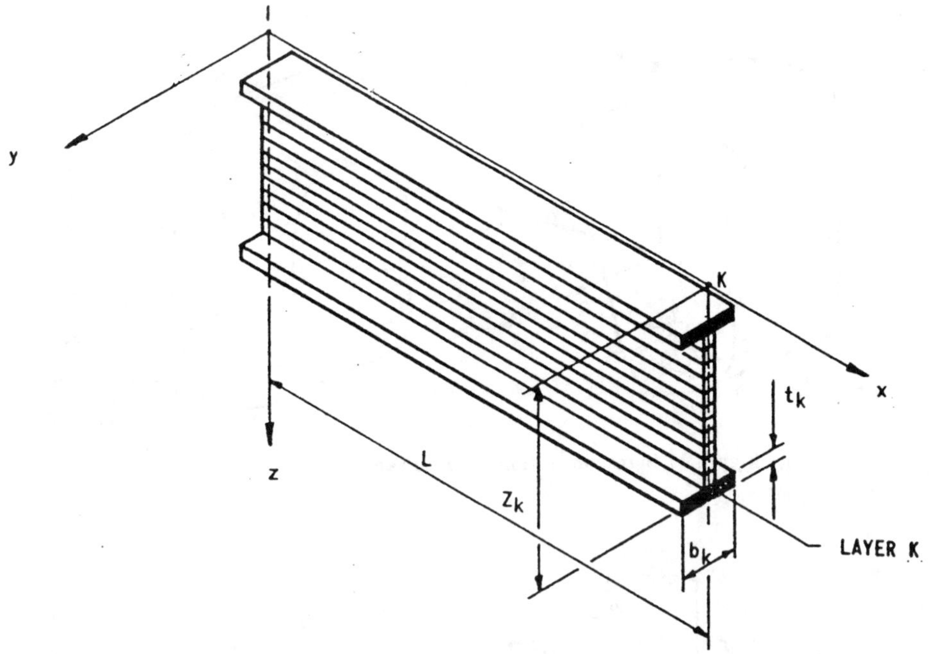

Figure 16 Layer discretization of a beam.

Also in (68)

$$\hat{\sigma}' = [N, M_{\bar{y}}, M_{\bar{z}}, T, Q_{\bar{y}}, Q_{\bar{z}}]^T \tag{70}$$

is the vector of local resultant stresses given by (see Figure 17)

$$[N, M_{\bar{y}}, M_{\bar{z}}, T, Q_{\bar{y}}, Q_{\bar{z}}]^T = \int\int_A [\sigma_{\bar{x}}, \bar{z}\sigma_{\bar{x}}, \bar{y}\sigma_{\bar{x}}, (\bar{y}\tau_{\bar{x}\bar{z}} - \bar{z}\tau_{\bar{x}\bar{y}}), \tau_{\bar{x}\bar{y}}, \tau_{\bar{x}\bar{z}}]^T dA \tag{71}$$

From (66) and (71), the relationship between resultant stresses and generalized strains can be written as

$$\hat{\sigma}' = \hat{\mathbf{D}}'\hat{\varepsilon}' \tag{72}$$

where

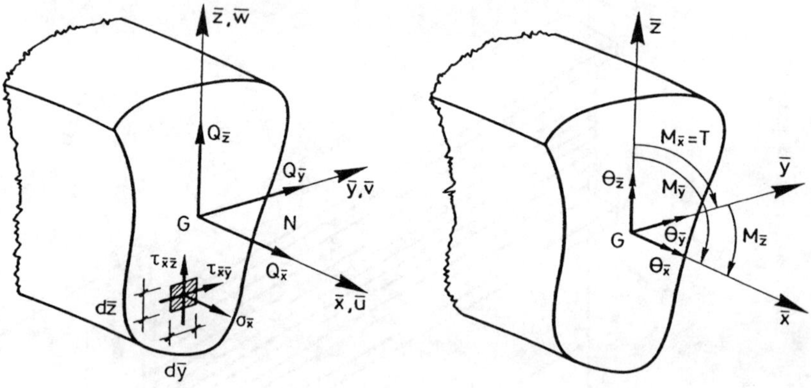

Figure 17 Sign convention for resultant stresses.

$$\hat{\mathbf{D}}' = \int\int_A \begin{bmatrix} E & & & & & \mathbf{0} \\ & \bar{z}^2E & & & & \\ & & \bar{y}^2E & & & \\ & & & (\bar{y}^2+\bar{z}^2)G & & \\ & & & & G & \\ \mathbf{0} & & & & & G \end{bmatrix} dA \tag{73}$$

In a layer formulation the integral in eq.(73) must be computed taking into account the material properties of each concrete or steel layer. For homogeneous material eq.(73) reduces to the well known expression

$$\hat{\mathbf{D}}' = \begin{bmatrix} EA & & & & & \mathbf{0} \\ & EI_{\bar{y}} & & & & \\ & & EI_{\bar{z}} & & & \\ & & & GJ & & \\ \mathbf{0} & & & & \alpha_{\bar{y}}GA & \\ & & & & & \alpha_{\bar{z}}GA \end{bmatrix} \tag{74}$$

where $I_{\bar{y}}$ and $I_{\bar{z}}$ are the principal moments inertia of the section, J is the torsional inertia ($J = I_{\bar{y}} + I_{\bar{z}}$) and $\alpha_{\bar{y}}$ and $\alpha_{\bar{z}}$ are the warping coefficients (for rectangular sections $\alpha_{\bar{y}} = \alpha_{\bar{z}} = 5/6$)

For non linear material analysis eq.(73) is written in an incremental form and the computation of $\hat{\mathbf{D}}'_{ep}$ involves now integration of the tangent material operator across the different layers.

6.2 Formulation of Timoshenko beam element

We consider a discretization of the beam in one dimensional straight C_o finite elements as the three node element shown in Figure 14.

The local displacements $\mathbf{u}'$ are interpolated as

$$\mathbf{u}' = \sum_{i=1}^{n} N_i \mathbf{I}_6 \mathbf{a}_i'^{(e)} \quad ; \quad \mathbf{a}_i'^{(e)} = [\bar{u}_i, \bar{v}_i, \bar{w}_i, \theta_{\bar{x}_i}, \theta_{\bar{y}_i}, \theta_{\bar{z}_i}]^T \tag{75}$$

From (69) and (75) we deduce

$$\hat{\varepsilon}' = \sum_{i=1}^{n} \mathbf{B}_i' \mathbf{a}_i'^{(e)} = \sum_{i=1}^{n}{}' \mathbf{B}_i' \mathbf{L}_i \mathbf{a}_i^{(e)} = \sum_{i=1}^{n} \mathbf{B}_i \mathbf{a}_i^{(e)} \tag{76}$$

where

$$\mathbf{B}_i = \mathbf{B}_i' \mathbf{L} \tag{77}$$

and

$$\mathbf{B}_i' = \begin{bmatrix} \frac{\partial N_i}{\partial s} & 0 & 0 & 0 & 0 & 0 \\ 0 & 0 & 0 & 0 & \frac{\partial N_i}{\partial s} & 0 \\ 0 & 0 & 0 & 0 & 0 & -\frac{\partial N_i}{\partial s} \\ 0 & 0 & 0 & \frac{\partial N_i}{\partial s} & 0 & 0 \\ 0 & \frac{\partial N_i}{\partial s} & 0 & 0 & 0 & -N_i \\ 0 & 0 & \frac{\partial N_i}{\partial s} & 0 & N_i & 0 \end{bmatrix} \tag{78}$$

is the local generalized strain matrix of node i.

The element stiffness matrix and the equivalent nodal load vector can be obtained in the standard manner as

$$\mathbf{K}_{ij}^{(e)} = \int_{l^{(e)}} \mathbf{B}_i^T \hat{\mathbf{D}}' \mathbf{B}_j ds \tag{79}$$

$$\mathbf{f}_i^{(e)} = \int_{l^{(e)}} \mathbf{N}_i^T \mathbf{t} ds + \mathbf{p}_i \tag{80}$$

The integrals along the element lenght are computed using numerical integration in the standard manner [14,15,25],

Timoshenko beam elements suffer from shear locking behaviour when used for slender beam situations. The simplest approach to avoid locking in this case is to use *reduced integration* for the shear terms in the stiffness matrix. In fact, it has been proved that a *uniform reduced integration* for all terms of $\mathbf{K}^{(e)}$ provides excellent results [15,25]. This simply requires to use of one, two and three point quadratures for the linear, quadratic and cubic beam elements, respectively.

Of particular practical interest is the simple two node linear (straight) beam element with a single integration point for which the stiffness matrix can be explicitely computed as

$$\mathbf{K}_{ij}^{(e)} = \bar{\mathbf{B}}_i^T \bar{\hat{\mathbf{D}}}' \bar{\mathbf{B}}_j l^{(e)} \tag{81}$$

where $(\bar{\cdot})$ denotes values at the element mid–point. The form of $\bar{\mathbf{B}}_i$ can be directly obtained form (78) by making $N_i = \frac{1}{2}$ and $\frac{\partial N_i}{\partial \bar{s}} = \frac{(-1)^i}{l^{(e)}}$.

6.3 Formulation of eccentric beam stiffener elements

Figure 18 shows a typical case of a shell element stiffed by an eccentric beam element. The shell element formulation could be based on any of the flat or degenerated layered shell elements presented in Lecture 1 and will not be repeated here. We will also assume for simplicity that the stiffner is based on the Timoshenko beam theory studied in previous section.

A further assumption will be that the beam stiffener is rigidly connected to one of the two shell surfaces along a nodal line. Therefore displacement compatibility conditions along that line require that the order of the beam and shell elements be the same. Also if both elements are rigidly connected the beam transverse section A_i has the same global rotations as the ith shell node to which it is connected. Finally we will assume that the normal nodal axe of the shell element $\vec{v}_{3_i}$ crosses the centroid of the beam section. Note, however, that it is not necessary that the directions of the nodal vectors in the shell coincide with the local axes of the beam section.

With these assumptions we can express the global displacements of the beam centroid G_i in terms of the displacements of the shell ith node as

$$\vec{u}_{G_i} = \vec{u}_i + \vec{\theta}_i \times \alpha_i \vec{v}_{3_i} \quad ; \quad \vec{\theta}_{G_i} = \vec{\theta}_i \tag{82}$$

or

$$\mathbf{u}_{G_i} = \mathbf{u}_i + \mathbf{A}_i \boldsymbol{\theta}_i \quad ; \quad \boldsymbol{\theta}_{G_i} = \boldsymbol{\theta}_i \tag{83}$$

with

$$\mathbf{u}_{G_i} = [u_{G_i}, v_{G_i}, w_{G_i}]^T$$

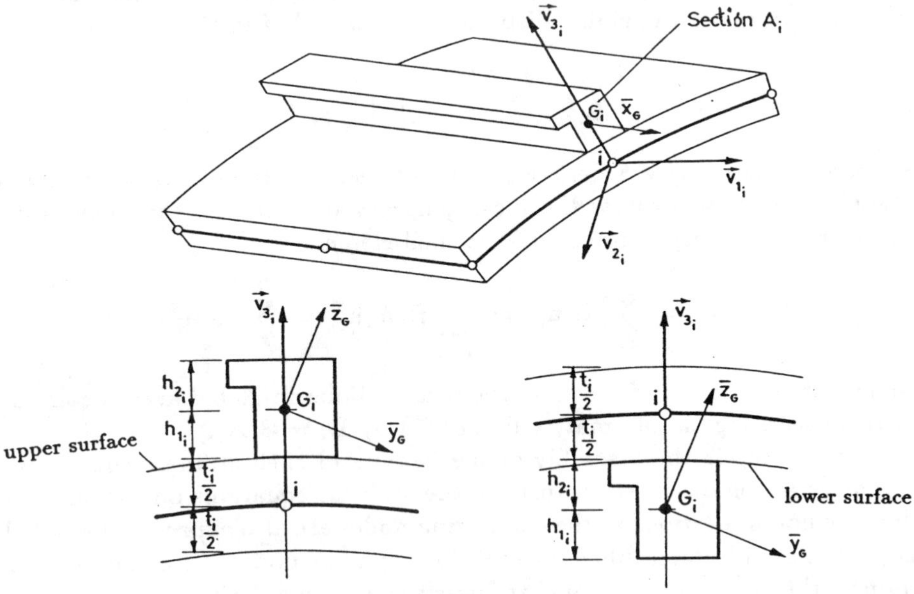

Figure 18 Eccentric beams element.

$$\mathbf{u}_i = [u_i, v_i, w_i]^T \quad ; \quad \mathbf{A}_i = \alpha_i \begin{bmatrix} 0 & v_{3_i}^z & -v_{3_i}^y \\ -v_{3_i}^z & 0 & v_{3_i}^x \\ v_{3_i}^y & -v_{3_i}^x & 0 \end{bmatrix} \tag{84}$$

$$\boldsymbol{\theta}_{G_i} = \boldsymbol{\theta}_i = [\theta_{x_i}, \theta_{y_i}, \theta_{z_i}]^T$$

and

$$\alpha_i = \begin{cases} \left(\dfrac{t_i}{2} + h_{1_c}\right) & \text{if the beam stiffner lays on the } \textit{upper} \text{ surface of the shell} \\ -\left(\dfrac{t_i}{2} + h_{2_c}\right) & \text{if the beam stiffner lays on the } \textit{lower} \text{ surface of the shell} \end{cases}$$

where t_i is the thickness of ith shell node and h_{1_i} and h_{2_i} are the distances of the beam centroid to the upper and lower shell surfaces, as shown in Figure 18. From

(83) we deduce that the global displacements vector of the beam centroid G_i can be expressed in terms of the global displacements of node i in the shell element as

$$\mathbf{a}_i^B = \begin{Bmatrix} \mathbf{u}_{G_i} \\ \boldsymbol{\theta}_{G_i} \end{Bmatrix} = \begin{bmatrix} \mathbf{I} & \mathbf{A}_i \\ \mathbf{0} & \mathbf{I} \end{bmatrix} \begin{Bmatrix} \mathbf{u}_i \\ \boldsymbol{\theta}_i \end{Bmatrix} = \hat{\mathbf{A}}_i \mathbf{a}_i^S \tag{85}$$

where super–indexes B y S denote beam and shell displacements, respectively.

Eq.(85) allows to express the local generalized strains in the beam stiffner in terms of the nodal displacements of the shell element as

$$\hat{\boldsymbol{\varepsilon}}' = \sum_{i=1}^{n} \mathbf{B}_i \mathbf{a}_i^B = \sum_{i=1}^{n} \mathbf{B}_i \hat{\mathbf{A}}_i \mathbf{a}_i^S = \sum_{i=1}^{n} \hat{\mathbf{B}}_i \mathbf{a}_i^S \tag{86}$$

The stiffness matrix of the eccentric beam stiffner in global axes is computed by eq.(79) substituting simply matrix $\mathbf{B}_i$ of (77) by $\hat{\mathbf{B}}_i = \mathbf{B}_i \hat{\mathbf{A}}_i$.

The final step is the assembly of the beam and shell stiffness equations. This may require in some cases to transform the stiffness equations on the shell element so that the nodal rotations at the connecting nodes are also expressed in global axes. This poses no additional difficulty even for coplanar nodes since the beam stiffner introduces the necessary rotational stiffness to avoid singularity of the global stiffness matrix [15].

In some cases the beam stiffner may have an arbitrary orientation with respect the shell mid–surface as shown in Figure 19. The only difference with the formulation described above is that now the relative position of each centroidal node in the beam with respect to the connecting ith shell node has to be precisely defined. Then the distance h_{G_i} between these two points and the unit vector $\vec{e}_{G_i}$ linking the two points must be computed. These two values replace now in (84) the distance α_i and the normal nodal vector $\vec{v}_{3_i}$, respectively. The rest of the formulation is identical to that presented above.

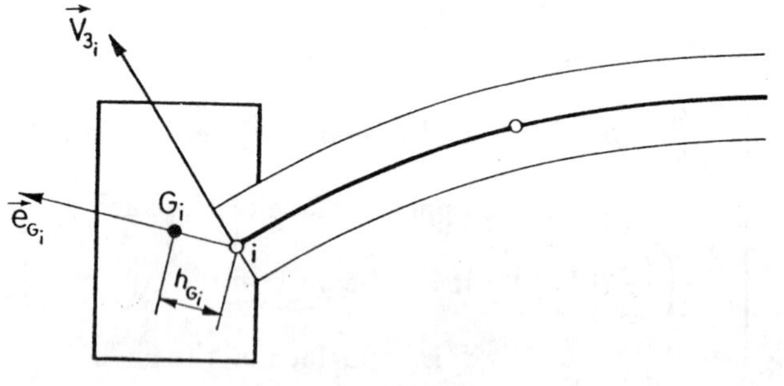

Figure 19 Beam stiffner arbitrary oriented with respect the shell mid surface.

6.4 Analysis of slab–beam bridges

Slab–beam bridges are a particular case of applications of the eccentric beam stiffener formulation described. For simplicity we will consider the bridge shown in Figure 20, formed by an assembly of a rectangular slab and straight beams of rectangular cross section. However, the case of inertia varying beams poses no greater difficulty.

In the more general case the slab will behave as a flat shell element and the beam element will require the 3D formulation as described earlier in this lecture. Again, for simplicity we will assume the same basic C_o finite element formulation for both type of elements (Reissner–Mindlin shell elements for the slab and Timoshenko beam elements for the beam).

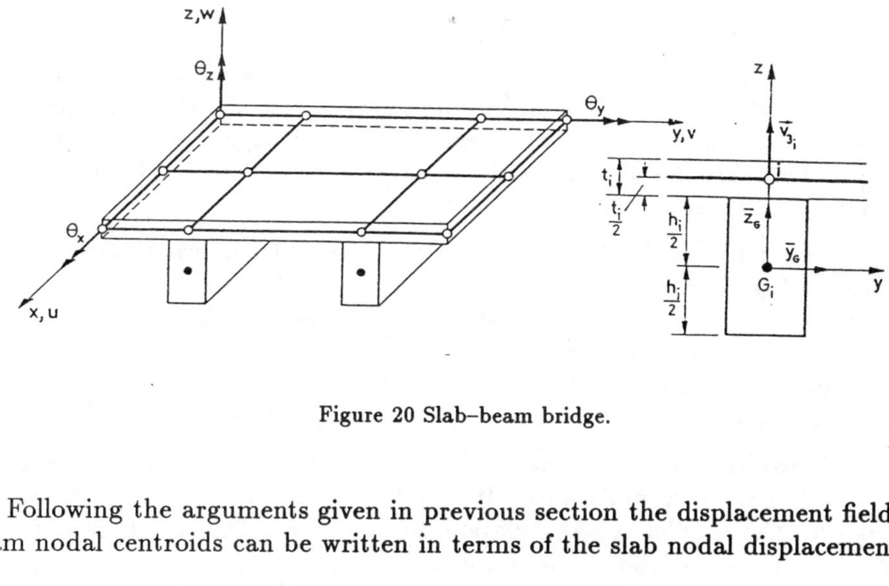

Figure 20 Slab–beam bridge.

Following the arguments given in previous section the displacement field of the beam nodal centroids can be written in terms of the slab nodal displacements as

$$\begin{aligned} u_{G_i} &= u_i + \frac{1}{2}(t_i + h_i)\theta_{y_i} \quad , \quad v_{G_i} = v_i - \frac{1}{2}(t_i + h_i)\theta_{x_i} \\ w_{G_i} &= w_i \quad , \quad \theta_{xG_i} = \theta_{x_i} \quad , \quad \theta_{yG_i} = \theta_{y_i} \quad , \quad \theta_{zG_i} = \theta_{z_i} \end{aligned} \tag{87}$$

Therefore, matrix $\mathbf{A}_i$ of (84) can now simply be written as

$$\mathbf{A}_i = \begin{bmatrix} 0 & \frac{1}{2}(t_i + h_i) & 0 \\ -\frac{1}{2}(t_i + h_i) & 0 & 0 \\ 0 & 0 & 0 \end{bmatrix} \tag{88}$$

The beam stiffness matrix is transformed for direct assembly with that of the slab as

$$\hat{\mathbf{K}}^b_{ij} = \hat{\mathbf{A}}^T_i \mathbf{K}^b_{ij} \hat{\mathbf{A}}_i \tag{89}$$

where $\mathbf{K}^b_{ij}$ is the standard beam stiffness matrix given by eq.(79) and $\hat{\mathbf{A}}_i$ is deduced from (85).

A simpler alternative to analyse slab beam bridges is to neglect the beam eccentricity effect. In this case the slab is modeled with simpler plate elements and the beams with standard 1D beam elements including torsional effects. In the next section we present an example where the different options to analyse a slab–beam bridge are compared.

6.4.1 *Example of linear analysis of a slab–beam bridge*

Figure 21 shows the geometry of a slab beam bridge simply supported in four points of each end section analyzed under unformly distributed loading of $q = 10KN/m^2$. A quarter of the structure has been analyzed due to symmetry as shown in Figure 21. For the slab and beams $E = 10^7 KN/m^2$ and $\nu = 0.3$. The analysis has been carried out with three different elements.

a) 4 node quadrilateral flat shell elements [24] for modelling both the slab and beams (Figure 22).
b) 4 node quadrilateral flat shell elements [24] for the slab and 2 node 3D Timoshenko beam element (Figure 22).
a) 4 node quadrilateral plate elements [24] for the slab and simple 1D Timoshenko beams including torsional effects (Figure 22).

Figure 22 shows the results for the normal stresses at the central section obtained with each of the three formulations. It can be noted that errors in the values computed with the simpler (c) formulation do not exceed 20% of those obtained with the more precise (a) and (b) assumptions.

7 NON LINEAR ANALYSIS OF SHELLS WITH ECCENTRIC BEAM STIFFNESS

The non linear analysis of concrete beam–shell assemblies follows precisely the lines explained for the shell case.Material non linearities in the beam and shell elements are modelled using a layered approach together with an adequate constitutive model for concrete and reinforcing steel as described in Section 3. The layered model allows to monitor onset and evolution of damage (i.e. cracking in concrete or plasticity in steel) at each individual layer in both the shell and beam stiffeners.

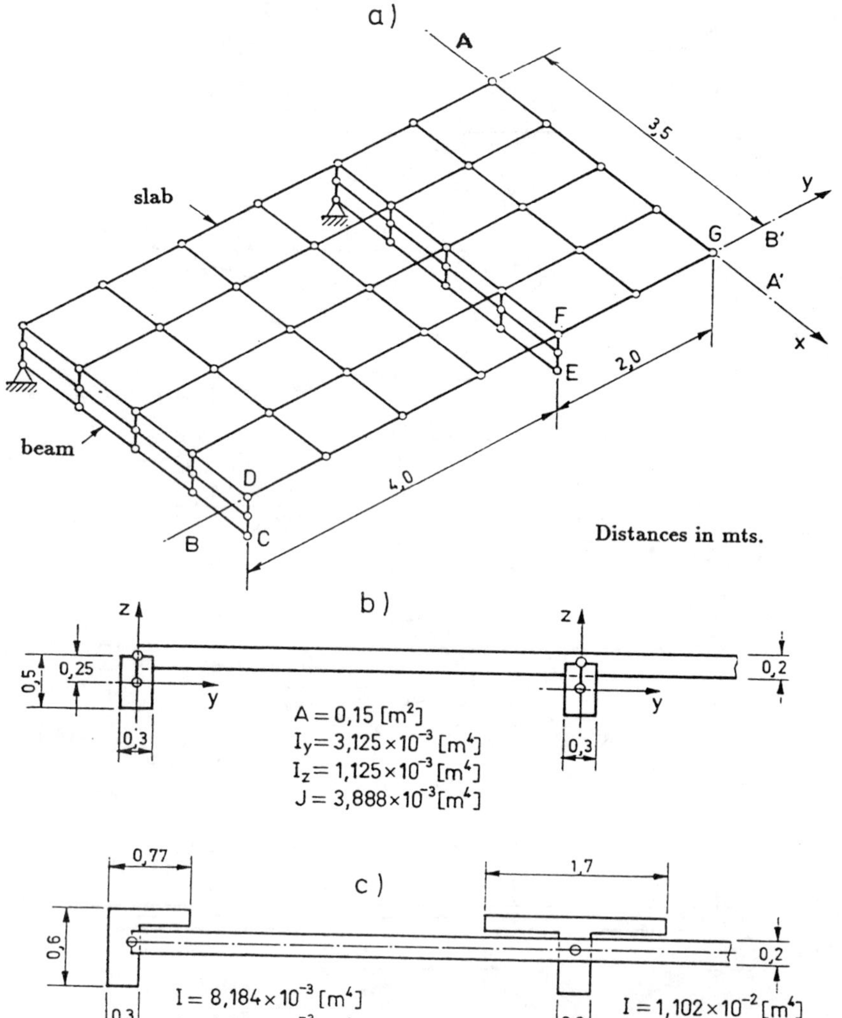

Figure 21 **Slab–beam bridge under uniformly distributed loading. (a) Discretization of slab and beams in flat shell elements (b) Discretization in flat elements (slab) and 2 node 3D Timoshenko beams, (c) Discretization in plate elements (slab) and 2 node 1D Timoshenko beams. 1/4 of the structure analized for symmetry.**

Geometric non linear effects can also be properly taken into account. Details of the adequate geometrically non linear formulations for the beam stiffeners can be found in [25,70].

Some examples of the non linear behaviour of beam–shell assemblies are given in next sections.

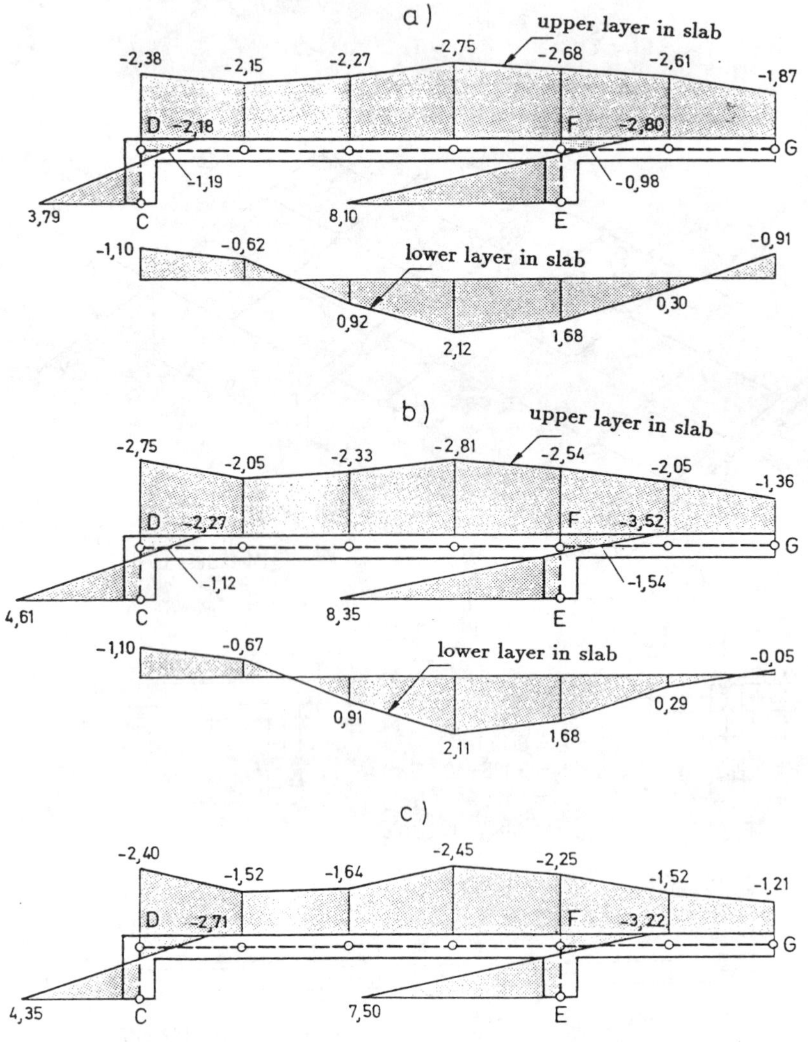

Figure 22 Slab–beam bridge under uniformly distributed loading. Normal stresses in central section for the formulations (a), (b) and (c) as described in Figure 21.

8 SOME EXAMPLES OF FINITE ELEMENT NON LINEAR ANALYSIS OF CONCRETE SHELLS

8.1 Example 1. bending test of a simply supported notched beam under central point load (fracture mode I)

Figure 23 shows the geometry and material properties of the beam and the finite element mesh of isoparametric 8 node Serendipity elements used [14]. The non linear behaviour of concrete has been modelled with the plastic–damage model described in Section 3.

Experimental results for the load–displacement curve for this example are shown in Figure 23 [71] together with the numerical results obtained in the analysis.

In Figure 24 the distribution of principal stresses for load points A,B, and C of Figure 23 is also shown. Note the relaxation of tensible stresses as cracking develops and the intensity of compressive stresses in the upper part ot the beam for the final state which requires adequate modelling of non linear compressive behaviour of concrete in these zones.

Figure 24 also shows an amplification of the notched zone for the final state (point C in the load–displacement curve of Figure 23). Note the strain localization in a band of elements modelling the progression of the crack towards the upper part of the beam.

Figure 24 finally shows the distribution of cracks for the maximum and final loads (points B and C in Figure 23). The length of each crack line represents the amount of plastic deformation in the corresponding ortogonal direction thus providing an indicator of the opening of the crack. The damaged elements zone has been amplified in Figure 24 showing the typical fracture mode I as expected. For further information see [74].

8.2 Example 2 analysis of a notched beam (mixed fracture mode)

This example is a reproduction of the experimental test performed by Arrea and Ingraffes [72]. The geometry of the notched beam, material data and loading conditions used to induce a mixed fracture mode (modes I and II) are shown in Figure 25. As it can be seen the steel beams, used to transmit the loads to the concrete beam, have also been considered in the analysis (assuming a linear behaviour) in order to take in account its rigidity. The numerical analysis was performed using eight–noded two dimensional finite elements, and the mesh used is shown in Figure 25. The *crack mounth sliding displacenent* (CMSD) at the notch tip (see Figure 25) was controlled using a spherical path technique [73]. Again the plastic–damage model for concrete described in Lecture 2 has been used.

Numerical results for the load–CMSD showing the points of onsetting of cracking (point A), instability (point B) and ultimate state analized (point C) have also been plotted in Figure 25. Good agreement with experimental results [72] also plotted in the same figure, is obtained.

In Figure 26 the cracking pattern at the peak (point B of Figure 25) and ultimate load (point C of Figure 25) are shown.It is interesting to note that cracking localizes in a narrow curved band after the peak load, for which all cracks are distributed almost vertically and form an angle of approximately 60° with the horizontal axis. Excellent agreement between the localized cracking band obtained numerically and experimentaly is achieved, as it can be seen in Figure 26.

The principal stress distributions at the onset of cracking (point A of Figure 25)

a)

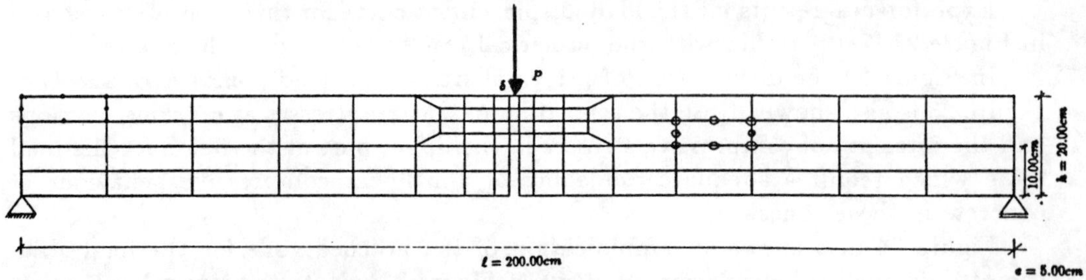

$\bar{\sigma}_e^{max} = 339.45\ kg/cm^2$

$\bar{\sigma}_e^{peak} = 360.00\ kg/cm^2$

$E_0 = 305810.40\ kg/cm^2$ $\quad\bar{\sigma}_e^{max} = 36.00\ kg/cm^2$

$\nu_0 = 0.20$ $\quad G_f = 0.126\ kg/cm$

$\phi_0 =\ cte. = 32°$ $\quad G_e = 12.600\ kg/cm$

$\psi_0 =\ cte. = 32°$ $\quad$ **Associated plasticity (no stiffness degradation)**

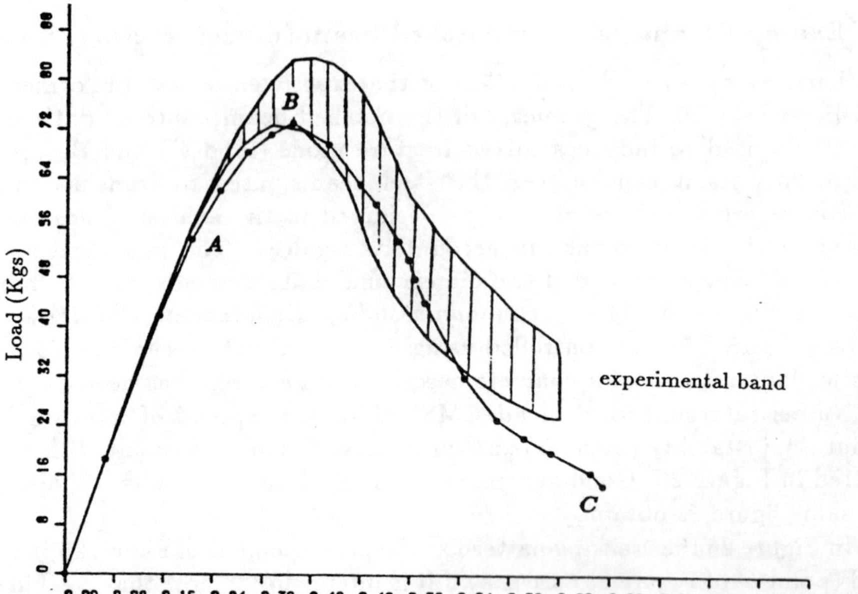

Figure 23 Simple supported notched beam. Load–displacement curve. Point *A*: onset of cracking; Point *B*: maximum load; Point *C*: Final state.

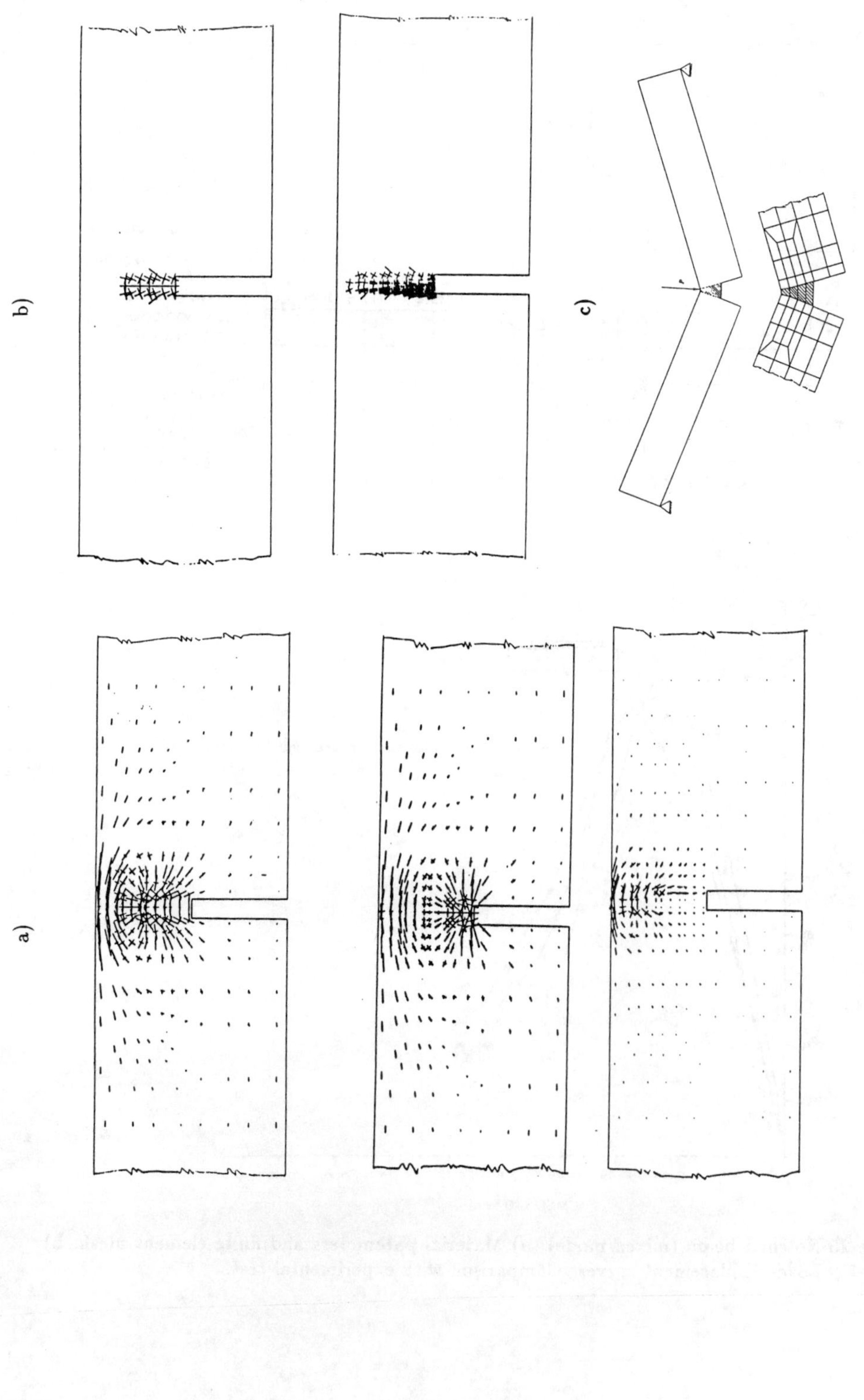

Figure 24 Simply supported notched beam. a) Principal stress distribution for points *A*, *B* and *C* of Figure 23. b) Cracking patterns for points B and C, c) Strain localization in the notched region.

a)

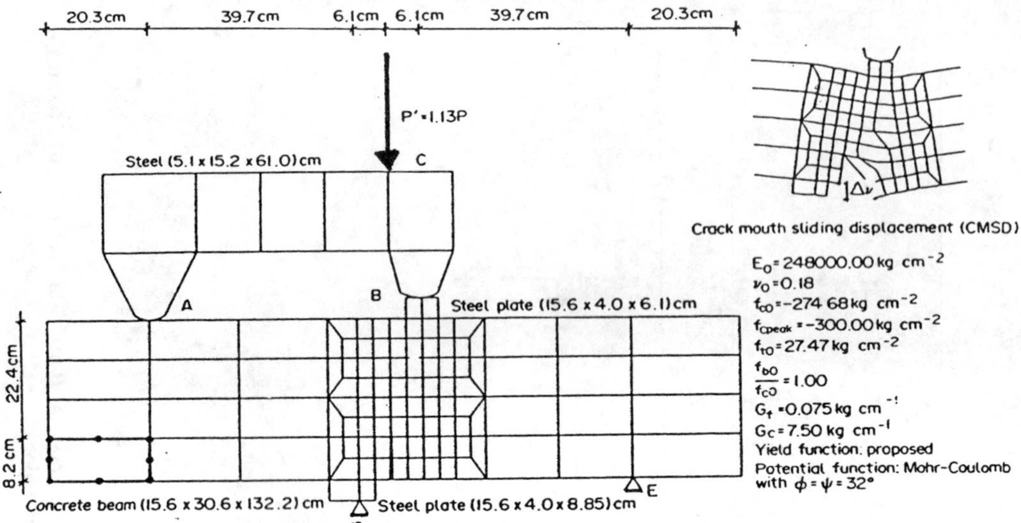

b)

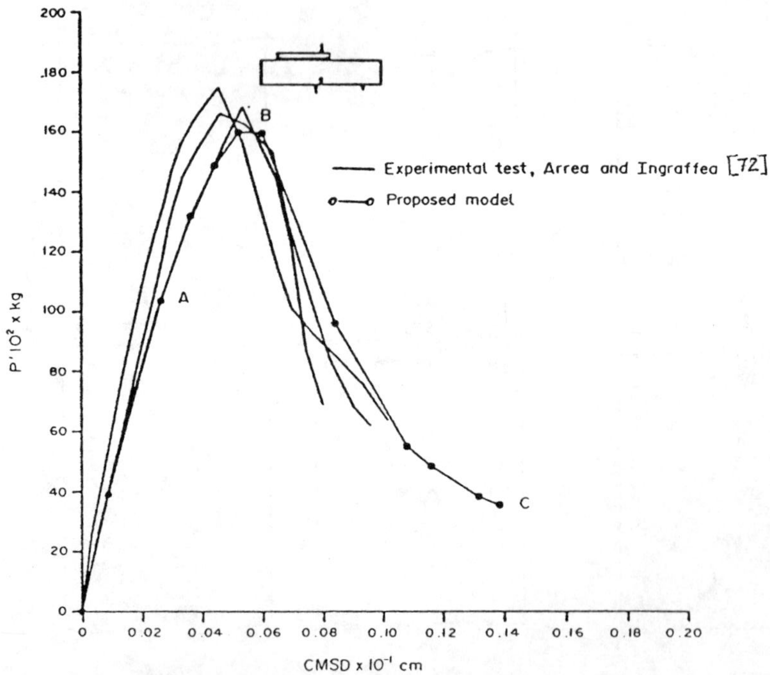

Figure 25 Notched beam (mixed mode). a) Material parameters and finite element mesh, b) Load–displacement curves. Comparison with experimental test.

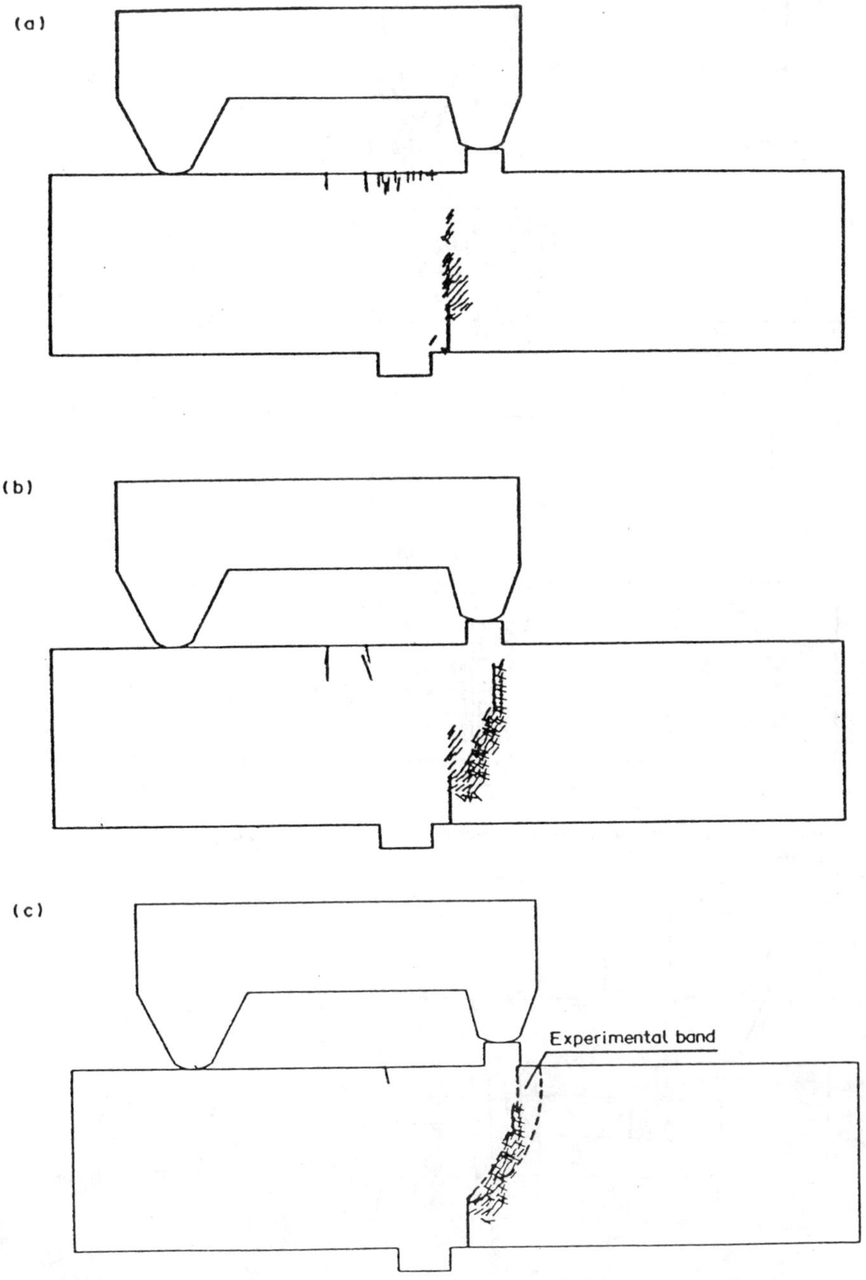

Figure 26 Notched beam (mixed mode). Distribution and localization of cracking a) at t he peak of stress (B) – All cracks; b) at the ultimate state (C) – Cracks greater that 3% of the maximum crack: c) at the ultimate state (C) – Cracks greater that 5% of the maximum crack.

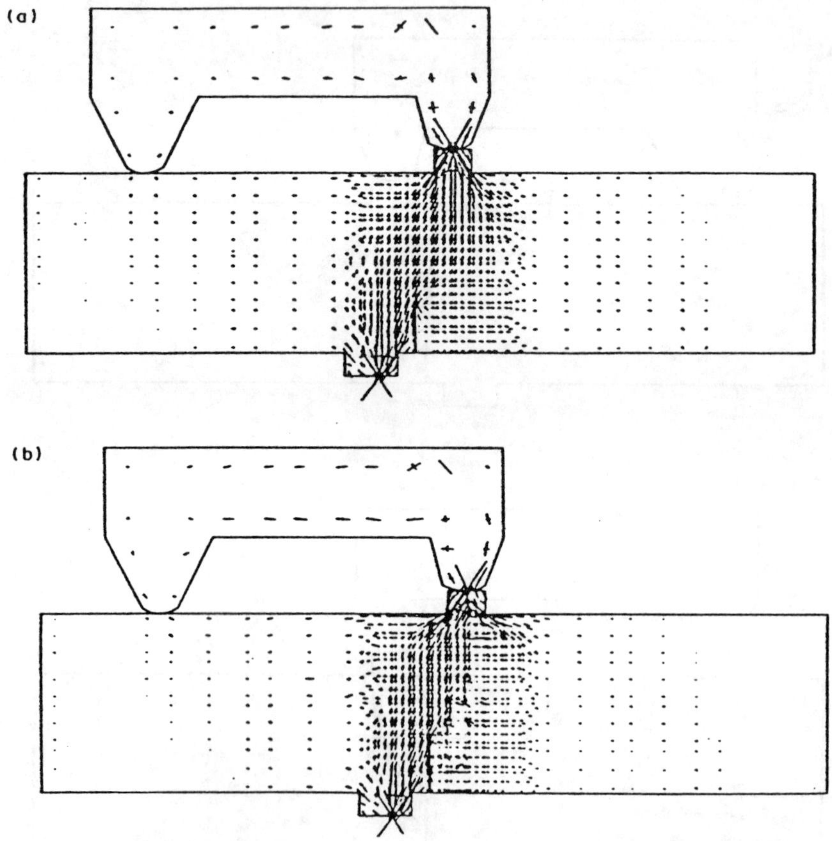

Figure 27 Notched beam mixed mode. Stress state: (a) at the elastic state (A); (b) at the ultimate state (C).

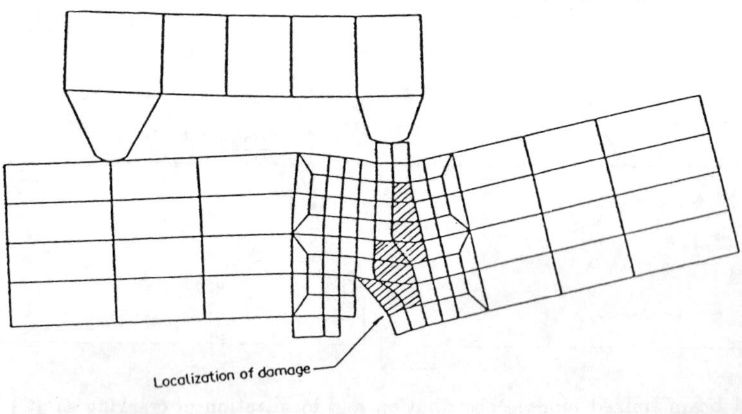

Figure 28 Notched beam mixed mode. Strain locatization for the final state (deformation are amplified 300 times).

and the ultimate state (point C of Figure 25) are shown in Figure 27. It is worth noting the stress relaxation in the zone where cracking localizes. This localization can also be clearly seen in Figure 28, where the deformed shape of the beam (amplified 300 times) at the end of the test is shown. Further details can be found in [64].

8.3 Example 3. Simply supported mixed beam-slab bridge

We consider here the analysis of a simply supported mixed steel beams-reinforced concrete slab bridge using the elasto-plastic-brittle model [16,17]. The geometry of the bridge is shown in Figure 29. Details of the cross section discretization in layers are also shown in Figure 29.

Some typical results of the analysis displaying the load-strain and load-deflection curves for two loading cases corresponding to four point loads acting on the midspan section are shown in Figure 30. Very good agreement of the measured midspan deflections with the numerical results obtained are also shown in Figure 30. For more information on this example see [75].

8.4 Example 4. Cylindrical concrete shell with edge beams

We present next the analysis of two reinforced and prestressed cylindrical shells studied by Roca [76]. The first analysis considers the shell and beam composed of reinforced concrete with geometrical and material properties given in Figure 31. The elasto-plastic-brittle model of ref [16] has again been used. The loading is the following: *Shell*: normal uniform pressure of 7 KPa. *Edge beams*: normal uniform loading of 1.41 KN/m.

Six degenerate shell elements and six Timoshenko beam elements have been used as shown in Figure 31b. The load-deflection curve obtained is shown in Figure 31c.

Figure 32 shows the crack distributions in the shell and beam with and without taking into account the effect of prestressing in the beam.

Finally, Figure 33 shows results for a similar problem considering now the prestressing of both the shell and edge beams. Further information can be obtained from [76].

8.5 Example 5. Cylindrical cryogenic reinforced concrete tank

The last example is the analysis of a cylindrical cryogenic reinforced concrete tank with an spherical dome for liquid gas storage [79]. The tank has been assumed to be fully clamped at the base, for simplicity. The geometry of the tank, material properties and the meshes of axisymetric solid and degenerate shell finite elements used for the thermal and structural problems can be seen in Figure 34. The tank has been analyzed for two different thermal conditions. The first one corresponds to the case of the tank filled with liquid gas at $-160°$C (due to the breakage of an

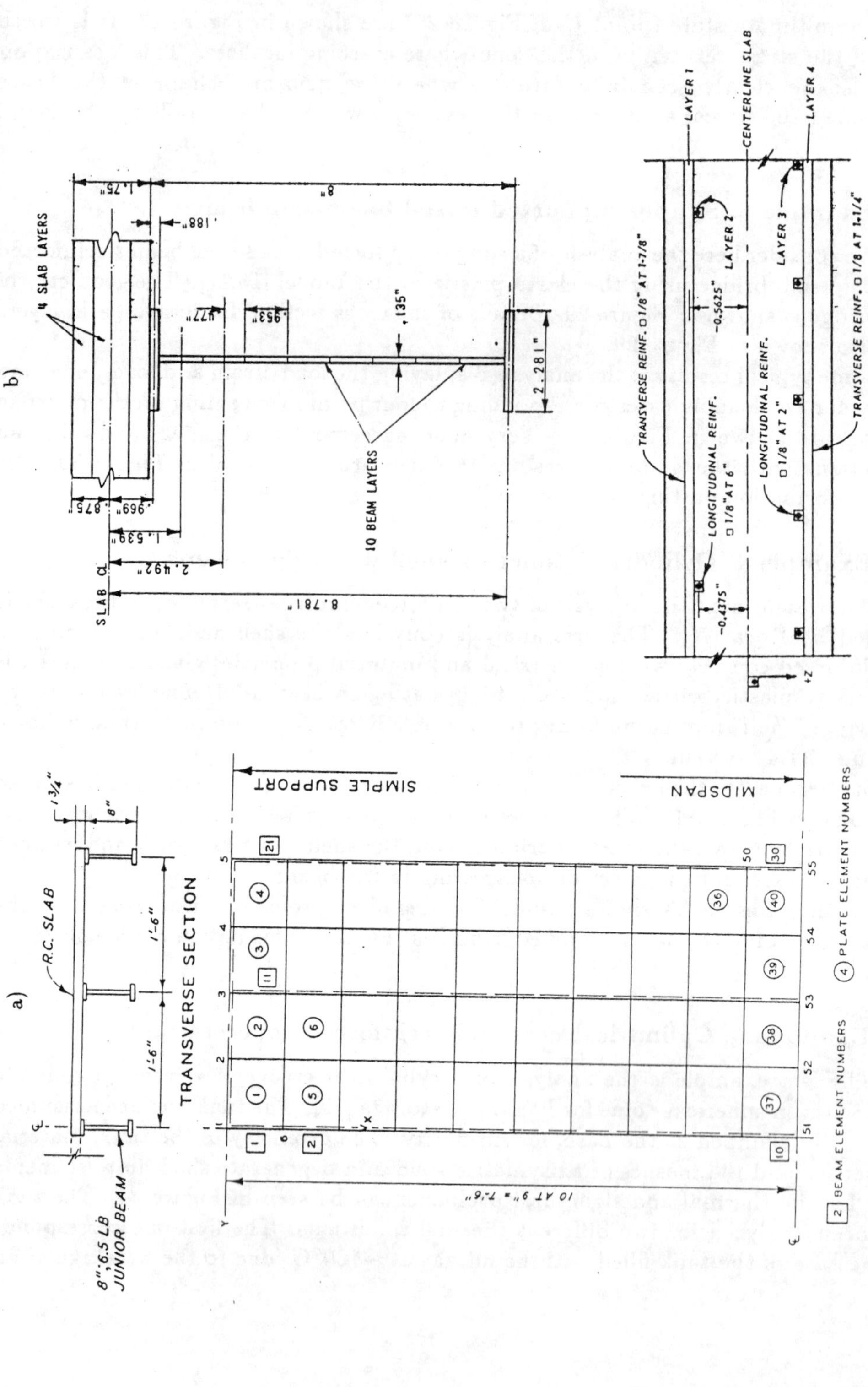

Figure 29 Discretization of simply supported mixed steel-concrete bridge.

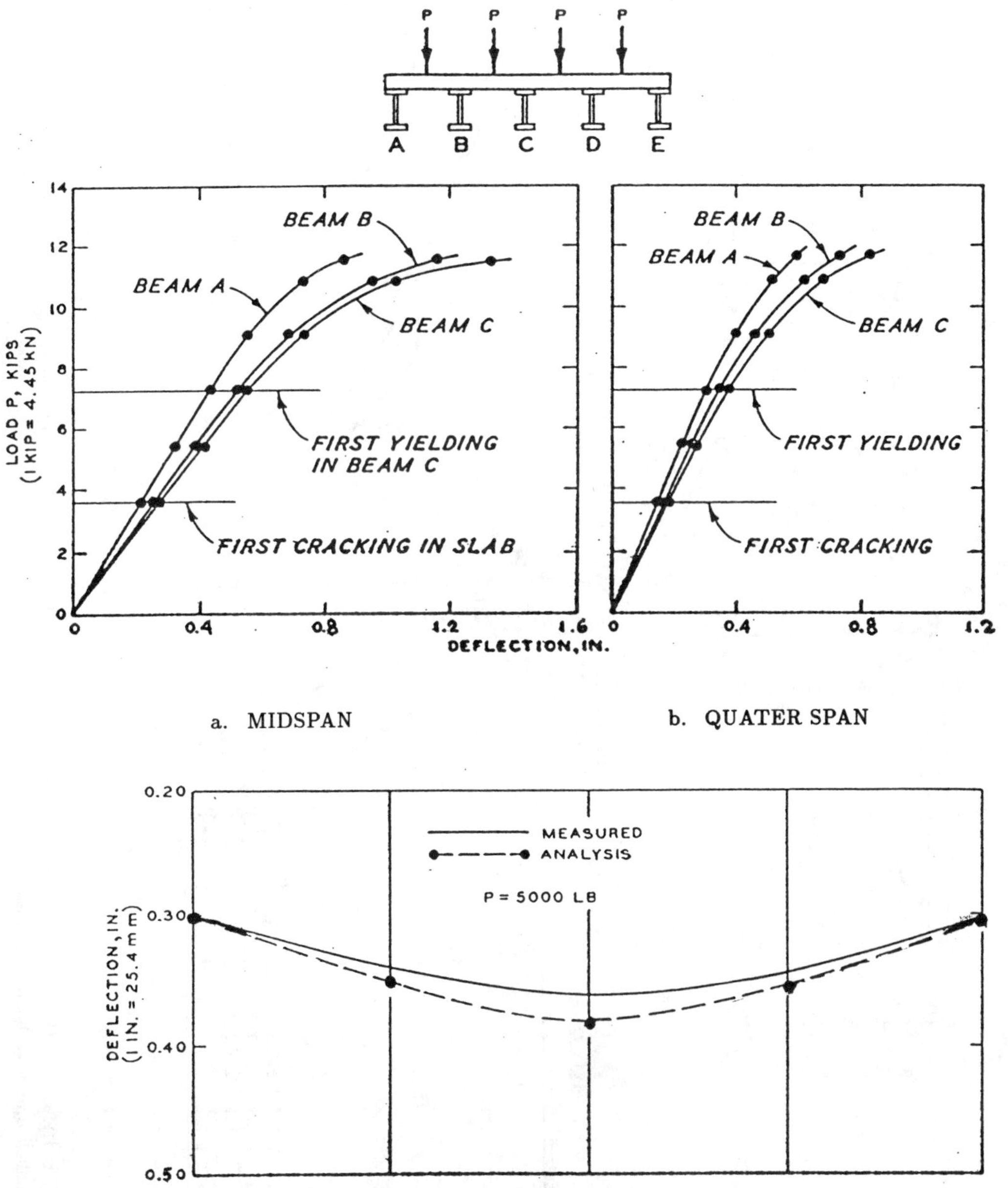

Figure 30 Simply supported mixed steel-concrete bridge. Load-deflection and midspan deflection curves for four central load case.

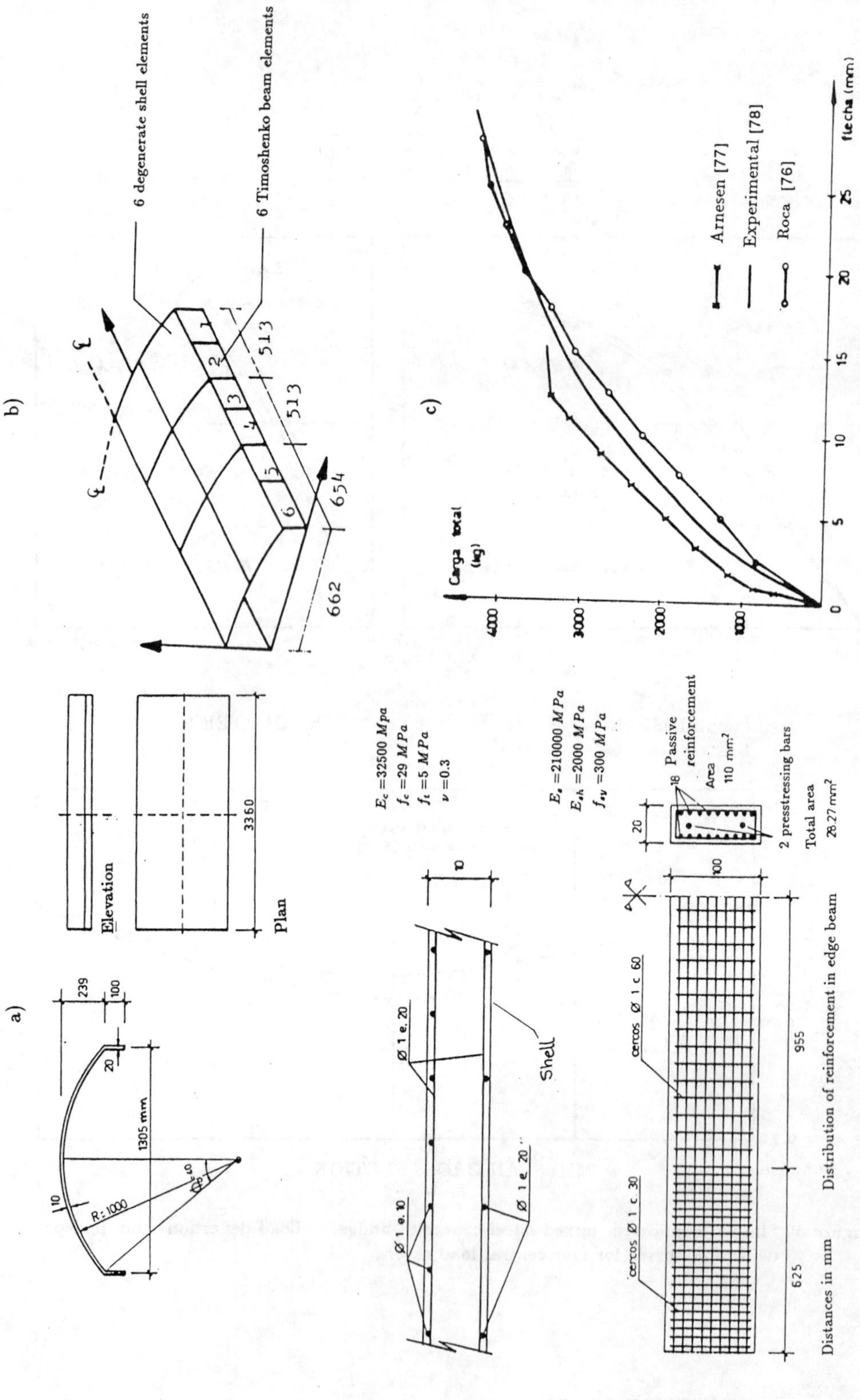

Figure 31 Cylindrical concrete shell. (a) Geometry and reinforcement steel in shell and edge beam, b) Finite element mesh of shell and edge beam, c) Load-displacement curve.

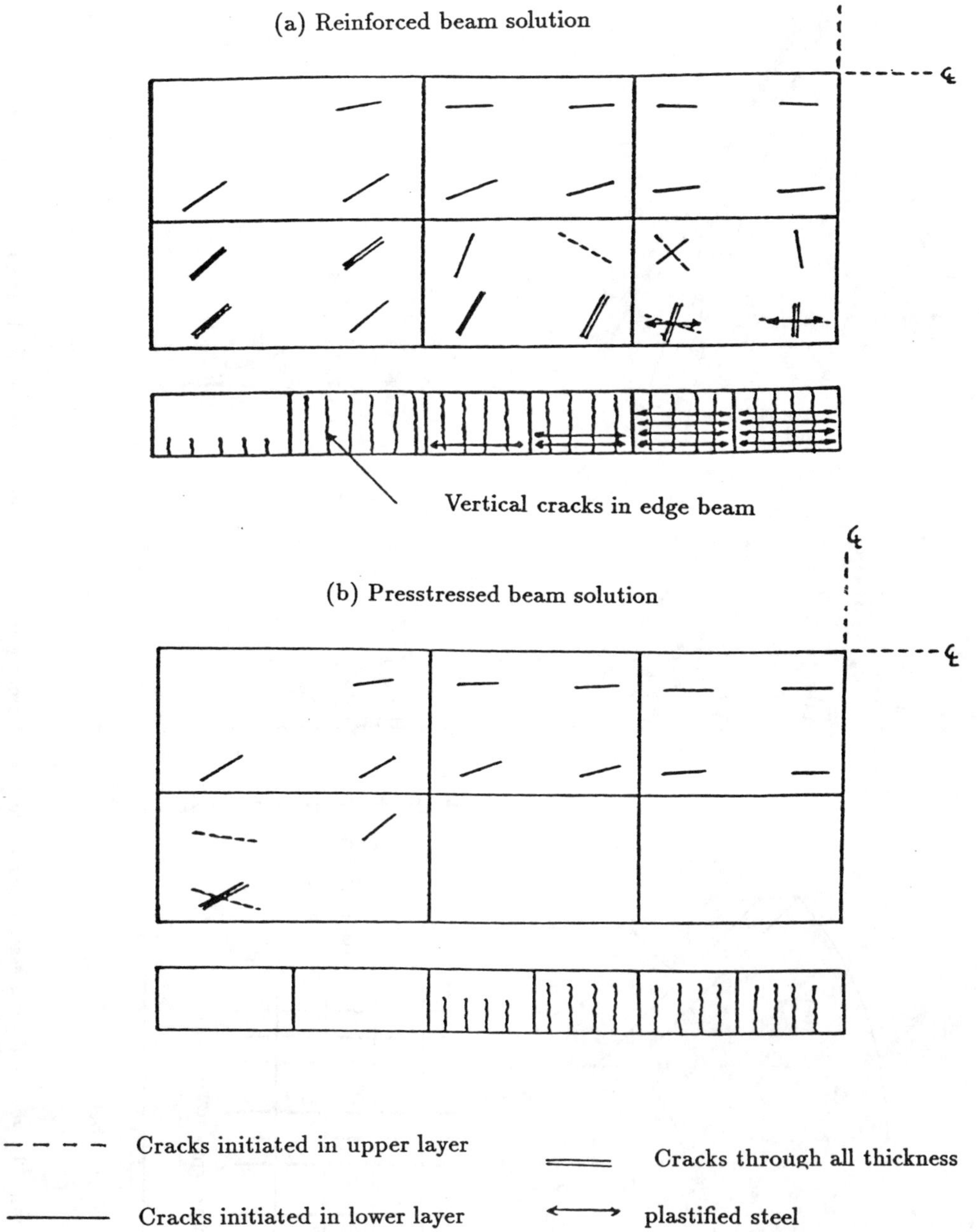

Figure 32 Cylindrical concrete shell. Crack patterns for reinforced steel and prestressing solutions.

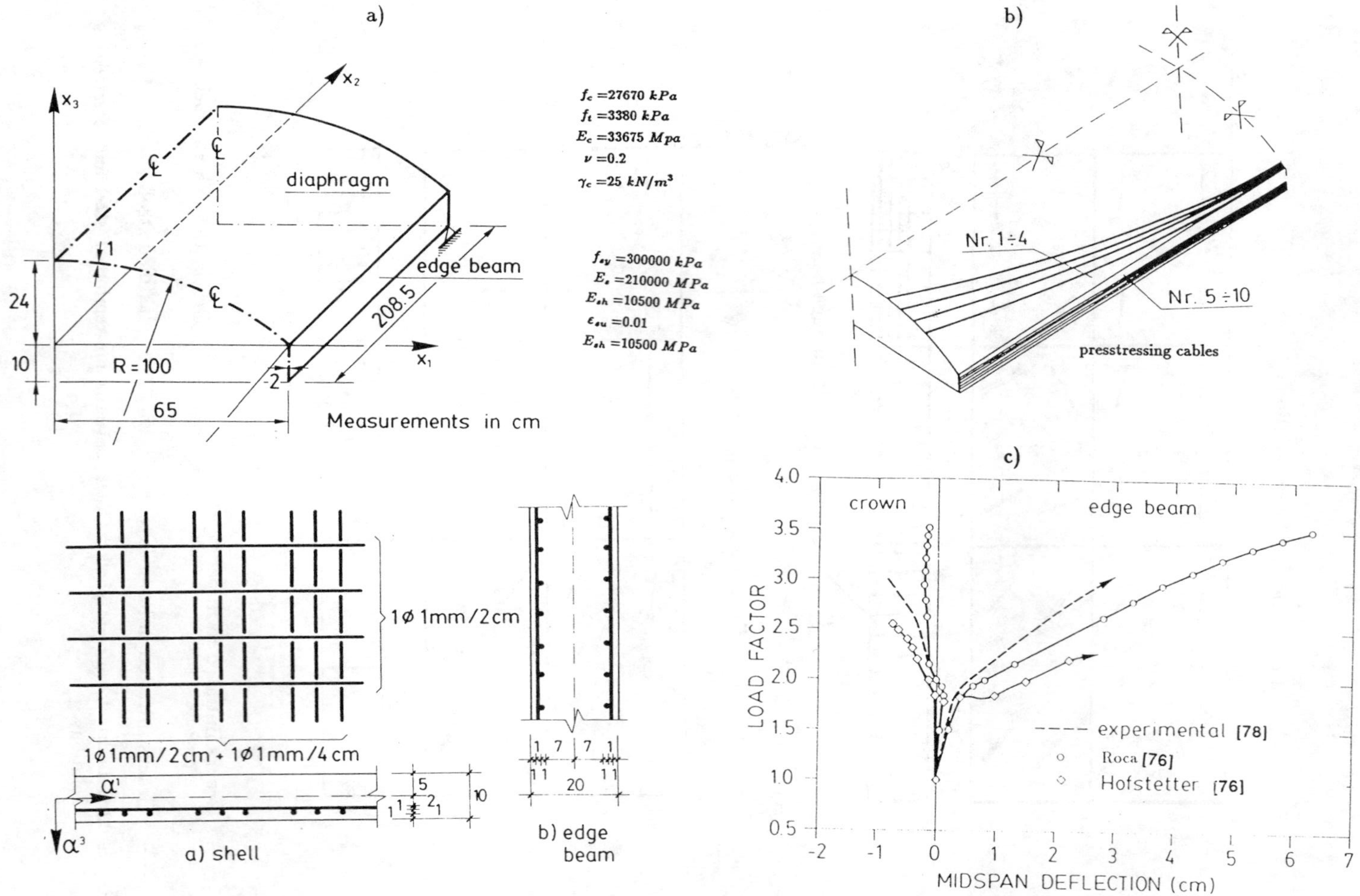

Figure 33 Cylindrical prestressed concrete shell. a) Geometry and reinforcement distribution, b) Presstressing distribution, c) Load–displacement curve

internal container). This problem has been studied in two steps. In the first step the non linear structural response of the tank for external loads such as vertical and circumferential presstressing and self weight has been analyzed. Then the non linear thermal structural behaviour for various temperature distributions obtained at different times has been studied. Results for the temperature contours and for radial and circumferential cracking zones in the tank at various times have been plotted in Figure 35. It can be seen that the initial external loads give place to very localized cracking zones only. The thermal loads, on the other hand, have an special effect on the amount of cracking in the upper part of the dome. The rest of the tank is not severely altered. Further detail can be found in [79].

REFERENCES

1. ASCE Committee on concrete and masonry structures, Task Committee on Finite Element Analysis of Reinforced Concrete Structures: A State-of-The-Art Report on Finite Element Analysis of Reinforced Concrete Structures, *ASCE Spec. Pub.*, 1981.
2. W.C. Schnobrich, "Behaviour of reinforced concrete structures predicted by the finite element method", *Computers and Structures*, Vol. **7**, pp. 365–376, 1977.
3. J.H. Argyris, G. Faust, J. Szimmat, E.P, Warnke and K.J. Willam, "Recent developments i the finite element analysis of stressed concrete reactor vessels", *Nuclear Engineering and Design*, Vol. **28**, pp. 42–75, 1974.
4. P.G. Bergan and I. Holand, "Nonlinear finite element analysis of concrete structures", *Computer Methods in Applied Mechanics and engineering*, Vol. **17/18**, pp. 443–467, 1979.
5. K.H. Gerstle, "Material modelling of reinforced concrete", Introductory Report, IABSE Colloqiuum on Advanced Mechanics of Reinforced Concrete, Delft. Volume–band 33, pp. 41–63, 1981.
6. D. Nod and A.C. Scordelis, "Finite element analysis of reinforced concrete beams", *American Concrete Institute Jounal*, Vol. **6**, No. 3, March, 1967.
7. L. Cedolin and S. Deipoli, "Finite element studies of shear–critical R/C beams, *ASCE Journal of th Engineering Mechanics Division*, Vol. **103**, No. EM3, pp. 395–410, Jue 1977.
8. M.N. Fardis and O. Buyukoztuk, "Shear stiffness of concrete by finite elements", *ASCE Journal pf the Structural Division*, Vol. **106**, No. ST6, pp. 1311–1327, June 1980.
9. Z.P. Bazant and L. Cedolin, "Fracture mechanics of reinforced concrete", *ASCE Journal of Eng. Mech. Division*, Vol. **106**, No.EM6, pp. 1287–13106, Decembe, 1980.
10. C.S. Lin, "Nonlinear analysis of reinforced concrete slabs and shells", PhD. Thesis, UC-SESM 73–7, university of California, April 1973.

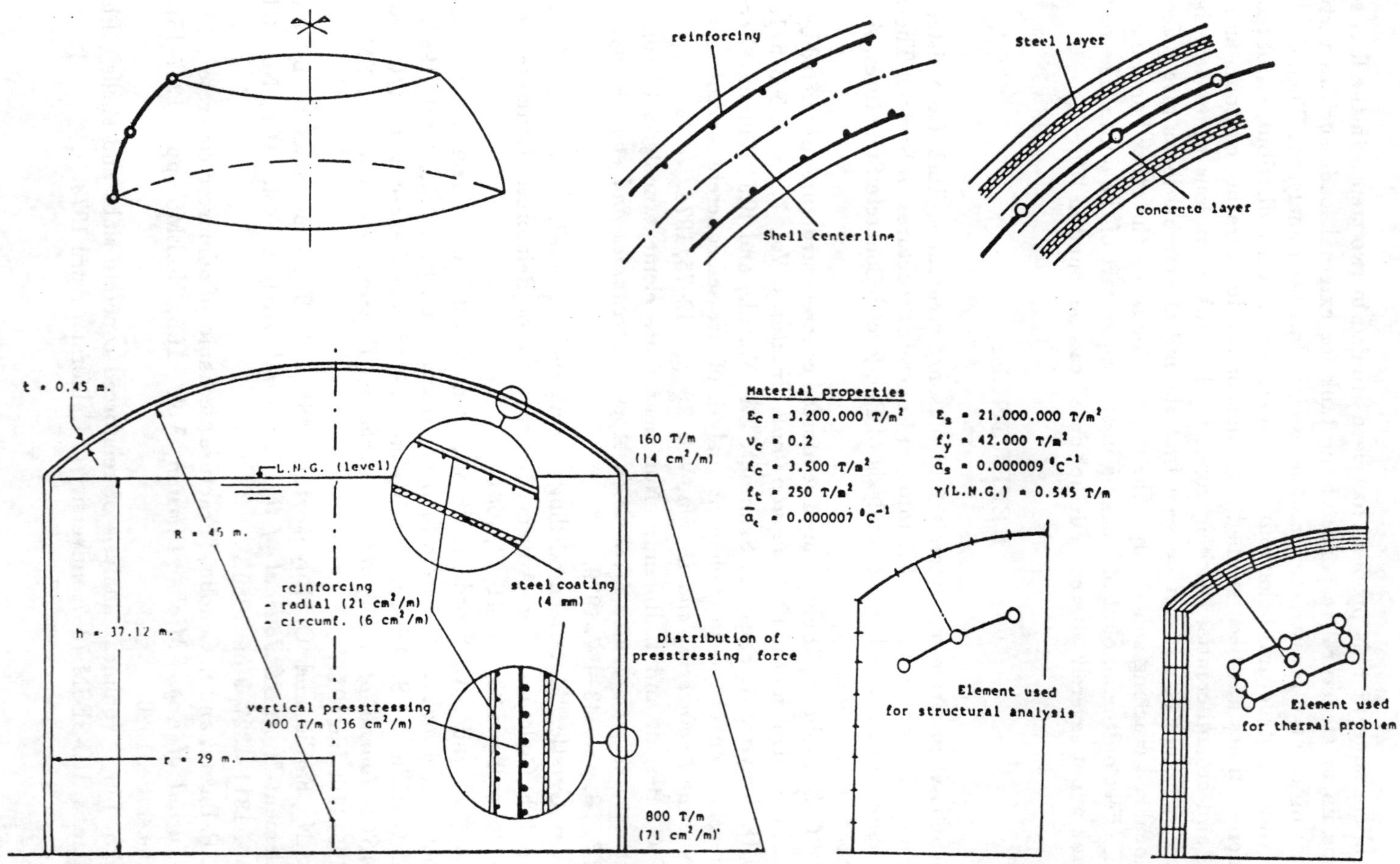

Figure 34 Cylindrical cryogenic concrete reinforcement distribution and layer discretization.

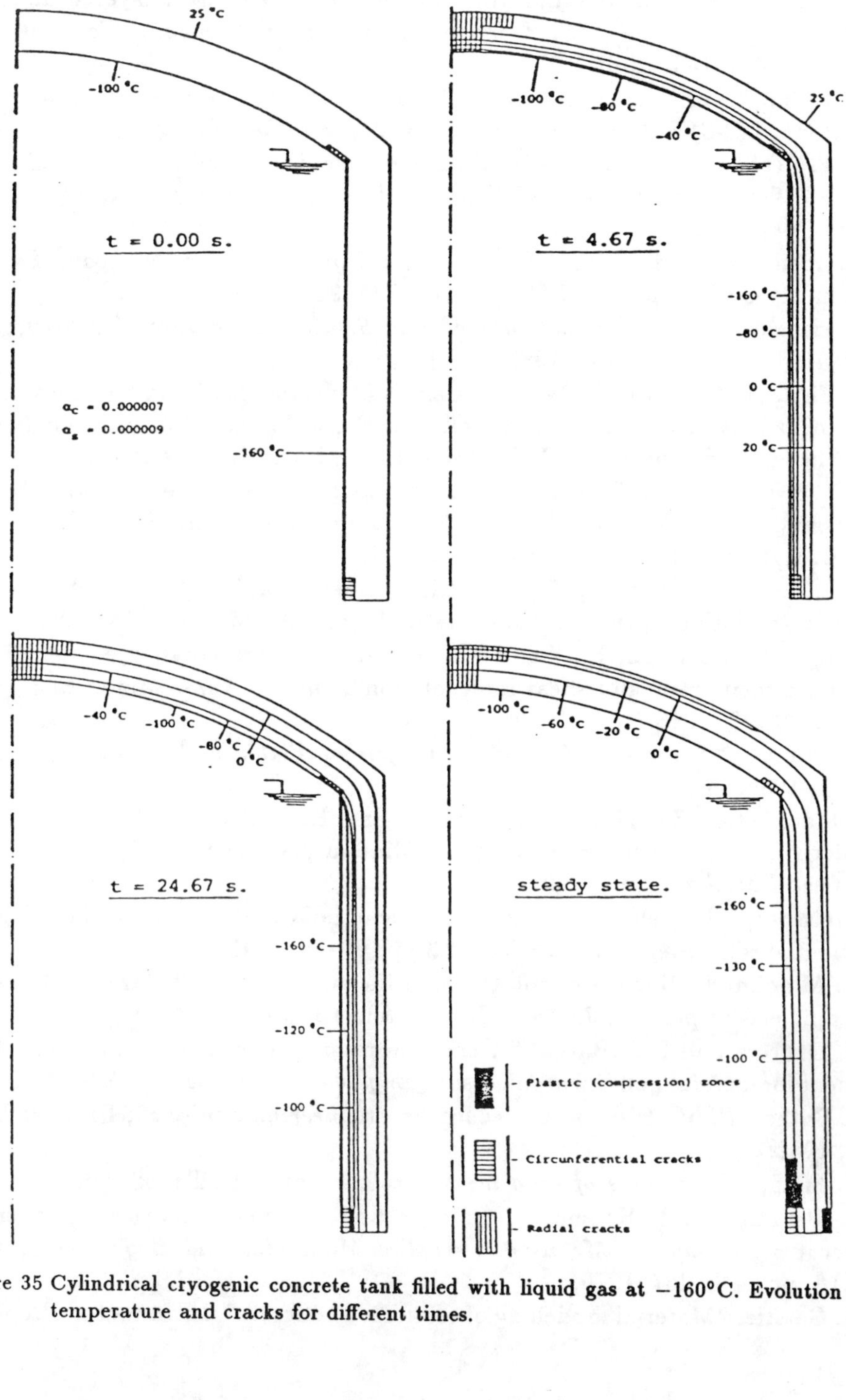

Figure 35 Cylindrical cryogenic concrete tank filled with liquid gas at −160°C. Evolution of temperature and cracks for different times.

11. F.R. Hand, D.A. Pecknold and W.C. Schnobrich, "Nonlinear layered analysis of RC Plates and Shels, *ASCE Journal of Structural Division,* Vol. **99**, No. ST7, pp. 1491–1503, July, 1973.
12. G. Mueller, "Numerical problems in nonlinear analysis of reinforcd concrete", report No. UC-SESM 77-5, University of California, Berkeley, September, 1977.
13. H.H.A. Rahman, "Computational models for the nonlinear of reinforced concrete flexural slab systems", Ph. D Thesis, C7Ph/66/82, University of Wales, May, 1982.
14. O.C. Zienkiewicz, and R.L. Taylor, "*The Finite Element Method*", Fourth Edtion Mc.Graw Hill, Vol. I (1989), Vol II (1991).
15. M. Crisfield, "*Finite Element Method and Solution Procedures for Structural Analysis*", Pineridge Press, 1986.
16. J.A. Figueiras and D.R.J. Owen, "Analysis of elasto–plastic and geometrically non linear anisotropic plates and shells" In Finite Element Software for Plates and Shells, E. Hinton and D.R.J. Owen (Eds.), Pineridge Press, 1984.
17. D.R.J. Owen and J.A. Figueiras, "Ultimate load analysis of reinforced concrete plates and shells", In Finite Element Software for Plates and Shells, E. Hinton and D.R.J. Owen (Eds.), Pineridge Press, 1984.
18. Zienkiewcz, O.C., Qu, S., Taylor, R.L. y Nakazawa,S., "The Patch test for mixed formulations", *Int. J. Num. Meth. Eng.*, Vol. **23**, pp. 1873–1883, 1986.
19. Huang, H.C. y Hinton, E., "A new nine node degenerated shell element with enhanced membrane and shear interpolation", *Int. J. Num. Meth. Eng.*, Vol. **22**, pp. 73–92, 1986.
20. H.C. Huang, "*Static and Dynamic Analysis of Plates and Shells*", Springer Verlag, 1989.
21. E. Oñate, O.C. Zienkiewicz, B. Suárez, and R.L. Taylor, "A methodology for deriving shear constrained Reissner-Mindlin plate elements", *Int. Journal Numerical Methods Eng.*, 1991.
22. E. Reissner, "The effect of transverse shear deformation on the bending of elestic plates", *J. Appl. Mech.*, Vol. **12** pp. 69–76, 1945.
23. R.D. Mindlin, "Influence of rotatory inertia and shear in flexural motions of isotropic elastic plates", *J. Appl. Mech.*, Vol. **18**, pp. 31–38, 1951.
24. E.N. Dvorkin and K.J. Bathe, "A continuum mechanics based four node shell element for general non-linear analysis", *Eng. Comp.*, Vol. **1**, pp. 77–88, 1984.
25. K.J. Bathe, '*Finite Element procedures in engineering analysis*", Prentice Hall Inc., 1982.
26. Y.C. Fung, "*Fundations of solid mechanics*", Prentice Hall Inc. 1965.
27. P.G. Bergan and I. Holand, "Nonlinear finite element analysis of concrete structures", *Computer Methods in Applied Mechanics and Engineering*, Vol. **17/18**, pp. 443–467, 1979.
28. K.H. Gerstle, "Material modelling of reinforced concrete", Introductory Report,

IABSE Colloqiuum on Advanced Mechanics of Reinforced Concrete, Delft. Volume–band 33, pp. 41–63, 1981.

29. H.A. Mang and H. Floegl, "Tension stiffening concept for reinforced concrete surface structres", IABSE Colloquium on Advanced Mechanics of Reinforced Concrete, final Report, pp. 351–369, Delft, June, 1981.
30. H. Kupfer, K.H. Hilsdorf and H. Rush, "Behaviour of concrete under biaxial stresses", Proceedings, American Concrete Institute, Vol. **66**, No. 8, pp. 656–666, August 1969.
31. L.L. Mills and R. M. Zimmerman, "Compressive strength under multiaxial loading conditions", *ACI Journal Proc.*, Vol. **67**, M–10, pp. 802–2087, Oct. 1970.
32. K.H. Gerstle et al. "Behaviour of concrete under multiaxial stress states", *ASCE Journal of Eng. Mech. Div.*, Vol. **106**. No. EM6, pp. 1383–1403, December 1980.
33. H.B. Kupfer and K.K. Gerstle, "Behaviour of concrete under biaxial stresses", *ASCE Journal of the Eng. Mech. Div.*, Vol. **99**, No. EM4, pp. 853–866, August 1973.
34. T.C.Y. Liu, A.H. Nilson and F.O. Slate, "Biaxial stress–strain relations for concrete", *ASCE Journal of the Structural Division*, Vol. **98**, No. ST5, pp. 1025–1034, May 1972.
35. L. Cedolin, Y.R.J. Crutzen and S. Deipoli, "Triaxial stress–strain relationship for concrete", *ASCE J. of the Eng. Mech. Div.*, Vol. **103**, No. EM3, pp. 423–439, June 1977.
36. K.H. Gerstle, "Simple formulation of biaxial concrete behaviour", *ACI Journal*, Vol. **78**, pp. 62–68, Feb. 1981.
37. C.S. Lin and A.C. Scordelis, "Nonlinear analysis of BC shells of general form", *ASCE Journal of the Structural Division*, Vol. **101**, No ST3, pp. 523–538, March 1975.
38. A.C.T. Chen and W.F. Chen, "Constitutive relations for concrete", *ASCE Journal of the Eng. Mech. Div.*Vol. **101**, No. EM4, pp. 465–481, August 1973.
39. O. Buyukozturk, "Nonlinear analysis of reinforced concrete structures", *Computer and Structures*, Vol. **7**, pp. 149–156, 1977.
40. Z.P. Bazant and P..D. Bhat, "Endochronic theory inelasticity and failure of concrete", *ASCE Journal of the Eng. Mech. Div.*, Vol. **102**, No. EM4, pp. 701–722, August, 1976.
41. Z.P. Bazant and S.S. Kim, "Plastic–Fracturing theory for concrete", *ASCE Journal of The Eng. Mech. Div.*, Vol. **105**, No. EM3, pp. 407–428., June 1979.
42. N.S. Ottosen, "A failure criterion for concrete", *ASCE Journal of Eng. MEch. Div.*, Vol. **103**, No. EM4, pp. 527–535, 1977.

43. J. Wastiels, "Behaviour of concrete under multiaxial stresses a leview", *Cement and Concrete Research*, Vol. **9**, pp. 35–44, 1979.
44. J. Wastiels, "Failure criteria for concrete subject to multiaxial stresses", Lecture held at Univ. of Illinois at Chicago Circle, Dept. of Materials Eng., 1981.
45. C.S. Lin, "Nonlinear analysis of reinforced concrete slabs and shells", PhD. Thesis, UC-SESM 73–7, university of California, April 1973.
46. W.F. Chen. "*Plasticity in Reinforced Concrete*", McGraw–Hill Book Company, 1982.
47. D.R.J. Owen and J.A. Figueiras, "Ultimate load analysis of reiforced concrete plates and shells", In *Finite Element Software for Plates and Shells*, E. Hinton and D.R.J. Owen (Eds.), Pineridge Press, 1984.
48. E. Andenaes, K. Gerstle and H.Y. Ko, "Response of mortar and concrete to biaxial compression", *ASCE Journal of Eng. Mech. Div.*, Vol. **103**, No. EM4, pp. 515–526, 1977.
49. D.R.J. Owen and E. Hinton, "*Finite Element in Plasticity*"Theory and Practice, Pineridge Press, Swansea, U.K., 1980.
50. O.C. Zienkiewicz and R.L. Taylor, "*The Finite Element Method*", Fourth Edition, Vol. **I**, 1989, Vol. **II**, 1991.
51. Y. Goto, " Cracks formed in concrete around deformed tension bars", *ACI Journal*, pp. 244–251, 1971.
52. H.H.A. Rahman, "Computatioal models for the nonlinear of reinforced concrete flexural slab systems", Ph. D Thesis, C7Ph/66/82, University of Wales, May, 1982.
53. R.I. Gilbert and R.F. Warner, " Tension stiffnening in reinforced concrete slabs", *ASCe Journal of the Structural Division*, Vol. **104**, No. ST12, pp. 1885–1900, 1978.
54. R.J. Cope, P.V. Rao, L.A. Clark and P. Norris, "Modelling of reinforced concrete behaviour for finite element analysis of bridge slabs", Numerical Meth. for Nonlinear Problems, Vol. **1**, Pineridge Press, Proceedings of the International Conference held at Univ. College of Swansea, pp. 457–470, 2–5 September, 1980.
55. A. Arnesen, S.L. Sorensen and P.G. Bergan, "Nonlinear analysis of Requirements for Reinforced Concrete", *ACI Standard* 318–77, Building Code, American Concrete Institute, Detroit, 1977.
56. R.C. Fenwick and T. paulay, "Mechanics of shear resistence of concrete beams", *ASCE Journal of the Structural Division*, Vol. **94**, No. ST10, pp. 2325–2350, October 1968.
57. J.A. Hofbeck, I.O. Ibrahim and A.H. Mattock, " Shear transfer in reinforced concrete", *ACI Journal*, pp. 119–128, 1969.
58. Y.D. Hamadi and P.E. Regan, "Bahaviour in shear of beams with flexural cracks", *Magazine of Concrete Research*, Vol. **32**, No. 111, pp. 67–78, 1980.

59. L. Cedolin and S. Deipoli, "Finite element studies of shear–critical R/C beams, *ASCE Journal of the Engineering Mechanics Division*, Vol. **103**, No. EM3, pp. 395–410, Jue 1977.
60. F.R. Hand, D.A. Pecknold and W.C. Schnobrich, "Nonlinear layered analysis of RC Plates and Shels, *ASCE Journal of Structural Division,* Vol. **99**, No. ST7, pp. 1491–1503, July, 1973.
61. Z. Bazant, "Mechanics of distributed cracking, *Appl. Mech. Rev.*, Vol. **39**, pp. 675–705, 1986.
62. E. Oñate, J. Oliver and G. Bugeda, "Finite element analysis of nonlinear response of concrete dams subject to internal loads"Europe–US Symposium on Finite Element Methods for Nonlinear Problems, (Edited by Bergan, Bathe and Wunderlich) Springer-Verlag, 1986.
63. S, Oller, "Un Modelo de daño continuo para materiales friccionales, Tesis Doctoral, Dept. De Estructuras, Univ. Politécnica de Catalunya, Barcelona, España, 1988.
64. J. Lubliner, S. Oller, J. Oliver and E. Oñate, "A plastic damage model for nonlinear analysis of concrete", *Int. J. Solids Struct.*, Vol. **25**, 3, pp. 299–326, 1989.
65. E. Oñate, S. Oller And J. Lubliner, "A constitutive model of concrete based on the incremental theory of plasticity", *Engng. Comput.*, Vol. **5**, 4, 1988.
66. E. Oñate, S. Oller, J. Oliver and J. Lubliner, "A fully elastoplastic constititutive model nonlinear analysis of concrete, Proc. Second Int. Conf. Advances Methods in Numerical Methods in Engineering, Theory and Applications–NUMETA (Edited by G. Pande and J. Middleton). Martinus Nijhoff, Swansea, 1987.
67. E. Oñate, S. Oller and J. Lubliner, "A constitutive model cracking of concrete based on the incremental theory of plasticity", in Proc. Int Conf. Computational Plasticity (Edited by D.R.J. Owen, E. Hinton and E. Oñate), Part 2, pp 1311-1327, Pineridge Press, Barcelona, 1987.
68. R. De Borst and P. Verneer, "Non associated plasticity for soils, concrete and rock", *Heron* 29, Delft, Netherlands, 1984.
69. J. Oliver, "A consisten characteristic length for smeared cracking models", *Commun, Appl. Num. Meth. in Engng.*
70. M. Crisfield, "Finite element methods for non linear structural analysis", J. Wiley, 1991.
71. J.G. Rots, P. Nauta, G. Kusters and J. Blaauwendraad, "Smeared crack approach and fracture localization in concrete", Heron, Univ. Delft, The Netherlands, Vol. **30**, 1985.
72. M. Arrea and A.R. Ingraffea, "Mixed mode crack propagation in mortar and concrete", *Cornell Univ., Dept. Struct. Engng Report 81-13*, Ithaka, New York, 1981.

73. M.A. Crisfeld, "A fast incremental iterative solution procedure that handles snap–through", *Comp. and Struct.*, Vol. **13**, pp. 55–62, 1981.
74. S. Oller, E. Oñate y J. Oliver, "Un modelo de fisuración del hormigón basado en la teoría incremental de la plasticidad", *II Simposium sobre Aplicaciones del Método de los Elementos Finitos en Ingeniería*, Universidad Politécnica de Cataluña, June 1986.
75. A. W. Wegmuller, "Overload behaviour of composite steel-concrete highway girder bridges", School of Engineering, University of Alabama, Birmingham, October 1986.
76. P. Roca, "Un modelo de análisis no lineal para el estudio del comportamiento de estructuras laminares de hormigón pretensado", *Ph. D. Thesis*, Universidad Politécnica de Cataluña, September 1988.
77. A. Arnesen, "Analysis of reinforced shells considering material and geometric non linearities", Report No. 79–1, *Division Struct. Mech.*, University Trandheim, Norway, July 1979.
78. A.L. Bouma, A.C. Van Riel, H. Van Koten and W.J. Beramek, "Investigations on models of eleven cylindrical shells mode on reinforced and prestressed concrete", *Proc. Symposium on Shell Research*, Delft, North Holland Pub. Co., 79–101, 1961.
79. E. Oñate, J. Oliver, R. Chueca, J. Peraire and R. Albareda, "Non linear finite element analysis of cryogenic concrete tanks under thermal actions", *II Int. Conf. on Cryogenic Concrete*, Amsterdam, 1983.

BASICS OF SHAPE OPTIMAL DESIGN

K.-U. Bletzinger and E. Ramm
University of Stuttgart, Stuttgart, Germany

ABSTRACT

At present optimum structural design is understood as a synthesis of several disciplines which individually are to a large extend developed [1]: (i) design modeling, (ii) structural analysis, (iii) behavior sensitivity analysis, (iv) mathematical programming, and (v) interactive computer graphics. Their interactions reflect the typical loop of each design process (Fig. 1). The art of structural optimization is to join the interdisciplinary dependencies in a clear, integrated overall model and to convert it into an efficient and practical computer code [2]. Today, all the leading software packages offer optimization modules, although of different quality and still mainly restricted to sizing.

A lot of special software and procedures have been developed and presented in the literature which deal with shape optimal design [3–6]. The related process is reviewed and accompanied by a number of excellent papers [7–9], proceedings [10–12], and books [13–15] which clearly show the evolution from the first beginnings with sizing of structures to the latest developments in shape optimal design. Fields of applications are steadily increasing. Especially in automotive [6] and aircraft industries [16] the methods found wide–spread recognition as instruments of numerical simulation to improve structural quality. Civil engineering is yet behind to use structural optimization as a regular tool. But also here the methods can be used as an extra design aid; e.g. for the shape design of free formed shells [17–20]. It is the intention of this contribution to give some insight into the basic ideas of the methods and the underlying modeling; the application to general shells is described in [21].

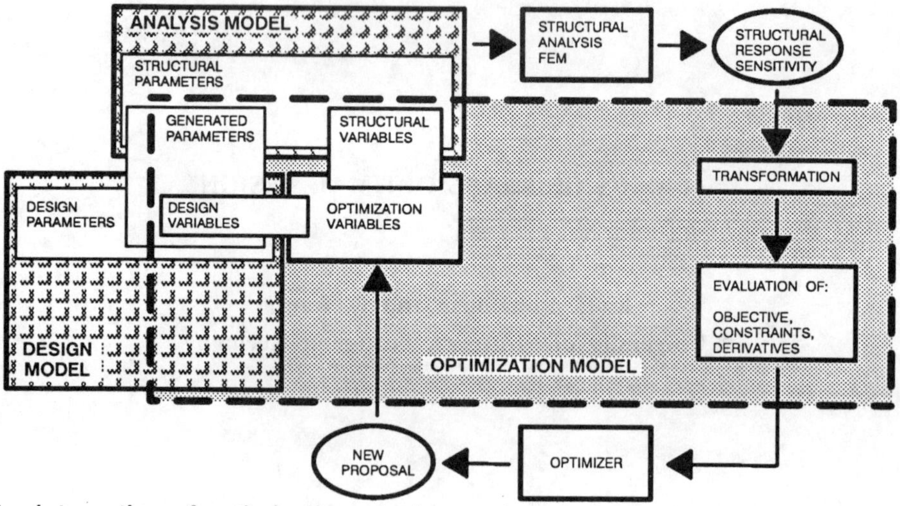

Fig. 1. Interaction of optimization, design and analysis models

1. THE OPTIMIZATION MODEL

A typical problem of structural optimization is characterized by an objective $f(\mathbf{x})$ and constraints $\mathbf{g}(\mathbf{x})$ and $\mathbf{h}(\mathbf{x})$ which are nonlinear functions of the optimization variables $\mathbf{x}$. In mathematical terms it can be stated as:

$$\begin{aligned} &\text{minimize:} && f(\mathbf{x}) \\ &\text{subject to:} && h_j(\mathbf{x}) = 0 \;;\quad j = 1,\ldots,m_e \\ & && g_j(\mathbf{x}) \leq 0 \;;\quad j = m_e + 1,\ldots,m \\ & && \mathbf{x}_L \leq \mathbf{x} \leq \mathbf{x}_U \;;\quad \mathbf{x} \in \mathbf{R}^n \end{aligned} \tag{1}$$

where $\mathbf{x}_L$ and $\mathbf{x}_U$ are lower and upper bounds on the variables respectively.

Within the optimization model all the above mentioned different disciplines are put together (Fig. 1). The structural responses (i.e. displacements, stresses, etc.) are calculated by any available structural analysis procedure. The methods of finite elements are widely used in this context. With the structural responses the objective function $f(\mathbf{x})$ and constraints $\mathbf{h}(\mathbf{x})$ and $\mathbf{g}(\mathbf{x})$ can be determined and fed back into the optimization model. In Fig. 2 a list of commonly used functions is given.

Methods to determine structural behavior sensitivities are tightly connected to the analysis method and follow usually the same analysis or structural model (e.g. the finite element mesh). Sensitivities serve as gradient informations in the optimization procedure. The geometry of the analysis model is controlled by a superior design model which may be defined in terms of the methods of Computer Aided Geometric Design (CAGD). Even complex shapes as well as structural thicknesses can be defined and controlled during the optimization procedure by some "natural" design parameters like characteristic lengths or heights of the structure or a few thickness parameters. They are independent of the chosen FE mesh. The set of optimization variables **x** is mainly recruited from these design model parameters. The solution of the optimization problem is determined by any method of non–linear mathematical programming. This is the most abstract stage of the whole procedure where all the methods mentioned above and their related models are in general used as black boxes. The efficiency is very much dependent on formulation and coding of their interactions. New values of the optimization variables are determined iteratively. They have to be brought back to design and analysis models to start the next optimization cycle. Variable linking rules have to be considered at this stage. These rules are used to define variable interdependences to get more realistic and problem oriented formulations. They are an important part of shape optimal design. Every step during optimization is accompanied by computer graphics to monitor precisely the process.

Design variables **x**:

$\mathbf{r}_d$, $\mathbf{r}_a$	Design and structural nodes	**A**	other cross section parameters
$\mathbf{t}_d$, $\mathbf{t}_a$	Design and structural thicknesses		

Objective functions f(**x**):		Constraint functions h(**x**), g(**x**):	
$W = \int \varrho dv$	Weight or volume	$h_W = \frac{W}{W_{all.}} - 1 = 0$	Weight or volume
$E = \frac{1}{2}\int \sigma\varepsilon dv$	Strain energy	$h_C = \frac{C}{C_{all.}} - 1 = 0$	Rotational inertia
$S = \int (\sigma - \sigma_{avg})^2 da$	Stress leveling	$g_u = \frac{u}{u_{all.}} - 1 \leq 0$	Displacement
$C = \int \varrho r^2 dv$	Rotational inertia	$g_\sigma = \frac{\sigma}{\sigma_{all.}} - 1 \leq 0$	Stresses

Fig. 2. Variables and functions of optimization model

2. COMPUTER AIDED GEOMETRIC DESIGN

The general methods of Computer Aided Geometrical Design (CAGD) are the basis of modern pre–processors to design structural geometries in two and three dimensions [22, 23]. Shapes are approximated piecewise by "design patches". Within each design patch the resulting shape $\mathbf{r}_a$ is parametrized in terms of shape functions H_i, patch parameters u, v, w, and design nodes $\mathbf{r}_{di}$ which describe the location of the patch in space:

$$\mathbf{r}_a(u, v, w) = \sum_{i=1}^{n} H_i(u, v, w)\ \mathbf{r}_{di} \tag{2}$$

There are many different shape functions available, e.g. Lagrangian interpolation, Coons transfinite interpolation, Bézier– and B–spline approximations.

Triangular and quadrilateral C_0–continuous domains with arbitrary edge descriptions are defined preferably by Coons patches. Typically, they are used to describe

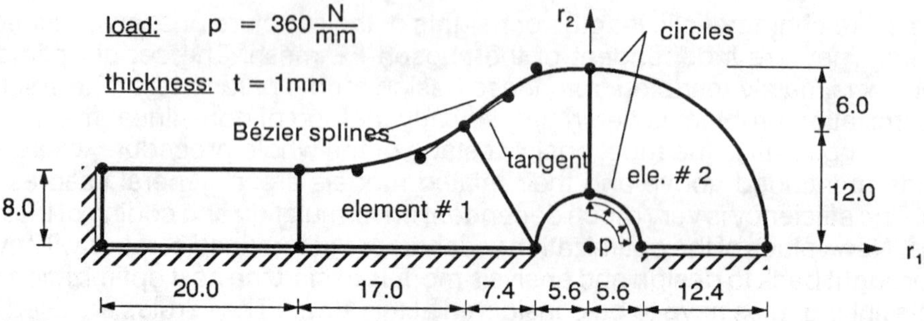

Fig. 3. Connecting rod example: design element discretization

plane problems, e.g. plane stress or plate bending problems where the final contour is to be determined. The geometry of the connecting rod (Fig. 3) is defined by in total four Coons patches. Within one design element different kinds of edges like Bézier spline (element no. 1) or circle (element no. 2) are combined with linear interpolations.

The shape function of a cubic Bézier patch is given by:

$$\mathbf{r}_a(u,v) = \sum_{i=0}^{3}\sum_{j=0}^{3} \mathbf{r}_{ij} \frac{3!}{(3-i)!i!} \frac{3!}{(3-j)!j!} u^i (1-u)^{3-i} v^j (1-v)^{3-j} \tag{3}$$

where $\mathbf{r}_{ij}$ are the position vectors of the 16 control nodes and u and v are parameters running from 0 to 1 (Fig. 4). Only the four vertices $\mathbf{r}_{00}$, $\mathbf{r}_{30}$, $\mathbf{r}_{03}$, $\mathbf{r}_{33}$ lie on the approximated surface.The remaining inner nodes are only approximating the shape which yields no differences in the results compared with equivalent Lagrange interpolation schemes but allows to construct very easily continuity conditions between adjacent design patches of continuous surfaces. This is demonstrated by an example. A shape is defined by four 16–noded Bézier patches as shown in Fig. 6. These patches are connected in a way that the generated shape is continuous in slopes across the common edges of adjacent patches. Consequently, corresponding design nodes of the involved patches have to stay on a common line during all following shape modifications. Across two adjacent edges this rule states as (Fig. 5):

$$\mathbf{r}_1^{(2)} = \mathbf{r}_0^{(2)} + \delta\,(\mathbf{r}_n^{(1)} - \mathbf{r}_{n-1}^{(1)}) \tag{4}$$

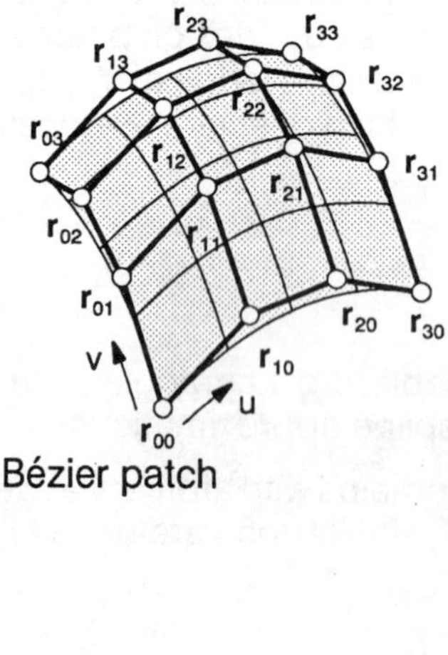

.Fig. 4. Bézier patch

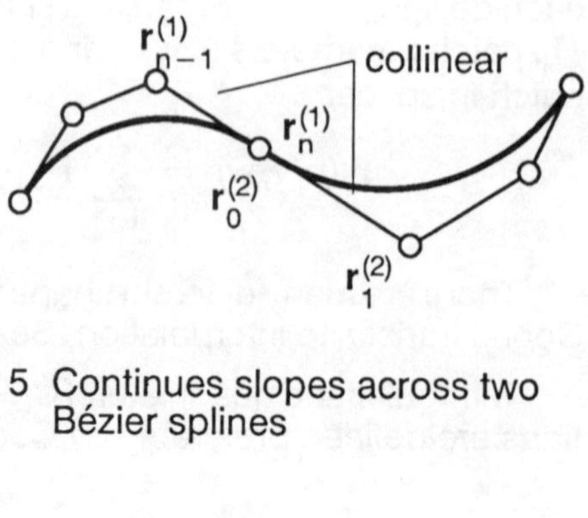

Fig. 5 Continues slopes across two Bézier splines

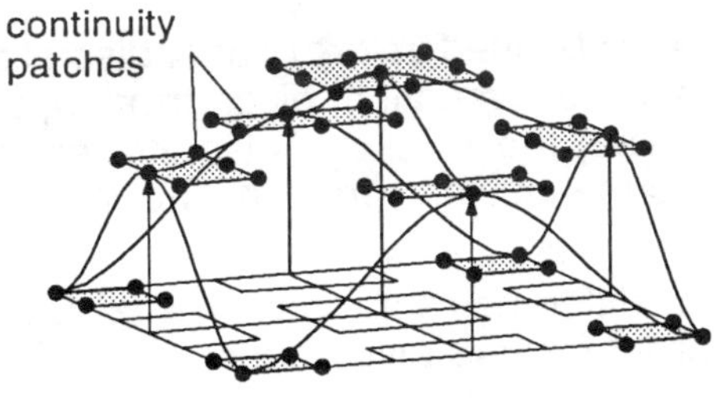

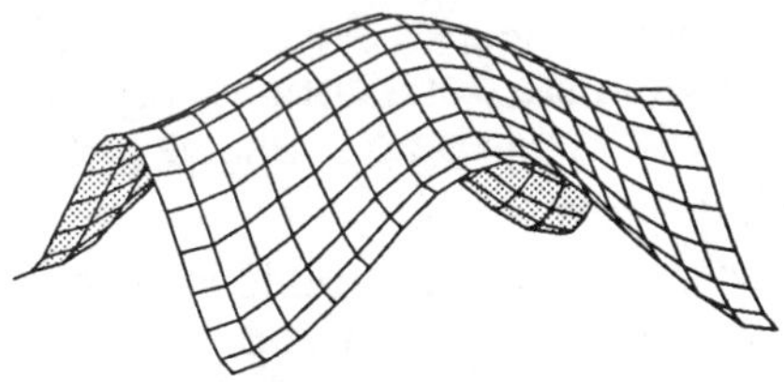

Fig. 6. Interactive surface modification; continuity patches connecting four Bézier patches

The tangent at the rod contour (Fig. 3) is defined that way. The same rule holds for the second dimension and the cross diagonal which leads to linear dependencies between at most nine design nodes. These topological relations are formulated in superimposed "continuity patches" [17]. They are generated automatically and preserved during manual user interactions and shape optimization. Fig. 6 shows different types of continuity patches depending on whether they are connecting two or four design patches or are defined at an isolated corner. In either case four nodes are independent and control the locations of the remaining nodes leading to a reduction of geometrical degree of freedom which is very welcome in structural optimization to stabilize the procedure.

3. STRUCTURAL ANALYSIS, FINITE ELEMENT METHOD

In the present study the well–known isoparametric degenerated solid elements are used for shell analysis applying an explicit integration across the thickness. These elements have been proven to contain the same mechanical content than elements based on a corresponding shear flexible shell theory; for an overview see [24, 25]. All kinds of bi–linear, bi–quadratic, and bi–cubic interpolated triangular and rectangular elements are available.

4. SENSITIVITY ANALYSIS

The sensitivity analysis supplies gradient informations on objective f($\mathbf{x}$) and constraints $\mathbf{h}(\mathbf{x})$ and $\mathbf{g}(\mathbf{x})$ with respect to optimization variables. They are not only important in the context of structural optimization together with gradient oriented optimizers, but also for system identification and statistical structural analysis. In general, any function t (objective or constraint) depends on optimization variables $\mathbf{x}$ and state variables $\mathbf{u}$, e.g. displacements. Thus, the total derivative of t with respect to $\mathbf{x}$ is given as:

$$\frac{dt}{d\mathbf{x}} = \frac{\partial t}{\partial \mathbf{u}}\frac{d\mathbf{u}}{d\mathbf{x}} + \frac{\partial t}{\partial \mathbf{x}} \tag{5}$$

where the determination of the response sensitivity $\frac{d\mathbf{u}}{d\mathbf{x}}$ is part of the job. It can be done by several different techniques. They can be divided into variational or discrete methods, dependent on whether the gradients are obtained before or after discretization. Nevertheless, the same results will be obtained if variation, discretization, and derivation are done consistently [18]. In the following the variants of discrete sensitivity analysis will be discussed [18, 26]. More informations on variational approaches may be found in [15, 27].

The easiest method to implement is the finite difference approach. The derivative of $\mathbf{u}$ at $\mathbf{x} = \mathbf{x}_0$ can be approximated by first–order forward differences:

$$\frac{du_j}{dx_i} = \frac{u_j(\mathbf{x}_0 + \Delta\mathbf{x}) - u_j(\mathbf{x})}{\Delta x_i} \tag{6}$$

This method suffers from possible truncation or condition errors, dependent on the step size D$\mathbf{x}$. Improvements may be achieved by central–difference approximations which, however, need two instead of one additional analyses for every variable x_i.

From the state equation $\mathbf{K}\,\mathbf{u} = \mathbf{R}$ (7)

the displacement derivatives can be derived analytically:

$$\frac{d\mathbf{K}}{dx}\mathbf{u} + \mathbf{K}\frac{d\mathbf{u}}{dx} = \frac{d\mathbf{R}}{dx} \tag{8}$$

$$\frac{d\mathbf{u}}{dx} = \mathbf{K}^{-1}\left(\frac{d\mathbf{R}}{dx} - \frac{d\mathbf{K}}{dx}\mathbf{u}\right) \tag{9}$$

This method is called discrete analytical sensitivity analysis. It is straightforward and consistent with the theory of structural optimization.

The major concern of this approach is the calculation of the pseudo load vector

$$\overline{\mathbf{R}} = \frac{d\mathbf{R}}{dx} - \frac{d\mathbf{K}}{dx}\mathbf{u} \tag{10}$$

which results in analytical derivations of the load vector $\mathbf{R}$ and the stiffness matrix $\mathbf{K}$. Especially for shape variables this procedure is cumbersome but worth–while if done so. In the special case of isoparametric finite elements the derivation of an element stiffness matrix yields:

$$\mathbf{k}_e = \int \mathbf{B}^T\mathbf{C}\,\mathbf{B}\,|\mathbf{J}|\,d\xi\,d\eta \tag{11}$$

$$\mathbf{k}_{e,x} = \int \mathbf{B}_{,x}^T\mathbf{C}\,\mathbf{B}\,|\mathbf{J}|\,d\xi\,d\eta + \int \mathbf{B}^T\,\mathbf{C}_{,x}\,\mathbf{B}\,|\mathbf{J}|\,d\xi\,d\eta + \int \mathbf{B}^T\mathbf{C}\,\mathbf{B}_{,x}\,|\mathbf{J}|\,d\xi\,d\eta + \int \mathbf{B}^T\,\mathbf{C}\,\mathbf{B}\,|\mathbf{J}_{,x}|\,d\xi\,d\eta \tag{12}$$

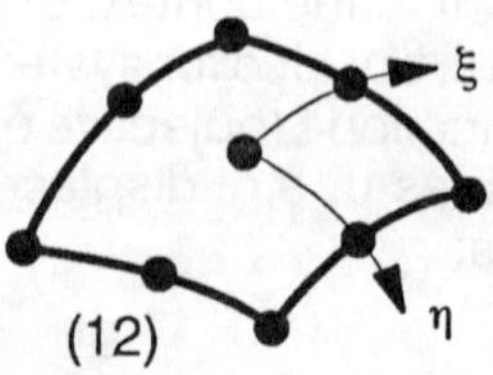

If the analytical derivations of the stiffness matrix and the element load vector are not available, the semi–analytical approach is an attractive alternative. It is based on a finite difference approximation of the pseudo load vector:

$$\overline{\mathbf{R}} = \frac{\Delta \mathbf{R}}{\Delta \mathbf{x}} - \frac{\Delta \mathbf{K}}{\Delta \mathbf{x}}\mathbf{u} \tag{13}$$

The semi–analytical method is easy to implement but suffers again from numerical errors. Additionally, this method is much more sensitive to truncation errors than the plain numerical approach. For bending structures this is explained by shear dominated sensitivity fields [28].

In contrast to (5) the derivative of t may be determined by an adjoint method. Again, there exist variational and discrete versions. If (9) and (10) are plugged into (5) one gets:

$$\frac{dt}{d\mathbf{x}} = \frac{\partial t}{\partial \mathbf{u}}\mathbf{K}^{-1}\,\overline{\mathbf{R}} + \frac{\partial t}{\partial \mathbf{x}} \tag{14}$$

Instead of determining first the displacement derivative, it may be more efficient to determine the adjoint variables **z** and to proceed with the final calculation of the function derivative:

$$\mathbf{z} = \mathbf{K}^{-1}\left(\frac{dt}{d\mathbf{u}}\right)^{T} \quad \text{and} \quad \frac{dt}{d\mathbf{x}} = \mathbf{z}^{T}\,\overline{\mathbf{R}} + \frac{\partial t}{\partial \mathbf{x}} \tag{15}$$

This version can be advantageous if the gradients of a small number of functions compared to the number of variables have to be evaluated. However, the main effort still remains with the calculation of the pseudo loads.

5. MATHEMATICAL PROGRAMMING

Most of sizing and shape optimization problems can be assumed to be continuous in gradients and curvature of the problem functions and are effectively solved by gradient oriented methods. A local solution of (1) is characterized as a stationary point of the corresponding Lagrangian function:

$$L(\mathbf{x},\mathbf{u},\mathbf{v}) = f(\mathbf{x}) + \mathbf{u}^{T}h(\mathbf{x}) + \mathbf{v}^{T}g(\mathbf{x}) \tag{16}$$

u and **v** are the Langrangian multipliers or dual variables. The necessary condition for the stationarity of L or the corresponding constrained minimum of f(**x**) is defined by the Kuhn–Tucker conditions. They give a set of nonlinear equations to determine the optimal solution **x***, **u***, **v***:

$$\frac{\partial L}{\partial \mathbf{x}}^{*} = \frac{\partial f}{\partial \mathbf{x}}^{*} + \left[\frac{\partial \mathbf{h}}{\partial \mathbf{x}}^{*}\right]^{T}\mathbf{u}^{*} + \left[\frac{\partial \mathbf{g}}{\partial \mathbf{x}}^{*}\right]^{T}\mathbf{v}^{*} = \mathbf{0} \tag{17}$$

$$\frac{\partial L}{\partial u_j}^{*} = h_j^{*} = 0 \quad ; \; j = 1,\ldots,m_e$$

$$\frac{\partial L}{\partial v_j}^{*} v_j^{*} = g_j^{*}\,v_j^{*} = 0 \quad ; \; j = m_e + 1,\ldots,m$$

$$v_j^{*} \geq 0$$

Where f^*, $\mathbf{h}^*$, $\mathbf{g}^*$ and L^* are the function values at the optimal solution. Constraints which are limiting the optimal solution are called active.

There exist several strategies of mathematical programming (MP) which can be distinguished by the type and number of independent variables they use. Due to Luenberger [29] they can be divided into (i) primal methods, (ii) penalty and barrier methods, (iii) dual methods, and (iv) Lagrange methods.

By now Lagrange methods are the most sophisticated optimization techniques. They are designed to solve the Kuhn–Tucker conditions (17) directly and are operating in the full space of primal and dual variables. Inequality and equality constraints can be considered simultaneously which allows flexible problem formulations. Iterative solution of eqns. (17) by subsequent linearization leads to a natural extension of the classical Newton/Raphson procedure which became known as SQP–method (sequential quadratic programming) [30].

In the k–th iteration step the corresponding quadratic subproblem states as:

$$
\begin{aligned}
&\text{minimize} && \tfrac{1}{2}\mathbf{d}^{(k)T}\mathbf{B}^{(k)}\mathbf{d}^{(k)} + \left(\frac{\partial f(\mathbf{x}^{(k)})}{\partial \mathbf{x}}\right)^T \mathbf{d}^{(k)} \\
&\text{subject to} && \frac{\partial \mathbf{h}(\mathbf{x}^{(k)})}{\partial \mathbf{x}}\mathbf{d}^{(k)} + \mathbf{h}(\mathbf{x}^{(k)}) = \mathbf{0} \\
& && \frac{\partial \mathbf{g}(\mathbf{x}^{(k)})}{\partial \mathbf{x}}\mathbf{d}^{(k)} + \mathbf{g}(\mathbf{x}^{(k)}) \le \mathbf{0} \\
& && \mathbf{d}^{(k)} \le \bar{\mathbf{x}} - \mathbf{x}^{(k)} \\
& && -\mathbf{d}^{(k)} \le \mathbf{x}^{(k)} - \underline{\mathbf{x}} \qquad ; \quad \mathbf{d}^{(k)} = \mathbf{x}^{(k+1)} - \mathbf{x}^{(k)}
\end{aligned}
\tag{18}
$$

$\mathbf{B}^{(k)}$ ... current Lagrangian second derivatives or their approximations in the case of quasi–Newton algorithms (BFGS, DFP)

Direct application of MP methods may turn out to be troublesome and time consuming because the internal approximation scheme of the chosen method (usually linear or quadratic) does not reflect to the physical properties of the problem and causes many iteration steps with a lot of expensive evaluations of functions and gradients. Performance can be improved if in an additional iteration cycle objective function and constraints are replaced by proper approximations. Usually, approximations are derived from first order linearizations with respect to generalized variables $\mathbf{y}$ as functions of $\mathbf{x}$:

$$
\begin{aligned}
\tilde{f}(\mathbf{y})^{(k)} &= f(\mathbf{y}^{(k)}) + \left(\frac{\partial f(\mathbf{y}^{(k)})}{\partial \mathbf{y}}\right)^T \left(\mathbf{y}^{(k+1)} - \mathbf{y}^{(k)}\right) \\
\tilde{f}(\mathbf{x})^{(k)} &= f(\mathbf{x}^{(k)}) + \left(\frac{\partial f(\mathbf{x}^{(k)})}{\partial \mathbf{x}}\right)^T \frac{\partial \mathbf{x}^{(k)}}{\partial \mathbf{y}} \left(\mathbf{y}(\mathbf{x}^{(k+1)}) - \mathbf{y}(\mathbf{x}^{(k)})\right)
\end{aligned}
\tag{19}
$$

$\mathbf{y}^{(k)} = \mathbf{y}(\mathbf{x}^{(k)})$ are chosen w.r.t. physical problem properties, e.g. $y_i^{(k)} = \frac{1}{x_i^{(k)}}$

The resulting approximated sub–problem still remains nonlinear in terms of the original variables **x** and has to be solved by suitable MP methods. Functions and gradients can be evaluated very efficiently since the substituted functions are known explicitly. Because of the linearization technique sub–problems show special properties like separability which can be exploited successfully by use of dual solution methods. Approximation techniques became known e.g. as convex linearization [31] or method of moving asymptotes [32]. Fleury [33] showed their relations to the methods of optimality criteria, the other important class of solution methods in structural optimization [34].

6. VARIABLE LINKING

The complexity of shape optimal design becomes obvious if we take a look at the variable interactions of the different models. A commonly used rule which links variables $\mathbf{r}_a$ of the analysis model via the design models with variables **x** of the optimization model is defined as [2]:

$$\mathbf{r}_a = \mathbf{r}_a^0 + \mathbf{L}_{ax}\mathbf{x} + \mathbf{H}_{ad}(\mathbf{r}_d) \tag{20}$$

with:
$$\mathbf{r}_d = \mathbf{r}_d^0 + \mathbf{L}_{dx}\mathbf{x} \tag{21}$$

In these relations $\mathbf{r}_a^0$ and $\mathbf{r}_d^0$ denote coordinates of analysis and design models, respectively, which remain constant during the optimization process. Linking matrices $\mathbf{L}_{ax}$ and $\mathbf{L}_{dx}$ describe linear relations between optimization variables **x** and variable coordinates of analysis and design model, respectively. $\mathbf{H}_{ad}$ denote nonlinear relations between design and analysis model and are identified as shape functions. In the simple case of Cartesian coordinates shape functions define linear relations. Then $\mathbf{H}_{ad}$ reduces to:

$$\mathbf{H}_{ad}(\mathbf{r}_d) = \mathbf{H}(u, v, w)\ \mathbf{r}_d \tag{22}$$

where u,v,w are parameters describing the shape (compare with eq. (2)).

The coefficients of linking matrix $\mathbf{L}_{dx}$ are defined by linking rules which help to fix typical and problem oriented degrees of freedom. Among them the following rules are useful in practical applications:

- prescribed move direction: $$\mathbf{r}_d = \mathbf{r}_d^0 + \mathbf{s}\ x_i \tag{23}$$
- linear combination: $$\mathbf{r}_d = \mathbf{r}_d^0 + \sum_i \alpha_i\ \mathbf{s}_i\ x_i \tag{24}$$
- change of basis: $$\mathbf{r}_d = \mathbf{r}_d^0 + \mathbf{s}_1\ x_1 + \mathbf{s}_2\ x_2 + \mathbf{s}_3\ x_3 \tag{25}$$

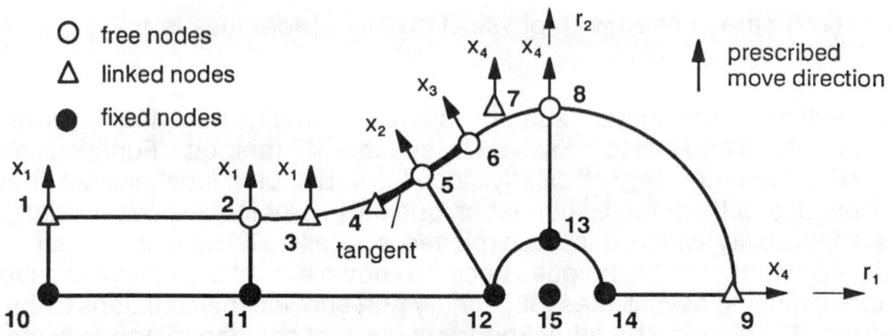

Fig. 7. Connecting rod example: linear linking rules

- symmetry: $$\mathbf{r}_d = \mathbf{r}_d^0 + \left(\mathbf{I} - 2\mathbf{n}\mathbf{n}^T\right) \mathbf{s}\, x_i \tag{26}$$

- projection: $$\mathbf{r}_d = \mathbf{r}_d^0 + \left(\mathbf{d}\mathbf{d}^T\right) \mathbf{s}\, x_i \tag{27}$$

$\mathbf{s}_i$ are the vectors of movement or base directions, respectively. $\mathbf{n}$ is the normal vector on the plane of symmetry, and $\mathbf{d}$ is the direction on which $\mathbf{s}$ is projected. Continuity conditions as they have been introduced before are another example of linear linking.

With eqs. (20) and (22) behavior sensitivity of any structural response g which is originally defined in terms of analysis variables $\mathbf{r}_a$ and state variables $\mathbf{u}$ is now defined as:

$$\frac{dg(\mathbf{u}, \mathbf{r}_a)}{d\mathbf{x}} = \left(\frac{\partial g}{\partial \mathbf{r}_a} + \frac{\partial g}{\partial \mathbf{u}} \frac{d\mathbf{u}}{d\mathbf{r}_a}\right) \frac{d\mathbf{r}_a}{d\mathbf{x}} \tag{28}$$

with: $$\frac{d\mathbf{r}_a}{d\mathbf{x}} = \mathbf{L}_{ax} + \mathbf{H}\,\mathbf{L}_{dx} \tag{29}$$

To give an example some linking rules are introduced for the connecting rod problem to reduce the number of variables to four (Fig. 7).

ACKNOWLEDGEMENTS

This work is part of the research project SFB 230 "Natural Structures – Light Weight Structures in Architecture and Nature" supported by the German Research Foundation (DFG) at the University of Stuttgart. The support is gratefully acknowledged.

REFERENCES:

1. Braibant, V. and Fleury, C.: Shape optimal design and CAD oriented formulation, Engng. with Comp., 1 (1986), 193–204.
2. Bletzinger, K.–U.; Kimmich, S. and Ramm, E.: Efficient modeling in shape optimal design, to appear in Computing Systems in Engineering, 1992.
3. Eschenauer, H.; Post, U. and Bremicker, M.: Einsatz der Optimierungsprozedur SAPOP zur Auslegung von Bauteilkomponenten, Bauingenieur, 63 (1989), 515–526.
4. Kimmich, S. and Ramm, E.: Structural optimization and analysis with program system CARAT, in: [12] 1989, 186–193.
5. Rasmussen, J.: The structural optimization system CAOS, Structural Optimization, 2 (1990), 109–115.
6. Chargin, M. K.; Raasch, I.; Bruns, R. and Deuermeyer, D.: General shape optimization capability, Finite Elements Anal. Des., 7 (1991), 343–354.
7. Schmit, L. A.: Structural synthesis – its genesis and development, AIAA Journal, 19 (1981), 1249–1263.
8. Vanderplaats, G. N.: Structural optimization – past, present and future, AIAA–Journal, 20 (1982), 992–1000.
9. Haftka, R. T. and Grandhi, R. V.: Structural shape optimization – a survey, Comp. Meth. Appl. Mech. Engng., 57 (1986), 91–106.
10. Bennett, J. A. and Botkin, M. E. (eds.): The Optimum Shape – Automated Structural Design, Plenum Press, New York, London, 1986.
11. Mota Soares, C. A. (ed.): Computer Aided Optimal Design – Structural and Mechanical Systems, NATO–ASI Series F: Computer and System Sciences, vol. 27, Springer, Berlin, Heidelberg, 1987.
12. Eschenauer, H. A. and Thierauf, G. (eds.): Discretization Methods and Structural Optimization – Procedures and Applications, Proc. GAMM–Seminar, Oct. 5– 7, 1988, Siegen, Lecture Notes in Engineering, Springer, 1989.
13. Vanderplaats, G. N.: Numerical Optimization Techniques for Engineering Design: With Applications, McGraw–Hill, New York, 1984.
14. Atrek, E.; Gallagher, R. H.;Ragsdell, K. M. and Zienkiewicz, O. C. (eds.): New Directions in Optimum structural Design, Wiley, Chichester, New York, 1984.
15. Haftka, R. T.; Gürdal, Z. and Kamat, M. P.: Elements of Structural Optimization, 2nd edition, Kluwer Academic Publishers, Dordrecht, 1990.
16. Petiau, C.: Structural optimization of aircraft, Thin Walled Struct., 11 (1991), 43–64.
17. Bletzinger, K.–U.: Formoptimierung von Flächentragwerken, Ph. D. Dissertation, Institut für Baustatik, Universität Stuttgart, 1990.
18. Kimmich, S.: Strukturoptimierung und Sensibilitätsanalyse mit finiten Elementen, Ph.D. Dissertation, Institut für Baustatik, Universität Stuttgart, 1990.
19. Ramm, E. and Mehlhorn, G.: On shape finding methods and ultimate load analyses of reinforced concrete shells, Engineering Structures, 13 (1991), 178–198.
20. Bletzinger, K.–U. and Ramm, E. Form finding of shells by structural optimization, to appear in Engng. with Comp., 1992.
21. Ramm, E.: Shape finding methods of shells, lecture notes CISM course Nonlinear Analysis of Shells by Finite Elements, Udine, June 24–28, 1991.
22. Böhm, W.; Farin, G. and Kahmann, J.: A survey of curve and surface methods in CAGD, Comp. Aided Geom. Des., 1 (1984), 1–60.
23. Faux, I. D. and Pratt, M. J.: Computational Geometry for Design and Manufacture, Ellis Horwood Publishers, Chichester, 1979.

24. Andelfinger, U: Untersuchungen zur Zuverlässigkeit hybrid–gemischter finiter Elemente für Flächentragwerke, Ph.D.–Dissertation, Institut für Baustatik, Universität Stuttgart, 1991.
25. Büchter, N. and Ramm, E.: Shell theory versus degeneration – a comparison in large rotation finite element analysis, submitted to Int. J. Num. Meth. Engng., 1990.
26. Haftka, R. T. and Adelman, H. M.: Recent developments in structural sensitivity analysis, Structural Optimization, 1 (1989), 137–151.
27. Dems, K. and Haftka, R. T.: Two approaches to sensitivity analysis for shape variation of structures, Mech. Struct. & Mach., 16 (1988), 501–522.
28. Barthelemy, B.; Chon, C. T. and Haftka, R. T.: Accuracy problems associated with semi–analytical derivatives of static response, Finite Elements Anal. Des., 4 (1988), 249–265.
29. Luenberger, D. G.: Linear and Nonlinear Programming, Addison–Wesley, Reading, 1984.
30. Schittkowski., K.: The nonlinear programming method of Wilson, Han and Powell with an augmented Lagrangian type line search function, Numerische Mathematik, 38 (1981), 83–114.
31. Fleury, C.: Structural optimization – a new dual method using mixed variables, Int. J. Num. Meth. Engng., 23 (1986), 409–428.
32. Svanberg, K.: The method of moving asymptotes – a new method for structural optimization, Int. J. Num. Meth. Engng., 24 (1987), 359–373.
33. Fleury, C.: Sructural weight optimization by dual methods of convex programming, Int. J. Num. Meth. Engng., 14 (1979), 1761–1783.
34. Berke, L. and Khot, N. S.: Structural optimization using optimality criteria, in: [11] 1987, 271–311.
35. Ding, Y.: Shape optimization of structures: a literature survey, Comp. & Struct., 24 (1986), 985–1004.
36. Esping, B. J. D.: A CAD approach to the minimum weight design problem, Int. J. Num. Meth. Engng., 21 (1985), 1049–1066.
37. Gallagher, R.H.: Fully stressed design, in: Optimum Structural Design – Theory and Applications (Eds. R. H. Gallagher and O. C. Zienkiewicz), J. Wiley, London, New York, 1973.
38. Lootsma, F. A. and Ragsdell, K. M.: State of the art in parallel nonlinear optimization, Parallel Computing, 6 (1988), 133–155.
39. Prasad, B.: Explicit constraint approximation forms in structural optimization, part 1: analyses and projections, Comp. Meth. Appl. Mech. Engng., 40 (1983), 1–26.
40. Ramm, E. and Schunck, E.: Heinz Isler – Schalen, Krämer, Stuttgart, 1986.
41. Santos, J. L. T.; Godse, M. M. and Chang, K.–H.: An interactive post–processor for structural design sensitivity analysis and optimization: sensitivity display and what–if study, Comp. & Struct., 35 (1990), 1–13.
42. Schwefel, H. P.: Numerical Optimization of Computer Models, J. Wiley & sons, Chichester, 1981.
43. Starnes J. H. and Haftka, R. T.: Preliminary design of composite wings for buckling, strength, and displacements constraints, J. Aircraft, 16 (1979), 564–570.
44. Thanedar, P. B.; Arora, J. S.; Tseng, C. H. et al.: Performance of some SQP algorithms on structural optimization problems, J. Num. Meth. Engng., 23 (1986), 2187–2203.
45. Adelman, H. M. and Haftka, R. T.: Sensitivity Analysis of discrete structural systems, AIAA–Journal, 24 (1988), 823–832.
46. Adeli, H. and Balasubramanyam, K. V.: A synergetic man–machine approach to shape optimization of structures, Comp. & Struct., 30 (1988), 553–561.
47. Arora, J. S. and Haug, E. J.: Methods of design sensitivity analysis in structural optimization, AIAA–Journal, 17 (1979), 970–974.

SHAPE FINDING METHODS OF SHELLS

E. Ramm
University of Stuttgart, Stuttgart, Germany

ABSTRACT

Shell structures are usually extremely thin and therefore heavily rely on an almost pure membrane stress state. In order to guarantee this stress state the shape of the shell plays a dominant role in the initial design of the structure. The present paper compares the classical geometric shapes with free form shells and discusses several shape finding methods. The most prominent procedure is the deforming flexible membrane under a controlling load case like pressure or dead load. The method has been applied experimentally as well as analytically. Other methods start with a prescribed stress state and take the surface geometry as unknown. The procedure may be based on membrane equilibrium equations or force–density equivalents. Finally, optimization schemes are applied using e.g. the strain energy as an objective and bounding the stress state to certain limits. Several examples for this approach are given.

1. OBJECTIVE

In the hierarchy of thin–walled structures shells are located right at the top. This rank is mainly due to their potential for elegance as well as structural efficiency. Shells may be extremely slender structures. Compared to classical domes having a radius to thickness ratio of up to 50 and natural eggs with a value between 50 and 100, modern concrete shell domes have been built with a ratio of up to 800. In such an extreme situation it is obvious that each material fiber has to participate in the load carrying process; in other words: the shell must be in an almost uniform membrane state. It is also clear that such an optimized structure may become extremely sensitive to slight deviations of the original design. This feature is the reason that in [1] the term 'prima donna' has been used for the shell. A shell which has been properly designed and built with extreme care will show excellent performance over the years. Prominent examples for this class are the concrete shells designed, built and monitored by the Swiss engineer H. Isler [1], [2]. On the other side, when certain basic rules for shell behavior are not taken into account, the "prima donna may be in a bad mood", i.e. the response of the shell will be rather poor. Typical examples are the dramatic change of the stress state in cooling towers when slight deviations of the ideal shape occur, the extreme sensitivity of buckling loads due to initial imperfections or the presence of unexpected inextensional deformations due to non–uniform settlements. Furthermore, an initial design with a considerable amount of bending may cause substantial cracking and creep over the time.

The beneficial effect of the curvature on the structural performance has been known and utilized already in ancient times. Being forced to rely on almost no tension materials, the early architects automatically learned by experience to shape arches and vaults so that they carry their loads by pure compression. Already very early analytical considerations about catenary, funicular systems and the best shape of vaults have been undertaken [3], Fig. 1. Later it was recognized that in contrast to a one–dimensional curved structure a shell is able to carry several smooth load cases by membrane action alone due to the 'double arch effect'. In so far, shells are à priori already optimized structures, provided certain basic membrane oriented design rules are satisfied. Since this is very often not the case, the important question of an ideal shape for shells and corresponding shape finding methods has to be raised. Frequently, a modification of the initial shape can substantially improve the original design, for example because of concentrated loading and support conditions, free edges, incompatibilities with adjacent structural elements. Furthermore, the objective of an almost fully stressed design may lead to a very specific shape of a shell, for example the drop–shaped echinodomes of on– and off–shore oil containments, Fig. 2. The question of an ideal shape has of course

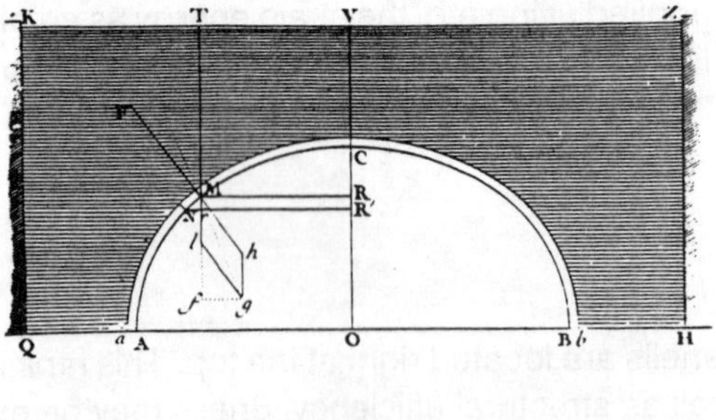

Fig. 1: Bossut's solution for the arch subject to hydrostatic pressure, taken from [3]

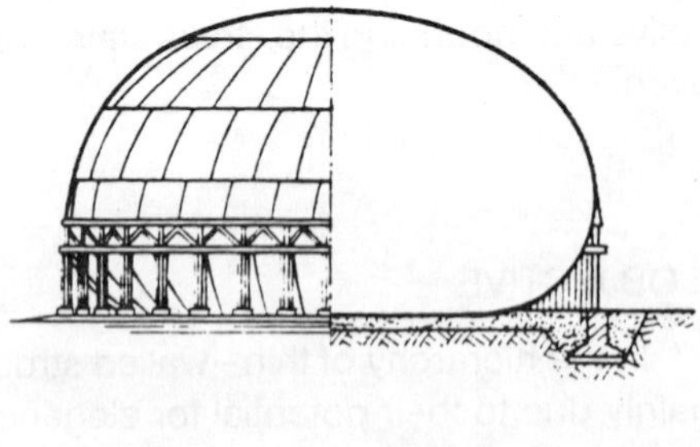

Fig. 2: Modern drop–shaped containment for oil storage, taken from [4]

to be answered in the context of the chosen material. Membrane tension structures have totally different requirements than reinforced concrete shells where tension and bending should be avoided to the greatest possible extend.

Medwadowski [5] describes the evolution of shell structures from the historical trial and error approach up to the modern era in which industrial techniques, new materials, analytical and computer analyses have revolutionized the design. He comes to the conclusion that "structural mechanics is not a leader in the development of form, it is a follower. Conversely, the development of the theories of shells was spurred by the practical problems of the forms important in practical application". It can be recognized that most shells are of regular shape, largely belonging to a few categories, often analytically defined. Despite a few exceptions, this tendency holds unfortunately to the present day.

This lack of variability is in clear contrast to the degree of sophistication reached in computer oriented analysis of shells in the meantime. Before the advent of the modern computer it was understandable that one avoided forms for which structural analyses were not available, called the 'play safe' design in [5]. It seems that this concept of using classical geometrical shapes has a lot of inertia, although numerical solution techniques have been available for general shells for many years.

The main purpose of the present study is to stimulate the discussion on free form shells. One fundamental step in the design of these shells is the form finding process for the initial shape, allowing the structure to be in an almost pure membrane state.

2. PRESENT SITUATION

The fact that almost all classical shell structures have a regular shape can probably be accounted for by a prescribed regular plan: domes on circular or polygonal plan, barrel, cloister and groin vaults spanning rectangular plans. This rule has certainly been followed up to the present day. The development of shell theory has added another reason: closed form solutions need an analytical definition of the shell geometry. A typical example is the simplicity of the membrane analysis of HP–shells which comes along with simple construction concepts (Candela). Not infrequently, the following design principle emerged: take an analytical defined surface, e.g. a sphere, cut out a certain segment and – since the amputated shell structure cannot exhibit anymore its ideal membrane state – add special stiffening elements at the boundaries. Typical examples are the cylindrical roof shell with edge beams and diaphragm walls at the ends or the spherical shells over a polygonal plan with heavy beams at the free edges (Kresge auditorium at MIT (1954), the opera in Dortmund (1964) or the convention hall in Frankfurt–Höchst (1963), see Fig. 3). The judgement is more a matter of taste and aesthetics than of structural reliability.

The alternative is the inverse formulation of the problem: given a few geometrical parameters like span and height, the load and a desired stress state, look for the natural shape of the shell. In many cases this concept allows one to avoid heavy edge beams leading to a more natural, elegant design of shells with free edges [1], [2], Fig. 4. In [6] these forms are denoted as structural shapes in contrast to the above mentioned geometric shapes.

3. SHAPE–FINDING METHODS OF FREE FORM SHELLS

3.1 Minimal Surface

The well–known equilibrium equation perpendicular to a membrane

$$\frac{n_1}{r_2} + \frac{n_2}{r_2} = p \tag{1}$$

with the pressure difference p, the principal radii r_i and the corresponding membrane forces n_i simplifies for a fully stressed situation $n_1 = n_2 = n$ to

$$\frac{1}{r_1} + \frac{1}{r_2} = \frac{p}{n} \tag{2}$$

The left side is related to the mean curvature $1/2\ (\varkappa_1 + \varkappa_2) = 1/2\ (1/r_2 + 1/r_2)$. It follows that equation (2) defines a surface of constant mean curvature, a bubble. The special case $p = 0$ or vanishing mean curvature is known as minimal surface and leads to the soap film solution [7]. Since then $\varkappa_1 = -\varkappa_2$ the shape is either flat or has negative Gaussian curvature $\varkappa_1 \varkappa_2 < 0$. Nowadays, these forms may be easily obtained by computer simulation, for example by shape optimization techniques [8]. Being a perfect shape for tension structures its quality for reinforced concrete shells is questionable.

3.2 The Flexible Membrane

A flexible membrane under one form generating load case assumes a shape entirely in tension and almost free of bending. The load can be uniform pressure acting on a clamped membrane leading to a pneumatic form (Isler calls them bubble shells), hydrostatic pressure or dead load in a hanging fabric model. The process of inverting this shape rendering pure compression is in particular appealing for no or low tension material like masonry or reinforced concrete.

The principle of inverting the catenary has been known for a long time, e.g. Giovanni Poleni's famous construction for the shape of St. Peter in Rome, 1748, or Gaudi's hanging model for the Colonia Güell chapel, 1908. The idea has been experimentally verified for concrete dams already in the 1950's. A rubber membrane was loaded downstream by water, dead load was simulated by extra weights. At the same time in India Ramaswamy [9] built shallow roof shells using a sagging fabric. The fabric is supported at all edges, loaded by fresh concrete and sags into a funicular shape. After the concrete hardens, the shells are inverted and used as units for roofs and floors. Usually these so–called funicular shells have a span of about one meter, but later single shells spanning almost 14 m have been built. In the meantime alternative load cases like hydrostatic pressure or concentrated loads also have been applied for liquid retaining structures or foundation footings.

The most successful application of this idea for large RC–shells goes back to Heinz Isler who has built over 1500 shells in the meantime. His experimental shape finding procedures include inflated membranes ("bubble shell"), shapes from flowing viscous fluid and the inverted hanging membrane [1], Fig. 5. Isler's shells do not only show structural elegance and efficiency but also have an excellent record of long term performance.

Fig. 3: Centennial hall in Frankfurt–Hoechst, (a) dome, (b) edge beam

Fig. 4: Shells by Heinz Isler in Switzer–land, (a) service station, Deitingen, (b), (c) Kilcher factory, Recherswil

In 1988 Mehlhorn and Kollegger [10] built and tested a reinforced concrete model shell with a plan of 5.5 x 5.5 m^2, a sag of 1 m, with free edges, only supported at the four corners, using the same principle of inversion, Fig. 6. The hanging reinforcement mat took the role of the fabric. The thickness of the concrete was 22 mm which was progressively increased to the corners to 60 mm. Finally, the shell was inverted as a whole and tested up to failure.

These shape generating procedures can of course be simulated by computer analyses based on geometrically nonlinear membrane or shell theory, using for example finite element methods [6], [11], [12]. For a given arbitrary plan a flat membrane is subjected to uniform or hydrostatic pressure, dead load, line or concentrated loads and the desired shape can be generated. This computer aided form finding process yields directly numerical data on the generated shape which can be further processed in a subsequent structural analysis.

The flexible membrane technique is an ideal procedure whenever one dominating load case is present. If this does not hold it has severe drawbacks. If the inversion is used, special care must be taken on areas where the membrane starts to wrinkle. The selected material for the shape finding process has not to be identical with that of the final structure; in the simplest case isotropic material data are introduced. In [6] anistropic material properties are used, allowing to trigger the final shape into a desired form. The idea reflects the situation if a real fabric is used in an experiment. By varying the material data or the load distribution the overall appearance or the local curvature, e.g. at the free edges, may be influenced. Since in general no correlation exists between the materials used for the shape finding and in reality only final shell analyses of the built structure can verify the quality of the chosen design.

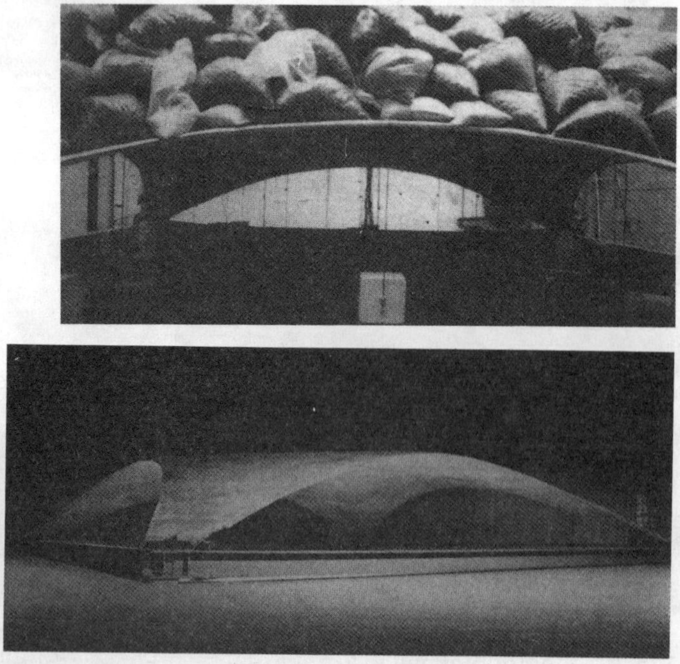

Fig. 5: Hanging model by Isler [1], [23] Fig. 6: Model shell [10]

3.3 The Inverse Membrane Problem

In general surface geometry and internal stress state show a highly nonlinear interdependence. In some situations it is of prime importance to control the stresses in a structure and take the geometry as unknown. This idea has been verified for prestressed cable structures by the so-called force density method [13]. In the nodal equilibrium equations each line cable element participates through the three components of the cable force S, for example in x direction $S\,(x_k - x_i)/l$; x_k, x_i being the coordinates of the adjacent nodes and taken as unknowns, l is the final cable length. Of course, such an expression is nonlinear, since S and l also depend on the deformed geometry. In the force density method the length is approximated by its unstressed value and S is the desired prestressing force. Thus the force density S/l is prescribed and the equilibrium equations are linearized, leading to rough values of the coordinates for the nodes. It is interesting to note that the force density is nothing else than the P–Δ–effect in frame analysis for tension, i.e. the geometric stiffness which geometrically "pulls in" the cables into the equilibrium position of the prestressed cable net. Consequently, the form–finding process reduces to solving the linear equation

$$\mathbf{K_g} \cdot \mathbf{x} = \mathbf{P} \tag{3}$$

with the geometric stiffness matrix $\mathbf{K_g}$ and the vectors of the nodal geometry $\mathbf{x}$ and nodal forces $\mathbf{P}$.

The extension to other elements like membranes and shells seems obvious, although the definition of a "force–density" based on a specified stress state is by far not straightforward. Therefore, in [14] the procedure was generalized introducing an iterative smoothing method. The designer specifies an initial trial shape and the desired stress state so that $\mathbf{K_g}$ can be computed. Based on equation (3) a new configuration is obtained. The iterative procedure is repeated until the equilibrium is satisfied. Alternatively least–square methods could also be used [14].

The inverse membrane problem has also been approached by Pucher's membrane equation [15].

$$z_{zz}\,F_{yy} - 2\,z_{xy}\,F_{xy} + z_{yy}\,F_{xx} + \overline{p}_z = 0 \tag{4}$$

$z = f(x,y)$ denotes the vertical coordinate of the shell surface defined in terms of the plane coordinates x, y; F(x,y) is a stress function related to the projections of the membrane forces ($\overline{N}_x = F_{yy}$, $\overline{N}_y = F_{xx}$, $\overline{N}_{xy} = F_{xy}$); $\overline{p}_z(x,y)$ is the vertical load. The indices define partial derivatives. For a given load and the corresponding stress state, for example for dead load, partial or uniform snow load, equation (4), supplemented by certain boundary conditions, can be solved numerically, for the unknown z coordinates, e.g. by a finite difference scheme [16], [17], [18]. In [19] a parametric study has been performed for different ratios of dead and uniform load, see Fig. 7. In [19] a comparison of shapes, generated either by this method or by large deformation finite element analyses, has been carried out resulting in negligible differences.

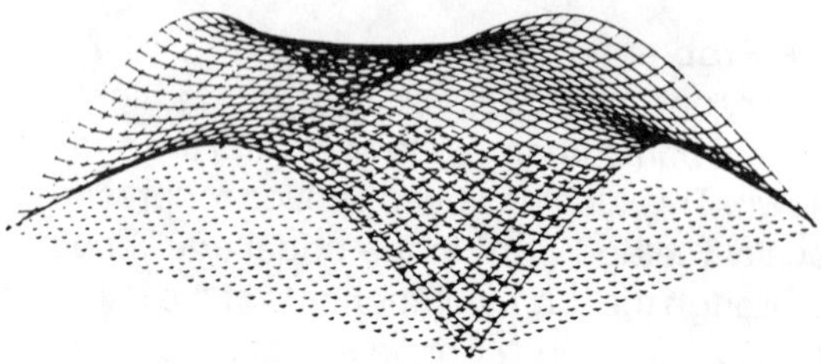

Fig. 7: Shell with free edges under dead and uniform load (ratio 8/3) [19]

3.4 Shape Optimization

The form finding of shells suggests to apply mathematical programming schemes. In [8], [20], [21]. [22] shape optimization techniques have been developed as a synthesis of geometrical design, structural analysis, sensitivity analyses and mathematical optimization. One basic feature is the design element [8] where CAGD–concepts are used to describe curved surfaces in space and their genesis by a few natural variables, i.e. coordinates of selected nodes. For example Coons–, Bézier– and B–spline interpolation schemes and related continuity patches are used. The principal question raised in connection of shells is concerned with a suitable objective. Obviously, minimum surface or weight make not always sense in this case. Therefore, a stress level function f_S or the strain energy f_E has been introduces [20].

$$f_S = \int_V (\sigma - \sigma_a)^2 \, dV$$
$$f_E = \frac{1}{2}\int_V \sigma\varepsilon \, dV \qquad (5)$$

σ_a is a prescribed allowable stress state.

Besides the surface geometry also the shell thickness may be introduced as optimization variables using the same kind of parametrization as for the surface. Again the intention is to get rid of the bending as much as possible, eventually leading to a pure membrane shell. The investigation can be controlled by selected constraints in stresses, displacements etc. This also allows to avoid tension areas (similar to wrinkles in the hanging fabric approach). An interesting feature of the method is that a compromise solution can be obtained including different load cases, individually weighted, in the objectives (see example below). Numerical experience so far confirms that there is no unique solution in every case; therefore, an interactive control device is very important for the designer. The extension of the method to include nonlinear and time–dependent phenomena like buckling or creep has not been done so far but is conceptually possible.

4. NUMERICAL EXAMPLES BASED ON SHAPE OPTIMIZATION

4.1 Shells of Revolution [8]

The example is used as benchmark demonstrating the principle capabilities of shape optimization in a qualitative way. The objective is either weight or strain energy. Starting from a spherical shell of uniform thickness (radius 10.0 m) with an apex hole of radius 2.5 m, the optimal shape is obtained for four different load cases: concentrated ring load $c = 255$ kN/m on the upper free edge, dead load ($\gamma = 78.5$ kN/m^3), uniform vertical load of 0.75 kN/m^2 (snow) and a uniform horizontal load of 0.72 kN/m^2 with pressure on one side and suction on the other side, roughly simulating wind. Either a fixed hinged or horizontal roller support is assumed. The material properties of steel are chosen: $E = 2.1 \cdot 10^5$ N/mm^2; $\nu = 0.3$. Stress constraints with a v. Mises stress limit of 160 N/mm^2 are considered.

Fig. 8 shows the optimal shapes for different loads and boundary conditions. Except for the "wind load", the final shape allows the shell to carry the load by almost pure membrane action. This holds for weight (concentrated and snow load) as well as strain energy (dead load, wind) minimization.

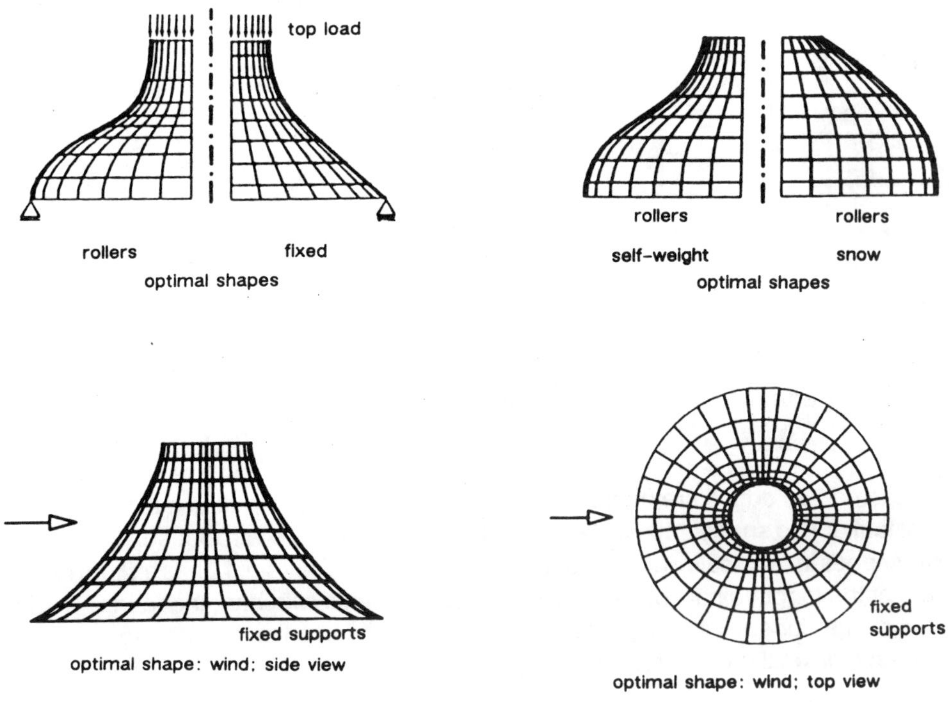

Fig. 8: Shape optimization of shell of revolution

4.2 Non–Regular Shell [20]

Two concrete shells with a uniform thickness t ($E = 3.4 \cdot 10^7$ kN/m², $\nu = 0.0$, $\gamma = 25$ kN/m²) have been investigated with respect to strain energy minimization. One load case of dead load and uniform life load p = 5 kN/m² is considered. The initial geometry of both shells with a thickness of t = 0.1 m is defined in Fig. 9 in which the height is plotted in a different scale.

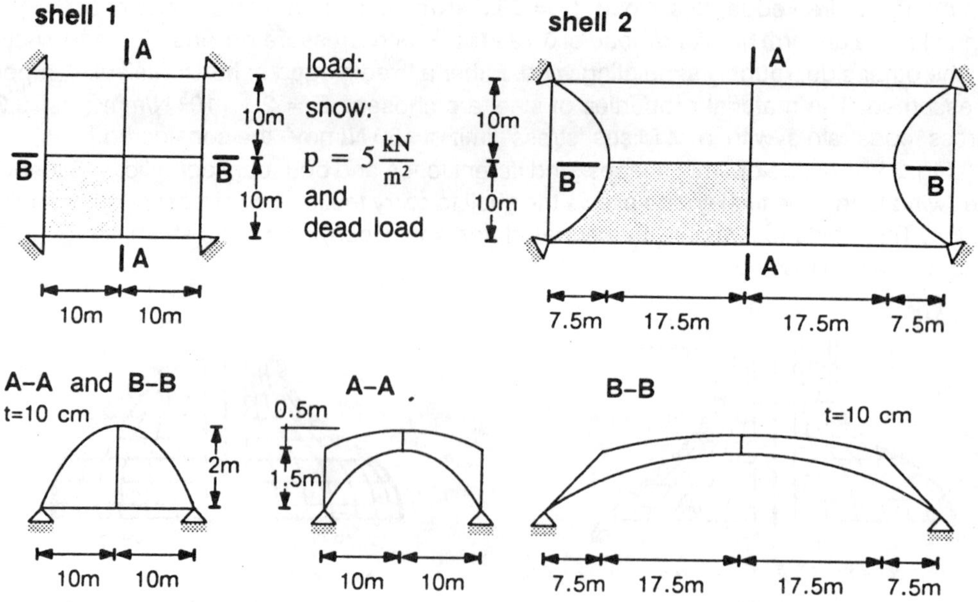

Fig. 9: Initial geometries of 4–point supported shells

The curved free edges are quadratic parabolas. Except that the total material volume is kept constant during shape optimization no further constraints are introduced. Fig. 10 shows the initial shapes in real scale and their optimal counterparts. Only shell 2 has a slight negative curvature at the longitudinal edges known from experiments with hanging fabric. The intermediate solution for shell 1 just before the optimum also indicates this phenomenon. Presumedly several local minima are close together.

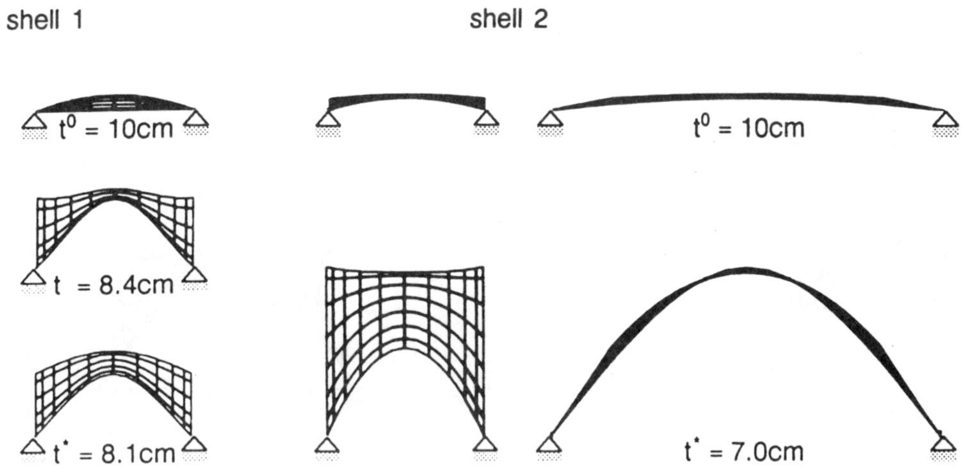

Fig. 10: Initial and optimal shape

The optimal shell 1 is further processed by thickness optimization based on weight minimization and a v. Mises stress limit of 2.1 N/mm^2. The thickness variation, defined by four independent variables, is given in Fig. 11. A subsequent shape optimization does hardly change the initial shape.

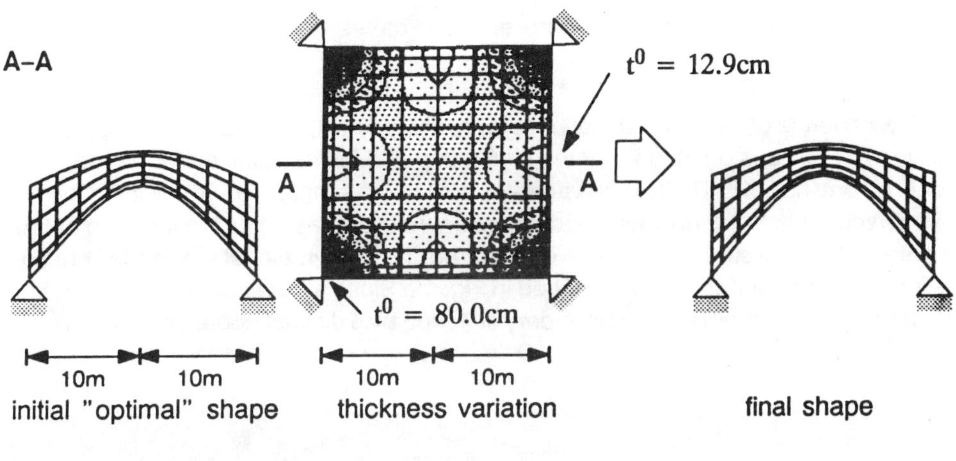

Fig. 11: Thickness and shape optimization

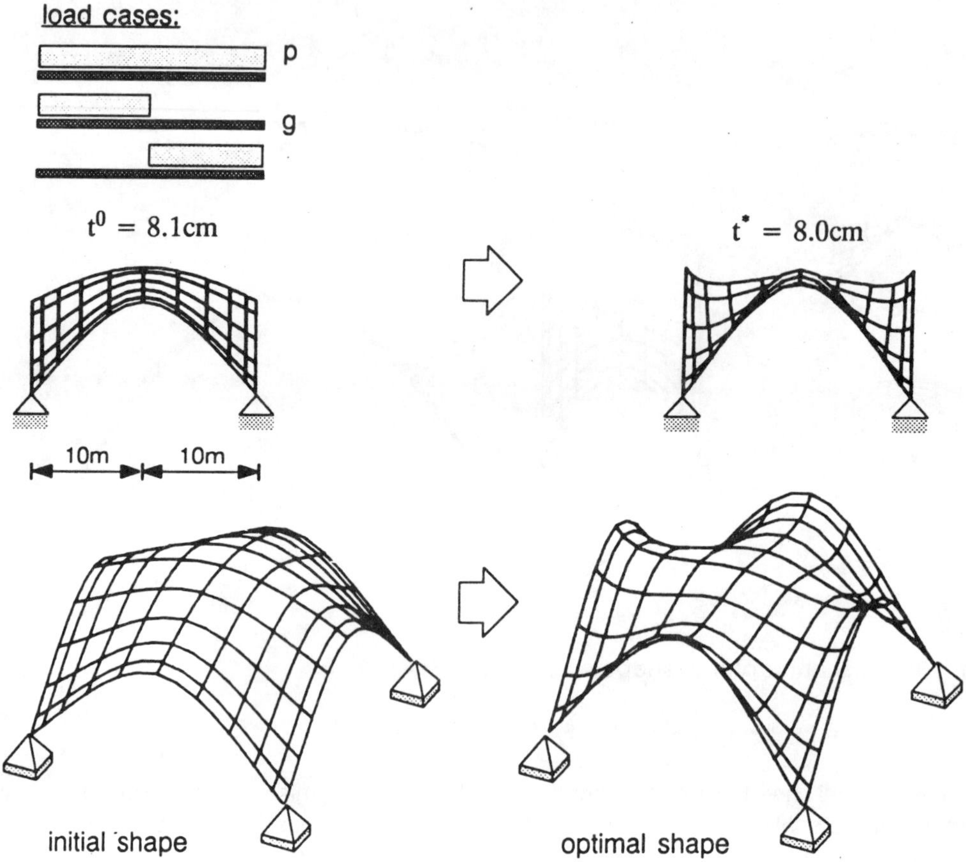

Fig. 12: Shape optimization with different load cases

The influence of several load cases is investigated next, again using the objective "strain energy" f_E. Besides dead load g the life load $p = 5\ kN/m^2$ acts either on the entire shell or only on one half of it. The design variables are linked, so that the symmetry of the structure is preserved. Therefore, only two load cases are relevant; as objective the sum of the non-weighted individual strain energies is considered. The total material volume is kept constant. The initial and optimal shapes are plotted in Fig. 12. Since a pure membrane state is not possible anymore, the shell needs a boundary stiffened by a distinct negative curvature (Fig. 12).

4.3 Free Form Concrete Shell [20]

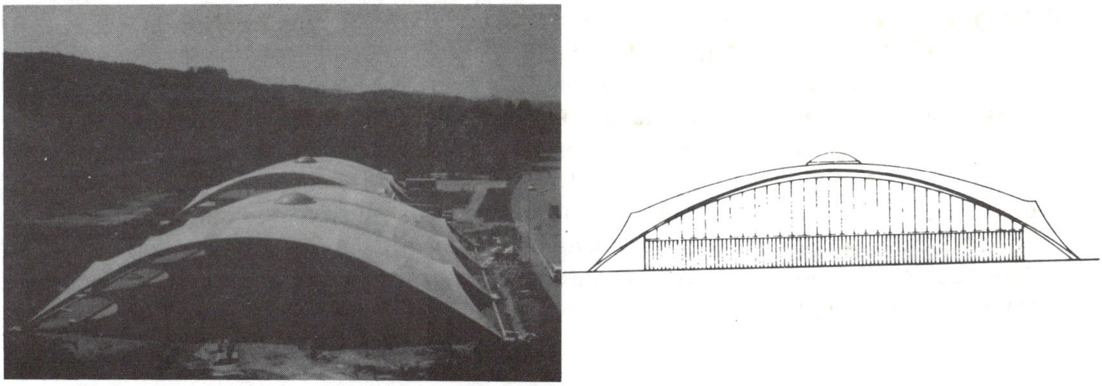

Fig. 13: Tennis center Solothurn, 1982, by Heinz Isler [1], [23]

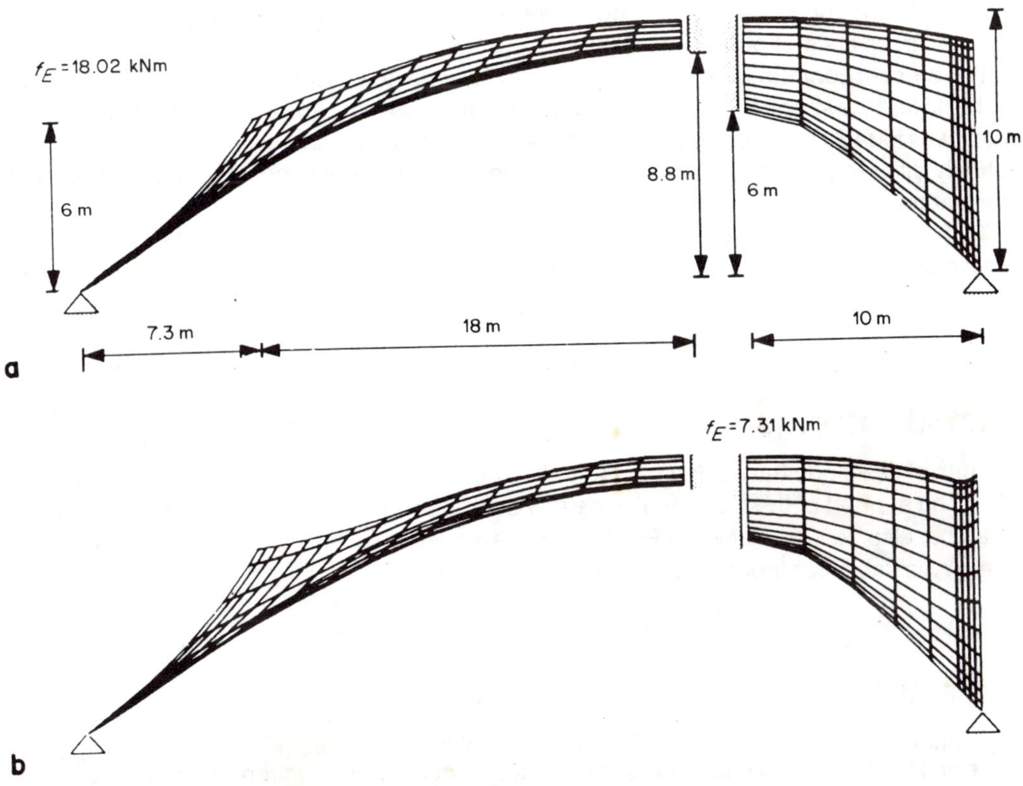

Fig. 14: (a) Initial and (b) optimal shape of free form shell

The shell in Fig. 14 is related to the reinforced concrete tennis hall designed by H. Isler [23], Fig. 13. The material data are $E = 3.1 \cdot 10^7$ kN/m^2, $\nu = 0.2$, $\gamma = 25$ kN/m^3. One load case of dead load plus uniform life load 5 kN/m^2 is considered. One quarter of the shell is idealized by 4 Bézier elements, linked by continuity patches and with 23 vertical nodal coordinates as design variables. 126 reduced integrated 8–node shell elements are used for a linear structural analysis. Again the strain energy is minimized. Since the material volume is kept constant, the uniform initial thickness $t^o = 0.10$ m slightly varies during optimization. It could be recognized that the shape near the free edges is sensitive and might even result in a sharp local curvature representing a kind of edge beam. If this is not tolerated by prescribed geometrical constraints, the optimal shape again shows a clear negative curvature near the free boundary (Fig. 14).

5. CONCLUSIONS

Shape sensitive structures like shells require high quality in design, analysis and manufacturing. Therefore, the main objective will be a membrane oriented design avoiding as fas as possible bending but also buckling phenomena. For RC shells the ideal situation is of course a stress state of pure compression.

The present contribution discusses different answers to this inverse formulation of question, i.e. to find the shape of the shell for a given ideal stress state and certain boundary conditions. It was found that several useful form finding methods are available which, although conceptually different, lead to similar results. It seems that the most versatile method is shape optimization.

To the author's knowledge all shape finding methods for shells so far are based on linear elastic material and exclude phenomena like material nonlinearities, creep and buckling. This is a challenging objective for further studies.

ACKNOWLEDGEMENTS

Parts of this study have been carried out in the research project SFB 230 'Natural Structures'. The support of the German Research Foundation (DFG) is gratefully acknowledged. The author also likes to thank his Ph.D. students Kai–Uwe Bletzinger and Stefan Kimmich for their valuable contributions.

REFERENCES:

1. Ramm, E. and Schunck, E.: Heinz Isler – Schalen". Karl Krämer Verlag, Stuttgart, 1986.
2. Isler, H.: New Shapes for Shells, Int. Colloquium on Construction, Processes of Shell Structures, IASS, Madrid, 1959.
3. Benvenuto, E.: An Introduction to the History of Structural Mechanics, Part II: Vaulted Structures and Elastic Systems, Springer, 1991.

4. Bramski,C.: Rotationssymmetrische tropfenförmige Behälter, W. Ernst & Sohn, 1976.
5. Medwadowski, S. J.: The Interrelation between the Theory and the Form of Shells, Symp. Shell and Spatial Structures: the Development of Form, Morgantown, West Virginia, 1978, IASS–Bulletin, 70, August 1979.
6. Hunter, I.S. and Billington, D.P.: Computational Form Finding for Concrete Shell Roofs, ACI – Spring Convention, Boston, MA, March 1991.
7. Hildebrandt, S. and Tromba, A.: Mathematics and Optimal Form, Scientific American Library, 1985.
8. Bletzinger, K.–U.: Formoptimierung von Flächentragwerken, Bericht 11 (1990), Institut für Baustatik, Universität Stuttgart, 1990.
9. Ramaswamy, G.S. and Rajasekaran, S.: Computer–Aided Form Generation of Funicular Shells, ACI – Spring Convention, Boston, MA, March 1991.
10. Kollegger, J. and Mehlhorn, G.: Analysis of a Free–Formed Reinforced Concrete Model Shell, in: Proceedings, 2nd Int. Conf. Computer Aided Analysis and Design of Concrete Structures. Pineridge Press, Swansea, UK, 1990.
11. Smith, P.G.: Membrane Shapes for Shell Structures, Ph.D. thesis, University of California, Berkeley, 1969.
12. Day, A.: A General Computer Technique for Form Finding for Tension Structures, Proceedings, IASS Symp. on Development of Form, Morgantown, West Virginia, 1978.
13. Scheck, H.J.: The Force Densities Method for Form Finding and Computation of General Networks, Comp. Meth. Appl. Mech. Engng., (1974), 115–134.
14. Haber, R.B. and Abel, J.F.: Initial Equilibrium Solution Methods for Cable Reinforced Membranes. Part I: Formulations; Part II: Implementation, Comp. Meth. Appl. Mech. Engng., 30 (1982), 263–284 and 285–306.
15. Pucher, A.: Über den Spannungszustand in doppelt gekrümmten Flächen, Beton & Eisen, 33 (1934), 298–304.
16. Korda, J.: Ribless Membrane Shells with Point Supports at the Corners, in: Int. Symp. on Shell Structures in Engineering Practice, Vol. I, Budapest, 1965, 179–190.
17. Csonka, P.: Point–Supported Shells with Free Boundary, Acta Technica Academiae Scientiarium Hungaricae, 75 (1973), 121–136.
18. Alpa, G., Bozza, E., Corsanego, A. and Del Grosso, A.: Shape Determination for Shell Structures on Pointlike Supports, IASS–Bulletin, 67, August 1978.
19. Berg, H.–G.: Tragverhalten und Formfindung versteifter Kuppelschalen über quadratischem Grundriss auf Einzelstützen, Bericht 79–2, Institut für Baustatik, Universität Stuttgart, 1979.
20. Kimmich, S.: Strukturoptimierung und Sensibilitätsanalyse mit finiten Elementen, Bericht 12 (1990), Institut für Baustatik, Universität Stuttgart, 1990.
21. Ramm, E., Bletzinger, K.–U. and Kimmich, S.: Trimming of Structures by Shape Optimization, in: Proceedings, 2nd Int. Conf. on Computer Aided Analysis and Design of Concrete Structures, Zell am See, Austria, April 1990, Pineridge Press, Swansea, U.K., 1990.
22. Ramm, E., Bletzinger, K.–U. and Kimmich, S.: Strategies in Shape Optimization of Free Form Shells". In: Nonlinear Computational Mechanics – a State of the Art – (Eds. P. Wriggers, W. Wagner), Springer, 1991.
23. Isler, H.: Elegante Modelle – die moderne Form des Schalenbaus, db, 7 (1990), 62–65.